NEW
CHRONOLOGICAL HARMONY
OF
HISTORY AND THE BIBLE

ALVIN F. SHOWALTER
B.A., M. DIV.

Alvin F. Showalter
Rom 11 : 2 5 - 32

NISAN JUBILEE YEAR 1988/89
120TH IN 6 000 YEAR CHRONOLOGY

Obtainable from:
Messianic Faith Ministries
P O Box 7345
Roggebaai
Cape Town
South Africa
8012

ISBN 0 86987 241 9

Printed in South Africa

"It is the Glory of God to conceals a thing: but the honour of kings is to search out a matter" Proverbs 25:2

"Men do not reject the Bible because it contradicts itself, but because it contradicts them" E. Paul Hovey

Dedicated
to the
Glory of the God of Abraham, Isaac,
and Jacob; to the Glory of Christ
the Messiah, (Emmanuel, God with us);
and to the Glory of the Holy Spirit
who reveals His eternal,unchangeable,
Holy Word.

Hopefully, the loyal, persistent, patient, and dedicated students of History and the Word will succeed in proving to themselves that this book is faithfully accurate as possible according to present criterions of Historic and Biblical ethics

"Ye shall not add unto the Word which I command you, neither shall ye diminish ought from it, that ye may keep the commandments of the Lord your God which I command you" Deut. 4:2

1

2

CONTENTS

CONTENTS

- - - - - - - - - - - - - - - -

BIBLE CHRONOLOGY - 4012 TO 600 BC

	SON'S BC BIRTH		
ADAM'S SIN	:4012:130+	4012-3082 BC Adam's sin & death	:130+800=930 G. 5:3
SETH	:3882:105+	3882-2970 BC Birth and death	:105+807=912 G. 5:6
ENOS	:3777: 90+	3777-2872 BC "Men called on God"	: 90+815=905 G. 5:9
CAINAN	:3687: 70+	3687-2777 BC Gen. 4:26	: 70+840=910 G. 5:12
MAHALALEEL	:3617: 65+	3617-2722 BC	: 65+830=895 G. 5:15
JARED	:3552:162+	3552-2590 BC (Inter-marriage)	:162+800=962 G. 5:18
ENOCH	:3390: 65+	3390-3025 BC	: 65+300=365 G. 5:21
METHUSELAH	:3325:187+	3325-2356 BC-lived until Flood	:187+782=969 G. 5:25
LAMECH	:3138:182+	3138-2361 BC Gen.	:182+595=777 G. 5:28
NOAH Gen.5:32	:2956:503+	2956-2006 BC 7:6;11:10	100+503+347=950 G. 9:28,29
Gen.10:21	:2800:	(500+100+350=950 G.9:28,29) F 500th yr of Noah	
	:2625:	Japheth born G.5:32	
SHEM	:2453:100+	2453-1853 BC	:100+500=600 G. 11:10
FLOOD BEGAN	:2356:		
FLOOD ENDS	:2355:		
ARPHAXAD	:2353: 35+	2353-1915 BC	: 35+403=438 G. 11:12
SALAH	:2318: 30+	2318-1885 BC	: 30+403=433 G. 11:14
EBER(HEBREWS)	:2288: 34+	2288-1824 BC	: 34+430=464 G. 11:16
PELEG (Earth-	:2254: 30+	redivided)+2254-2015(G.10:25)	: 30+209=239 G. 11:18
REU	:2224: 32+	2224-1985 BC	: 32+207=239 G. 11:20
SERUG	:2192: 30+	2192-1962 BC	: 30+200=230 G. 11:22
NAHOR	:2162: 29+	2162-2014 BC	: 29+119=148 G. 11:24
TERAH Acts 7:4	:2133:130+	G. 12:4+2133-1928 BC	:130+ 75=205 G. 11:32
ABRAHAM his call	:2003:100+	2003-1828 BC	:100+ 75=175 G. 17:1
ISAAC	:1903: 60+	1903-1723 BC	: 60+120=180 G. 25:26
JACOB(Age to-	:1843:130+	Egypt)+1843-1696 BC	:130+ 17=147 G. 47:9,28
ISRAEL-EGYPT	:1713:215 years	1713-1498BC	:215: Ex.6:16;Gal.3:17
SIN UNINPUTED	:1518:	:N.32:11:1518 BC	:Num.32:11;New Gen.
EXODUS (MOSES)	:1498: 40+	1498 BC	:40:Deut.2:7;2:14;8:2
JOSHUA	:1458: 25+		: 25:Josephus V. 1-29
NEW GENERATION	:1433: 15+	Adjustable	:15:Jud. 2:10;Num.32:11
MESPOTAMIA)	:1418: 8+	New Gen.	: 8:Jud.3:8 MESPOTAMIA
OTHNIEL	:1410: 40+		: 40: Jud. 3:10
EGLON (MOAB)	:1370: 18:	El Amarna	: 18:Jud. 3:13 MOABITE'S
EHUD : 80	:1352: 20:	1352 BC :80 yrs total-	: 20:Jud.3:30 -rest 80
JABIN: years	:1332: 20:	Shamgar	: 20:Jud.3:31;4:2;5:6 yr:
BARAK: total	:1312: 40+	1272 BC : 1313 BC	: 40:Jud.5:31 rest:
MIDIANITES)	:1272: 7:	: 1274 BC	: 7:Jud.6:2 MIDIANITES
GIDEON	:1265: 40:	Egyptian	: 40:Jud. 8:28
ABIMELECH)	:1225:	Conoubine : 1216 BC	: 3:Jud. 8:31 EGYPTIAN
TOLLA	:1222: 23:		: 23:Jud. 10:2
JAIR	:1199: 22:		: 22:Jud. 10:3
PERIOD OF SIN	:1177: 1:	Adjustable : 1188 BC	:1:Jud.10:6;11:26 300
PHILISTINES20)	:1176: 18:	Ammonites : 1176 BC	: 18:Jud. 10:8,IS.7:2;13;14
JEPHTHAH	:1158: 6:	:1158	: 6:Jud.12:7;11:26 300
IBZAN	:1152: 7:		: 7:Jud. 12:8,9
ELON	:1145: 10:		: 10:Jud. 12:11
ABDON	:1135: 8:		: 8:Jud. 12:14
SAMSON	:1127: 20:		: 20:Jud. 15:20
NO RULER	:1107: 5:	Adjustable	: 5:J.17:6;IK.6:1 480
SAUL	:1102: 40:		: 40:IS.10:1;Acts13:21
DAVID	:1062: 40:	1022 BC :1062 BC :1065 BC	: 40:IIS.5:4;IK2:11;IC.
SOLOMON	:1022: 40+	1018 BC :1022 BC :F:accession	40:IK.11:42;IIC.9:30
REHOBOAM	: 982: 17:	982 BC : 982 BC : Left	17:IK.14:21;IIC.12:13
ABIJAM	: 965: 3:3:	: 965 BC : correct	3:IK.15: 2;IIC.13: 2
ASA	: 962: 41:	: 962 BC :date of	41:IK.15:10;IIC.16:15
JEHOSHAPHAT	: 921: 25:d:	This King's : 920 BC : kings:	25:IK.22:42;IIC.20:31
JEHORAM	: 896: 8:	chronology : 895 BC :prorated	8:IIK.8:17;IIC.21:5
AHAZIAH	: 888: 1:	incorrect by : 887 BC : extra	1:IIK.8:26;IIC.22:2
ATHALIAH	: 887: 6:	1 or 2 years : 886 BC :9 months:	6:IIK.11:3;IIC.22:12
JOASH	: 881: 40:	without : 880 BC : for	40:IIK.12:1;IIC.24:1
AMAZIAH	: 841: 29:	prorating : 840 BC : both	29:IIK.14:2;IIC.25:1
UZZIAH	: 812: 52:	extra : 811 BC : Asa	52:IIK.15:2;IIC.26:3
JOTHAM	: 760: 16:	months : 759 BC : and	16:IIK.15:33;IIC.27:8
AHAZ	: 744: 16:	allowed. : 743 BC :4:Hezekiah:	16:IIK.16:2;IIC.28:1
HEZEKIAH	: 728: 29:	See pages : 727 BC :7: for	29:IIK.18:2;IIC.29:1
MANASSEH	: 699: 55:	35 - 46 : 697 BC :y: year	55:IIK.21:1;IIC.33:1
AMON	: 644: 2:	: 642 BC :s: allowed	2:IIK.21:19;IIC.33:21
JOSIAH	: 642: 31:	: 640 BC :J: for	31:IIK.22:1;IIC.34:1
JEHOAHAZ	: 611: 1:	1 yr for 3mos : 609 BC :Jehoahaz:	1:IIK.23:3;IIC.36:2
JEHOIAKIM	: 610: 11:	: 609 BC : and	11:IIK.23:36 IIC.36:5
JEHOIACHIN	: 599: 1:	1 yr for 3mos : 598 BC : Coniah	1:IIK.24:8;IIC.36:9
ZEDEKIAH	: 598: 6:	: 597 BC : 6: to 5th Jehoiachin	
5th yr Captiv.	: 592:	592 BC : 592 BC :Captivity	
		:586 BC : 586 BC - Jerusalem fell-Babylon	

PERIOD OF THE KINGS DATED

JUDAH	REIGN	CO-REGENCY	ISRAEL	REIGN	CO-REGENCY
	B.C.			B.C.	
1. Rehoboam	– 982–965		1. Jeroboam	– 982–961	
2. Abijam	– 965–962				
3. Asa	– 962–920		2. Nadab	– 961–960	
			3. Baasha	– 960–936	
			4. Elah	– 936–935	
(9 extra months prorated–Asa's Reign)			5. Zimri	– 935	(7 days)
			6. Omri – Tibni		935–931
			7. Omri	– 931–924	
			8. Ahab	– 924–902/1*	
4. Jehoshaphat	920–895	923–920 – – –	Asa diseased		
			9. Ahaziah	– 901–899	904–901
			10. Jehoram	– 899–887	903–899
5. Jehoram	– 895–887	899–895			
6. Ahaziah	– 887–886	888	11. Jehu	– 887–859	
7. Athaliah	– 886–880				
8. Joash	– 880–840				
			12. Jehoahaz	– 859–842	
			13. Jehoash	– 842–826	844–842
9. Amaziah	– 840–811	843–840 – – –	Joash diseased		
10. Uzziah	– 811–759		14. Jeroboam II	– 826–785	
Co-regent with Uzziah – – – – – – –			Jeroboam II		799–785
Earthquake	– 785**		15. Uzziah	– 785–775***	
Olympiads	– 776 B.C.		16. Zachariah	– 774	
			17. Shallum	– 773	
			18. Menahem	– 773–763	
			19. Pekahiah	– 763–761	
11. Jotham	– 759–743	775–759	20. Pekah	– 761–741	
(Prisoner	– Assyria)	743–740			
12. Ahaz	– 743–727		21. Hoshea(Gov.)	– 741–732****	
			Hoshea(Gov.)	– 732–730 – freedom	
13. Hezekiah	– 727–697(9 mos. prorated)		Hoshea(King)	– 730–722 – Israel	
14. Manasseh	– 697–642		**(CAPTIVITY**	**– 721 B.C.)**	
15. Amon	– 642–640				
16. Josiah	– 640–609				
17. Jehoahaz	– 609	(3 months)			
18. Jehoiakim	– 609–598				
19. Jehoiachin	– 598/7 B.C. (the last 3 months of Nisan year)				
20. Zekekiah	– 597/6 – 586/5 – Jerusalem fell – **Babylonian Captivity**				

* Ahab dies in the year of the Battle of Karkar – 902/1 B.C.
** Jeroboam II probably died the year of Uzziah's Earthquake – 785 B.C.
*** Uzziah ruled both kingdoms ten years – II K.14:28;15:1;II C.26:10
**** Hoshea ruled as an Assyrian puppet for nine years and then regained
   freedom for two years before becoming King in Israel

PREFACE

The problem of correlating and dating Biblical and Ancient Historical events is an inherited legacy faithfully transferred from generation to generation and from ages past, which should be solved in an era in which "knowledge shall be increased", and during "the time of the end of this present age." The Holy inspired Word, which is the only true basis for any new chronological revelation, states, "That thou, O Daniel, shut up the Words; and seal the book, even to the time of the end; many shall run to and fro; and knowledge shall be increased" Daniel 12:4.

The final book of God's precious revelation to all mankind adds, "And I saw in the right hand of Him that sat on the throne a book written within and on the backside, sealed with seven seals. And I saw a strong angel proclaiming with a loud voice, who is worthy to open the book, and loose the seals thereof? And I beheld . . . and in the midst of the elders stood a Lamb (Messiah) as it had been slain . . and they sung a new song, saying, Thou art worthy to take the book, and to open the seals thereof: for thou wast slain, and hast redeemed us to God by thy blood out of every kindred, and tongue, and people, and nation" Rev. 5:1,2,6a,9.

The above scriptural quotations reminds us first that the Word of God was sealed only to the time of the end; that only the Jewish Messiah, Jesus Christ, can unlock, by His Spirit, the knowledge hid in His Holy Word, and that John's book of Revelation completed the revelation of the Canon of scripture opening the final seals to a closed book. The first quotation written by Daniel before 500 B.C. confirms that the Word was not complete at that time. The last quotation by John over 600 years later testifies that the revelation and blue print of God has been revealed and completed. Since that time until this, knowledge and understanding of the scriptures and of the blue print of God has gradually increased. Each generation of believers, through the guidance of the Spirit of God, has built on the knowledge of their predecessors. Today, as knowledge in every realm and sphere of life is increasing, we can expect a break through in the realm of the spiritual, as well as in the scholastical realm of the revelation of God's Word.

Our prayer is that some part of this research gathered in this book may be an additional step in vindicating the Holy Word of God before the nations; that a true chronological system may be established, and that all who read may be struck by an irrevocable conviction that simplicity

7

of faith in God's Word is the sure key to light and spiritual revelation.

In our study of the writings of archaeologists, historians, and theologians, we have been often impressed that they lacked a definite yardstick for a sure chronological framework on which to hang their findings and conclusions involving Ancient History. By this we mean that they had no definite established continuous chronology on which to work or to interlock their historical discoveries. Many times, with a stroke of a pen, they passed over hundreds of years, not fearing lest they contradict other historical facts. In fact they had no need to fear because there had been no well established, definite, generally accepted chronological structure. Could it not be possible to establish a simple Bible chronological framework and then prove it by correlating Historical facts? Were there gaps in chronology, or do History and the Bible weave themselves into a beautiful finished pattern?

In our first attempts, we will have to admit that the way seemed rather confused and long. Yet, as time passed (22 years), and our material mounted, a conviction was growing that Biblical chronological study was the key to insurmountable problems which have faced the historians when correlating the dates of events from the Bible with those in Ancient History. Besides, when we weigh the thousands of Historical and Biblical events available and correlate them with known dates, there is no room for long chronological gaps, except for the Noahic Flood.

We are now convinced that Daniel's prophetic seventy weeks are based on a 360 day year and this unlocks the books of Daniel, and Revelation; while the years of the Judges, Kings, and the ages of the patriarchial fathers, are based on a full astronomical year of 365 1/4 days. With this in mind we felt that another effort to establish a true Biblical Chronology should be attempted so that all Biblical and Historical events with acceptable dates might possibly fall into their proper places in these last days. This, as you will see, occurs so frequently and so supernaturally so that it is impossible to deny the implications.

I am indebted to **Mr.** D.I. Cole, former lecturer at the planetarium in Cape Town, and for information gained from his conversation and notes in the monthly journal of the Astronomical Society of Southern Africa; for the excellent advice of Noel Roland of the Cape who read, corrected, and made important suggestions concerning the manuscript; for suggestions of various friends who have read the whole or portions of the manuscript; for the kind permission for the excellent quotations, from publishers and authors of books, magazines, articles, and notes, are greatly appreciated. These quotations from these eminent scholars correlated with Bible quotes builds an incontrovertible bulwark of proof before the nations in these last days; and last, but not least of all, for my wife, Nina Lois, who prevailed in love through the years while I compiled this manuscript, and for her kindness in reading the manuscript and compiling the excellent index at the back of the book.

8

- - - - - - - - - - -

INTRODUCTION

The chronology of the Holy Bible will yet gain its rightful laurels and position in the esteem of historians, pastors, archaeologists, anthropologists, and teachers alike. It will yet prove the only true yard-stick pin-pointing and correlating historical events. Even though, at this present time, there is a tendency among many of those who honor the massoretic text to deny the possibility of a true Biblical Chronology.

The à priori argument, after all, is whether a certain portion of Bible Chronology can be proven correct by providing a suitable framework for historical events? Then, the only logical conclusion would be that, the remainder will stand the stringent test of time and meticulous analysis. If the year of one phase of history proves to be 365 and one fourth days in length and fits the proven pattern of historical facts and archaeological survey, then there is no logical reason to count other phases as unreliable. For example, if historical and archaeological surveys fix the date of Jacob's entrance into Egypt as 1710 B.C., and simple Bible Chronology dates it 1713 B.C., then we cast our lot on the side of the Word of God. Especially so, when we find that 1713 B.C. (Sethos, Salitos or Rameses II's greatest glory) is exactly four hundred years before the "second year" of King Seti I (1313 B.C.) who, that year, commemorated the four hundred year Stella at Tanus.[1] The same event was also celebrated one hundred years later during the reign of Seti II (1213 B.C.), at which time an interesting traditional story which compares to Joseph's was told. These coincidences do not dampen but increase faith in Bible Chronology. Also, we are encouraged when other events begin to fall into place like ten pins. Otherwise, if we had found that Bible Chronology had contradicted all known historical facts then we would have given up our research long ago and readily accepted the present day theory of long gaps in chronology.

It could be said that, "simplicity is the mother of invention." Thus, we are setting forth a simple Biblical chronological framework or blue print which has itself been hid throughout the centuries within the pages of the Word of God, which may yet stand the test and criticism of both the evangelicals and critics alike. Personally, this chronological framework has never disappointed us and has only led us on and on to unlock other secrets heretofore hidden from the natural eye. Therefore, we boldly publish this so that others may have access to this historical pattern; thereby, adding their accumulated knowledge, so this last day historical jigsaw puzzle may fall into its rightful place and form its rightful format for greater future revelations.

It is said, "There is nothing new under the sun". Therefore, all these steps may have been suggested at some time in the historical past, yet the combining of them in this present relationship may not have been attempted. Also, the need of a framework (Biblical) is indisputably important. If the period of the Kings fit within a certain span of time,

---

1. Wilson, John A., "The Culture of Ancient Egypt", p. 240; Publisher: University of Chicago Press.

then it is relatively simple to coordinate the events. This, as we will see, simplifies the entire procedure and makes us wonder why the chronological key had not been discovered long ago. This also guards against overlapping reigns according to the imaginations of the mind, which only leads to more confusion with multiplied problems, and brings chronological study into disrepute.

Further, this chronology has only to do with this age (the last 6000 years) in which we are living. Adam was the first in the genealogical line beginning this age. The analysis of the six days of creation or the possibility of ages before is not to our interest. We are only considering and unfolding the secrets of this age, since Adam sinned and began to age, which we believe is unlocked by means of the Book of Books. We shall let the analysis of the Pre-Adamic ages to the keen insight and imagination of the anthropologists, and geologists. Adam's life began during the sixth day (Gen.1:24-28) of recreation (Psa.104:30). What that means as far as the ages past, we will not consider? We do believe that that point of Adam's sin and expulsion from Eden was in the year 4012 B.C. as we later vindicate.

To the general reader we would exhort to presist beyond the rather heavy introductory reading (first 66 pages) of the "Two Kingdoms Correlated" to obtain the true treasures found in this book. These first few pages of chronological computation are absolutely necessary to establish our foundation date (592/1 B.C.) of our whole historical framework, which inturn, is interwoven with Israel's explusion from Jerusalem and their captivity at Babylon. Also, a good many years of research were involved in compiling and harmonizing Biblical and Historical data found at the back of this book, especially the charts, "Bible Chronology and History Correlated", pages 211-235, which can not be fully expounded within the context of this book. Yet, it is a treasure house of information for all, and the "first" successful work to fully bring History and the Bible into one accord. Therefore, it is suggested that this book can not be lightly read to truly absorb its astounding importance, and Historic and Biblical value.

May we politely suggest that we have been a "gadabout" long enough with the GAD (General Accepted Date) chronological theory, and that it is about time to "draw nigh" to GOD (God's Official Date) closely advocated by John J. Bimson, David Livingston, Hans Goedicke, William H. Shea, Bietak, and Donald Redford. It has been suggested that these authors have problems, and their Exodus dates are subject to variation, which we, providentially, have been led, over these last 22 years, to slowly iron out; therefore establishing a satisfactory date for the Exodus, which correlates with both History, the Bible, and with these eminent men's research.

- - - - - - -

Recognition of the following seven chronological steps in the ancient Biblical plan helps to simplify the correlation of events and limits the number of events to be co-ordinated in each phase. Two of the steps (most ancient) are based on genealogical information given in the Bible, while the remaining five steps are confirmed by Biblical measurements or spans of time given clearly in the Word of God. Therefore, since we believe in the plenary, verbal inspiration of the eternal Word, historical events should fit into this God given pattern without force or connivance.

The foundation year of our chronological framework is based on the well known date of the "fifth year" of Jehoiachin's Captivity (592/1 B.C. - Ezekiel 1:2)[1]. Clear confirmation concerning the validity of this year will be presented establishing this date as a solid foundation on which to build our chronological structure. Each step or phase of this chronological framework will be considered in detail as we progress through the book.

We believe that when the Word of God speaks of a year in the Biblical genealogy or in reference to the reigns of kings, it refers to a full astronomical year of three hundred and sixty five and one fourth days. The only exception is found in dealing with the prophetic plan of God in the books of Daniel and Revelation where the Word refers to a prophetic year of 360 days. Thus, we follow this simple rule in the following seven steps of our chronological plan (One other exception is the Biblical length of the Flood which acts as a prophetic study key).

Now that we have the seven steps of our Bible Chronology before us, (page 12) let us consider the synchronism of events and dates preceding our foundation date (592/1 B.C.). This year is a well known date and is clearly confirmed by data gleaned both from the Bible, and from historical records.

These years, which we are now to consider, involve the last five Judean kings before the destruction of Jerusalem in 586 B.C.. By correlating this data before us, we can conclusively prove to ourselves what year the Bible is referring to when it mentions our foundation date, the fifth year of "Jehoiachin's Captivity" (Ezekiel 1:1,2). The data for our first chronological step, which precedes 0 B.C. to our foundation date, is based on Ezekiel's prophecy chapter four verse five, and is dated in Ezekiel chapter one verse two. All of the first seven chapters of Ezekiel were prophesied in the same year, the "fifth" year of Jehoiachin's Captivity; therefore, Ezekiel chapters eight to nineteen were prophesied in the following, "sixth", year of "Jehoiachin's Captivity" (Ezekiel 8:2) etc.. Thus, this method continues on through the book of Ezekiel with all chapters dates based on a certain year of Ezekiel's prophecy during the time of "Jehoiachin's Captivity".

1. Not the "fifth year" of Jehoiachin's "reign" which would be 593/2 BC.

11

SEVEN STEPS IN OUR NEW BIBLE CHRONOLOGY

PALAEOLITHIC AGE - before 4012 B.C. time was not calculated and man did not age. Adam & Eve had perfect fellowship with God(Heaven on earth). 4012 BC - **NEOLITHIC AGE** - Adam & Eve sinned and began to age with expul- : sion from Eden during the 6th day of recreation (Psa.104:30).

: - 1656 years to Noah's Flood - Gen.5:3-31;7:6;9:28,29;11:10,11.

:   2476 BC - Noah began to build the Ark for 120 year testimony.

: - - - -2356 B.C. - Noahic Flood (Most logical historical time).

:   "First Intermediate Period" - "History Almost Silent"

: -428 years to Abram's call (427 years plus 1 year for Flood).

:   2003 B.C. - The birth of Abraham.

: - - - - - 1928 B.C. - Call of Abraham (Left Haran - Gen.12:1-3):
                                                                    :
:   4 gener. - Abram, Isaac, Jacob & Levi-215 years.:
                                                                    :
Gal.3:17 - : - 430 years (Jacob to Egypt - Half of Egyptian Captivity):
1713 BC   :
:   4 gener. - Kohath, Amram, Moses & Gershon-215 yrs:
Ex.12:40,41;   Gen. 15:16   :
14:4;18:30; : - 1498 BC (**EXODUS**) Thutmose II dies, Hatshepsut reigns - :
15:19.   . :(Deut.2:14;Num.1:1;20:1 -2nd yr,1st mo.Kadesh-1496- :
                                                              BC):
    . :   "And it came to pass in the (Canaan - 1458 BC:
                476   480th year after the children Deut.2:7;8:2
480 years - - . - : -——- of Israel were come out of
                yrs   the land of Egypt, in the 4th
    . :   year of Solomon's reign over
Solomon's   Israel" I Kings 6:1
1st year - - -. - : - - 1022 B.C. - The first year of Solomon's reign.
& 4th yr - 1018 BC   :
                40   "And thou shall bear iniquity of
Evil - I K.11:6,9-12 -: -——- the house of Judah 40 days: I have
                yrs   appointed thee each day a year"
             :   Ezekiel 4:6
             - - 982 B.C. - Rehoboam & Jeroboam reigns.
             :   "For I have laid upon thee the years
                   of their iniquity, according to the
         390 yrs - : - number of the days, three hundred and
                   ninety so shalt thou bear the iniquity
             :   of the house of Israel" Ezekiel 4:5
Ezek. 1:1,2;4:5 - Fifth year Jehoiachin's "Captivity"*- July 5, 592/1 BC
             : - 592 yrs from 0 B.C. to the 5th year of
                   Jehoiachin's Captivity.
             : - - 0 B.C.

* The fifth year of Jehoiachin's **reign** would be 593/2 B.C.

12

The well known book, Halley's "Pocket Bible Handbook", agrees and confirms our theory when it states, "Since Ezekiel was as meticulous in dating his visions, even to the exact day, it is assumed that all that follows a given date belongs to that date, till the next date is mentioned".[1] This is clearly portrayed in our "Ezekiel Chart" soon to follow.

Another point which we must clarify is that when Ezekiel refers to the fifth year, he means that the fifth year has been completed, and the months following are included in the sixth year. This will be clearly illustrated later. We should also remember that it is the "fifth year" of his "Captivity" and not of his "reign" (His three month reign involved the last three months of the previous "Nisan" year).

Now, let us begin to correlate the events leading up to our foundation date. Please follow along in the corresponding years in our chart "Israel's and Judah's Chronology Correlated" starting with Nisan year 641/0 B.C. (page 45). That date was the final year of King Amon's reign over Judah, and the accession year of Josiah. Therefore, Josiah's first regnal year was 640/39 B.C.. His thirteenth year was 628/7 B.C., which was also the beginning of Jeremiah's prophesies (Jer. 25:1-3). In 627/6 B.C. Jeremiah completed his first year of prophecy, therefore, his 23rd year of prophecy would have been completed in year 605/4 B.C. (Jer. 25:3). This 23rd year (605 B.C.) is synchronized (Jer. 25:1) with the 4th year of Jehoiakim, which in turn, was also the first year of Nebuchadnezzar's reign (accession year - his 7th year was 598/7 B.C. - Jer. 52:28). "The British Museum's tablet number 21946 confirms that Nebuchadnezzar's first eleven years, including his accession year, dates from 605 to 594 B.C.".[2]

We also know that Josiah reigned 31 years (II Kings 22:1), and if he began his rule in 740 B.C., his reign would have ended sometime in 609 B.C., Jehoiakim's accession year (accession year is the year the previous king died).

Now, let us continue with the Biblical text of the inspired Word found in II Kings 23:29, "In his (Josiah's) days Pharaoh Necho II, King of Egypt, went up toward (not against Assyria) the King of Assyria to the river Euphrates (against Babylon): and Josiah went against him; and he slew him (Josiah) at Megiddo, when he had seen him".

Babylon was soon to be Judah's greatest enemy, yet Josiah was going out against Egypt who was opposing Babylon. What a dilemma? Josiah justly reaped the harvest. The following Biblical text in II Chron. 35:20-24 is confirmation: "After all this, when Josiah had prepared the temple (during revival - II Kings 22:3; II Chron. 34:14) Necho II King of Egypt came up to fight at (not against) Carchemish by Euphrates: and Josiah went out against him. But he (Necho II) sent ambassadors to him, saying; what have I to do with thee, thou King of Judah? I come not against thee this day, but against the house wherewith I have war (Babylonia, Josiah's enemy): for God commanded me to make haste (Egypt

1. Halley's. "Pocket Bible Handbook", 18th Edition (1943) p. 289
2. Thiele, E.R., "The Mysterious Numbers of the Hebrew Kings",1965,p.163
   Wm. B. Eerdmans Publishing Co.

13

was too late to help Assyria); forbear thee from meddling with God, who is with me, that he destroy thee not. Nevertheless, Josiah would not turn his face from him, but disguised himself, that he might fight with him, and hearkened not unto the words of Necho from the mouth of God, (Why? Egypt was God's instrument to help the Assyrians destroy Josiah's enemy, Babylon, at Carchemish, but because of Josiah's interference, Egypt arrived too late to help Assyria) - - - and came to fight in the valley of Megiddo. And the archers shot at King Josiah, and the king said to his servants, 'Have me away; for I am sore wounded' - - and he died - - ".

Additional historical sequence of this Egyptian attack involving Josiah is supplemented by a badly damaged fragment of cuneiform text in the British Museum, and translated by C.I. Gadd years after it was dug up in Mesopotamia. It read as follows: "In the month of Du'uz (Duzu or Tammuz - June or July 609 B.C.) the king of Assyria procured a large Egyptian army and marched against Harran to conquer it - - - till the month of Ululu (Elul - August or September) he fought against the city but accom-plished nothing".[1]

Carchemish was a place of decisive battles. The Bible text and general history verifies this especially by referring often to the phrase "the Battle of Carchemish". The Assyrians had lost their capital at Nineveh to Babylon in 612 B.C.. Then Haran fell to them in 610/9 B.C., there-fore, the Assyrian capital was set up at Carchemish which sets the scene of our historical events. Providentially, Pharaoh Necho II was sent by God to help Assyria fight and protect Carchemish against the Babylonian attack. Babylon would soon become Judah's greatest enemy, yet Josiah went out to interfere with the army of Necho II. Josiah rightly lost his life in delaying Necho II at the pass of Megiddo (II Chron. 35:20). Egypt was, no doubt, also delayed at Riblah (II Kings 23:33) and arrived at the city of Carchemish too late. Nebuchadnezzar had surprised the As-syrians and had captured the city of Carchemish. He in turn defeated Necho and his forces in a hand to hand battle and pursued them to Hamath. The story relates that "not a man escaped to his own country", which seems like a rather doubtful statement. As a result of this battle in 609 B.C. the Assyrian Empire was completely destroyed and Egypt sur-vived but became a secondary power. Babylon soon gained ascendancy throughout the fertile cresent and attacked Judah in 606 B.C..

After the death of Josiah in either June or July, 609 B.C., Jehoahaz was made king for three months (II Kings 23:31) during the very time Necho II was at Carchemish to oppose the Babylonian forces. Necho II, after his retreat, returned three months later and put Jehoahaz in prison in Riblah in the land of Hamath (II Kings 23:33,34). Jehoahaz later died in Egypt. No doubt Egypt was able to keep a military force that far north temporarily facing the Babylonian army to protect her yearly tribute from Judah (II Kings 23:33). Necho II then took Josiah's other son, Eliakim, and changed his name to Jehoiakim, and made him king over Judah in the month of September, 609 B.C. (II Kings 23:34; II Chron. 36:5). Jehoiakim reigned eleven years until December 6, 598/7 B.C. (Jehoi-

---

1. Keller, W., "The Bible as History", 1967, p. 275, Econ Verlagsgruppe, West Germany   14

achin, his son, followed reigning 3 months and ten days - II Chron. 36: 9,10 - to the end of the Nisan year, March 16, when he was taken "Captive" and to Babylon in the Nisan year April 22,597/6 B.C.).

In the second year (accession) of Jehoiakim, and during his first regnal year (608/7 B.C.) when his son, Jehoiachin, was eight years old (II Chron. 36:9), Jehoiakim made Jehoiachin co-regent with himself for the next ten years. Jehoiakim had made this decision since the Medes and the Neo-Babylonians under Nabopolasser had gained these new victories in the north. The Medes had annexed the north and the north east part of the territory, while Babylon kept the south and south-west. Thus, Nebuchadnezzar had been sent by the aging Nabopolasser to possess Syria and Palestine. Therefore, Jehoiakim saw the need of a co-regent in the face of these repeated ominous threats from the north. (This is the simple answer to the problem of two ages given to Jehoiachin when he began to reign - II Chron. 36:9; II Kings 24:8). Two years later in 606/5 B.C. Nebuchadnezzar began his attack on Judah, and according to Daniel 1:1; 2:1, Daniel and his friends with others, were either captured or surrendered immediately (The Jewish people were released from Babylon exactly 70 years later - 536/5 B.C. - Jer. 25:11; Dan. 9:2). His first attack was in the third year of Jehoiakim's reign (Dan. 1:1 - 606/5 "regnal year") when Daniel and his friends were put on a three year diet before they stood before the king (Dan. 1:5). They were called before the king in his, Nebuchadnezzar's, second regnal year (Dan. 2:1 - 603/2 B.C. (See Chart).

For further information concerning the prelude to this attack, Jeremiah 46:1-2 fills in the details: "The Word of the Lord which came to Jeremiah the prophet against the Gentiles; against Egypt; against the army of Pharaoh Necho II, King of Egypt, which was by the river Euphrates in Carchemish, which Nebuchadnezzar, King of Babylon, smote in the fourth year of Jehoiakim (Accession year would also be 606/5 B.C.) the son of Josiah King of Judah". Carchemish was mentioned in the attack by Necho II in 609 B.C. (II Chron. 35:20). Now, Necho II is back again facing Babylon at the same location in the year 606/5 B.C.. The Egyptians under Necho attacked Kimuhu and after a four month siege captured the city. No doubt, during this time, Carchemish was once again involved according to Jer. 46:1-2. Nabopolasser responded by establishing a garrison at Quramati on the east bank of the Euphrates. The Egyptians crossed the Euphrates and drove the Babylonians from Quramati.[1]

Because of Nabopolasser's age, his son, Nebuchadnezzar was sent to deal with this matter. Egypt was finally defeated at Carchemish, and the remnant was dispersed at Hamath (Jeremiah 46:2). Thus, "Nebuchadnezzar conquered the whole area of the Hatti-country" (Syria). This took place in the year 606/5 B.C. during the fourth year of Jehoiakim's reign (Accession year). This conquest would soon include all territory south to the border of Egypt (Jeremiah 46:14). As Nebuchadnezzar

1.Thiele, E.R."The Mysterious Numbers of the Hebrew Kings"1965, page 165
  Wm. B. Eerdmans, Publishing Co.

15

moved on south, the first group of captives, along with Daniel and his friends, were carried away to Babylon. Later, during the final clean up operation and during the attack on Jerusalem, Nebuchadnezzar was suddenly notified of the death of his father, King Nabopolassar (August 16, 605 B.C.). An interesting resume of the following events are found in Josephus, against Apion, i.19, 136-139; Antiquities X. 221-224 until Nebuchadnezzar took the throne twenty-one days later on September 7, 605/4 B.C..[1]

It was also in the fourth year of Jehoiakim that Jeremiah, while in prison (Jeremiah 36:1-5), composed a roll of a book containing God's words of warning to the nation. The following year in December (9th month - a fire was in the hearth - Jer. 36:22) of the fifth year (604/3 B.C.) of Jehoiakim's reign (Jeremiah 36:9) the roll was read before Jehoiakim, after which, he burnt the book. Jeremiah rewrote the roll and warned that "the King of Babylon shall certainly come and destroy this land, and shall cause to cease from thence man and beast. Therefore, thus saith the Lord of Jehoiakim, King of Judah; he shall have none to sit upon the throne of David (see curse on his son Jehoiachin - Jeremiah 22: 24-30): and his dead body shall be cast out in the day to the heat (see Jeremiah 22:17-19; 52:27,28), and in the night to the frost" (Jeremiah 36:29b,30). Therefore, as a forewarning of these promises, Nebuchadnezzar, after taking the throne, arrived again in the Hatti territory (Syria), and later received tribute from all the kings. This was in the year 604/3 B.C., which was the fifth year of Jehoiakim's reign. This was also the year, because of his sinfulness, Jehoiakim was put under tribute and "became a servant (to Nebuchadnezzar) three years (604/3 - 601/0 B.C.): then he turned and rebelled against him" (II Kings 24:1). The three years of servitude was brought to a sudden end for a short period in 601/0 B.C. when Babylon and Egypt met in an all-out battle. They smote each other and Nebuchadnezzar returned to his land and stayed there the following year (600/599 B.C.). Therefore, Jehoiakim continuously "rebelled against him", because of Nebuchadnezzar's defeat.

The following year (599/8 B.C.), which was the tenth year of Jehoiakim, and the sixth year of Nebuchadnezzar, "the Babylonian king once more campaigned in Hatti-land and plundered the desert Arabs".

The next year (598/7 B.C.), which was Nebuchadnezzar's seventh year, he gathered his forces to revenge the rebellion of Jehoiakim (see chart page 46). The Bible states this concerning the attack: "And the king of Babylon smote them, and put them to death in Riblah in the land of Hamath. Thus, Judah was carried away Captive (2nd group) out of their own land. This is the (number of) people whom Nebuchadnezzar carried away captive: in the seventh year (of Nebuchadnezzar) three thousand Jews and three and twenty" (Jeremiah 52:27,28 - 598/7 B.C.).

In 1955 D.J. Wiseman was deciphering old Babylonian cuneiform tablets which are now about 2575 years old. This is what he read confirming the

1. Thiele, E.R., "The Mysterious Numbers of the Hebrew Kings" 1965, page 165,167 , Eerdmans Publishing Co., Grand Rapids, Michigan

above Biblical text: "In the seventh year (of Nebuchadnezzar) in the month Chislev (Dec. 6, 598/7 B.C.), the King assembled his army and advanced on Hatti-land (Syria). He encamped over against the city of the Judeans (dethroned Jehoiachim) and conquered it (after Jehoiachin's 3 months and 10 days reign) on the 'second day' of Adar (March 16, 597 B.C. - New Nisan year 597/6 B.C.). He took the King (Jehoiachin) prisoner, and appointed in his stead a king after his own heart (Zedekiah). He exacted heavy tribute and had it brought to Babylon"[1] (Nisan year April 22, 597/6 B.C.).

Therefore, the "seventh year" of Nebuchadnezzar was also the "eleventh" and last year of Jehoiakim's rule. The month Chislev (December) was the final month of his reign. Nebuchadnezzar took him captive and allowed his co-regent, Jehoiachin, to continue to rule as an "eighteen year old" for only three months and ten days. The Bible text found in II Chronicles 36:5-7, 8b,9,10 continues with the story: "Jehoaikim was twenty and five years old when he began to reign, and he reigned eleven years in Jerusalem (Sept. 7, 609 to Dec 6, 598 B.C.): and he did that which was evil in the sight of the Lord his God (Destroyed God's Word). Against him came up Nebuchadnezzar King of Babylon, and bound him in fetters, to carry him to Babylon. Nebuchadnezzar also carried of the vessels of the house of the Lord to Babylon, and put them in his temple at Babylon. - - - : and Jehoiachin his son reigned in his stead. Jehoiachin was eight years old when he began to (co-) reign (II Chron. 36:9), reigned three months and ten days (when 18 years old - II Kings 24:8) in Jerusalem: and he did that which was evil in the sight of the Lord. And when the year (Nisan year 598/7 B.C.) was expired (on March 14th), King Nebuchadnezzar sent (2nd day of the new Nisan year - Mar. 16th - some of his forces) and brought him to Babylon (during the new Nisan year April 22, 597/6 B.C.) with the goodly vessels of the house of the Lord, and made Zedekiah, his brother, King over Judah and Jerusalem".

Thus, we reiterate, the attack against Judah by Babylon came on December 6th of the seventh (Nisan) year of Nebuchadnezzar (598/7 B.C.) dethroning Jehoiakim, and Jehoiachin, his son, was also carried away captive to Babylon after three months and ten days reign (II Chron. 36:9 - Mar. 16) during the beginning of Nebuchadnezzar's eighth (Nisan) year (see II Kings 24:12), whose "Captivity" took place during the "new" Nisan year on April 22, 597/6 B.C. (II Chron. 36:10), "when the (old) year was expired" (Old Nisan year 598/7 B.C.).

Also, when "the year was expired", the twenty-one year old Zedekiah was placed (Mar. 16) on the throne of Judah by the King of Babylon, and Zedekiah then ruled eleven years (597/6 - 586/5 B.C. - Jer. 52:1; 39:2; II Chron. 36:11; II Kings 24:17,18). These same eleven years extended from the "eighth" to the "nineteenth" year of Nebuchadnezzar. In his nineteenth year he destroyed Jerusalem (II Kings 25:8,9). This is a perfect correlation of God's Word with history.

It might be interesting to add that some early commentators state that

1. Keller, W., "The Bible as History", 1967, p. 280, Econ Verlagsgruppe West Germany

17

Nebuchadnezzar was in continuous siege against Tyre during this whole period from his seventh year until his twentieth year.[1] If so, the two final attacks on Jerusalem were divergent from the main battle at Tyre. This view is now doubtful. Certainly, if Nebuchadnezzar was not in continuous attack, he had isolated Tyre by by-passing her in his campaigns. Ezek. 26:1-14 warns Tyre in the eleventh year of "Jehoiachin's Captivity" (586/5 B.C.), which was the 19th year of Nebuchadnezzar, of the coming judgment which was partially fulfilled upon Tyre during the following years in 585 B.C. and in 572 B.C. (Nebuchadnezzar's 20th and 23rd years). It was completely fulfilled many years later by Alexander the Great when the ruins of the coastal city of Tyre was cast into the sea to make a causeway for Alexander's attack against the island stronghold. The former theory is now questioned, and the present date for the final attacks on Tyre are in years 585 and 572 B.C. (see chart).

Tyre's pride is rebuked by God through Ezekiel's prophecy in Ezekiel 26: 2-4: " - - - Because Tyrus hath said against Jerusalem; Aha, she is broken that was the gates of the people: she is turned unto me: I shall be replenished, now she is laid waste: Therefore, thus saith the Lord God; Behold, I am against thee, as the sea causeth his waves to come up. And they shall destroy the walls of Tyrus, and break down her towers (Nebuchadnezzar): I will also scrape her dust from her, and make her like the top of a rock" (the powerful forces of the great army of Alexander the Great fulfilled this prophecy many years later).

Zedekiah became King over Judah in the first month (March 16) of the Nisan year 597/6 B.C., while Jehoiachin was taken into "Captivity" to Babylon in the second month, April 22, 597/6 B.C.. Therefore, the first regnal year of Zedekiah was completed in the same month of the Nisan year 596/5 B.C., while Jehoiachin's first year of "Captivity" ended practically at the same time one month later. Therefore, the eleventh year of Zedekiah's "reign" (II Kings 25:2,8), and Jehoiachin's eleventh year of "Captivity" (Ezekiel 26:1,2) fell in the same year in which Jerusalem was destroyed (586/5 B.C.). Thus, Ezekiel, the prophet, introduces the phrase the year of "Jehoiachin's Captivity" to date the prophecies in his book. This is where many chronologists make a mistake by dating the first year of "Jehoiachin's Captivity" one year too early (not his reign). Therefore, Jehoiachin's "first" year of "Captivity" was completed in March/April in the second month of the new Nisan year 596/5 B.C.. His second year was completed in 595/4 B.C., which, in turn, was the year that Psammetichus II, Pharaoh of Egypt, ascended the throne following Necho II (610 - 595 B.C.). Jehoiachin's third year of "Captivity" (594/3 B.C.), no doubt, included many of Jeremiah's prophecies against all the countries of the middle east (Jer. 46:13 to 51:59) including the destroyer (Babylon).

In the 4th year of Zedekiah (593/2 B.C. - Jer. 51:59), which was also

1. Intern. Standard Bible Encyclopaedia, 1949, Vol. 4, p. 2128

the <u>fourth</u> year of "Jehoiachin's Captivity", Jeremiah sends a warning to Babylon by the hand of the Prince Seraiah who accompanied Zedekiah, the King of Judah, to Babylon. This is the year that Zedekiah, no doubt, made his covenant with Nebuchadnezzar (Ezekiel 17:13 - <u>4th year</u> - Jer. 51:59). Zedekiah sealed the destiny of Jerusalem by this sinful decision. Therefore, Ezekiel, a "Captive" in Babylon, began his prophetic ministry the following year (Ezekiel 1:2) by warning Judah during the <u>last seven years</u> before the destruction of Jerusalem (592/1 - 586/5 B.C. inclusive). Ezekiel begins his book "Ezekiel" by dating his first prophecy during the year of our basic "Biblical Chronological Foundation Date" in the <u>fifth year of "Jehoiachin's Captivity"</u> (Ezekiel 1:2;4:4,5,6 - Nisan year 592/1 B.C.).

A rather interesting archaeological confirmation of this very important date in our chronological plan comes to us through the Assyriologist, E.F. Weidner in the year 1933 while translating writings on tablets and sherds in the basement of the Kaiser-Friedrick Museum. He found four different receipts for supplies issued to the recipient "Ja'-u-kinu (Jehoiachin) King of the (land of) Judah". "The Babylonian clay receipts moreover bear the date of the <u>thirteenth year</u> of the reign of King Nebuchadnezzar". That means year 592/1 B.C. (The very year Ezekiel began his prophecy) fell <u>five years</u> (fifth year of "Jehoiachin's Captivity") after the surrender of Jerusalem and the second deportation (April 22, 597/6 B.C.). "In addition the Babylonian steward of the commisarist has mentioned in three cases five of the King's sons, who were in charge of a servant with the Jewish name of 'Kenaiah'".[1]

Therefore, our basic "Chronological Foundation Date" 592/1 B.C. (<u>the 5th year of "Jehoiachin's Captivity" to Babylon</u>) correlates perfectly with the <u>13th regnal</u> year of Nebuchadnezzar (see chart on page 46). The provisions of Jehoiachin and his party wonderfully verifies the veracity of God's Holy Word: "And for his diet, there was a continual diet given him (Jehoiachin) of the king of Babylon, every day a portion until the day of his death, all the days of his life" Jeremiah 52:34.

We can now clearly portray and verify the dates in chart form by the means of the fir.t eight years of Ezekiel's prophecy following (Page 20).

During these same "seven years" before the Babylonian Captivity (592/1 B.C. - 586/5 B.C. inclusive), Ezekiel describes the Glory of God in symbolic language which compares with the words used by John in his book of Revelation. John is also portraying and speaking concerning the last seven years before the close of this present age which ends with the Tribulation Period (Jacob's Troubles - Jeremiah 30:7), and the Battle of Armageddon (Rev. 16:16). The Word of God states that "half of the city (Jerusalem) shall go forth into captivity, and the residue of the people shall not be cut off from the city" Zech. 14:2b. The similarity as to the number of years concerning which both Ezekiel and John prophecied and the common terms which both used should have future historical significance. Listed following our Ezekiel Chart, these following comparisons should make an interesting study for some at this age end.

1. Keller, W., "The Bible as History", page 279, **Econ Verlagsgruppe,West Germany**   19

"EZEKIEL'S CHART"

```
                                              982 BC-Division of the
Ezekiel's prophecy (year)                   :   Kingdom
   :     Accession of Zedekiah."Jehoiachin's Captivity"April 22,597/6BC
   :     1st yr "Jehoiachin's Captivity" & Zedekiah   :     596/5BC
   :     2nd yr "Jehoiachin's Captivity"              :     595/4BC
   :     3rd yr "Jehoiachin's Captivity"              :     594/3BC
   :     4th yr "Jehoiachin's Captivity"              :     593/2BC
1st yr. - 5th yr & 1st yr prophecy  Ez.1:2-4th mo(July):5th day  592/1BC
   :        Ezekiel's vision dated - Ez.1:2- - -:     :
   :        5th yr                   Ez.2           :-          :
   :        5th yr                   Ez.3           :- 390 yrs spans Israel's
   :        Ezekiel vision's length- Ez.4:5- - -:                  sins
   :        5th yr                   Ez.5              (Nisan year)
   :        5th yr                   Ez.6
   :        5th yr                   Ez.7
2nd yr. - 6th yr & 2nd yr prophecy  Ez.8:2 6th mo.(Sept) 5th day 591/0BC
   :        6th                      Ezekiel chapters 8 to 19
3rd yr. - 7th yr & 3rd yr prophecy  Ez.20:1 5th mo.(Aug)10th day 590/9BC
   :        7th yr                   Ezekiel chapters 20 to 23
4th yr. - 8th yr & 4th(No prophecy)                             589/8BC
5th yr. - 9th yr & 5th yr prophecy  Ez.24:1 10th mo.(Jan) 10th   588/73C
   :        9th yr                   Ezekiel chapters 24 to 28
6th yr. -10th yr & 6th yr prophecy  Ez.29:1 10th mo(Jan)12th day 587/6BC
7th yr. -11th yr & 7th (To Egypt)   Ez.30:20 1st mo.(Apr) 7th  Jerusalem
Fall    -11th yr (Against Egypt)    Ez.31:1 3rd mo.(Jun)1st  fell 586/5BC
Fall    -11th(Jer.29:2;52:5-12)TyrusEz.26:1(4,5th Jul-Aug)10th IIK25:1-8
8th yr. -12th yr "Report of fall"   Ez.33:21 10th mo(Jun)5th day 585 B C
         12th yr  Against Egypt      Ez.32:1 12th mo(Aug)1st day Tishri-
         12th yr  Against Egypt      Ez.32:17 12th Mo. (Aug) 15th day
```

TERMS IN "EZEKIEL" & "REVELATION" COMPARED

(In Revelation the "Glory of God" also reveals the Church Age)

EZEKIEL (7 YRS) (Judgment - Captivity)	"THE GLORY OF GOD"	REVELATION (7 YRS) (Judgment- purification)	
1. Whirlwind (Cloud)	Ez.1:4	"In the Spirit"	Rev. 4:2
2. Living creatures	Ez.1:5	Four beasts	Rev. 4:6
3. Man like	Ez.1:5,26	Man like Rev.4:10, 11;1:13	
4. Faces	Ez.1:6,10	Faces	Rev. 4:7
5. 4 wings (Cherubims)	Ez.1:6;10:21	6 wings (Seraphim)	Rev. 4:8
(Judgment)	**Ez.10:2**	(Purified)	Isa. 6:2
6. Man,Lion,Ox,Eagle	Ez.1:10	Man,Lion,Ox,Eagle	Rev. 4:7
7. Lamp	Ez.1:13	Lamp	Rev. 4:5
8. Lightning	Ez.1:13	Lightning	Rev. 4:5
9. Rainbow	Ez.1:28	Rainbow	Rev. 4:3
10. Eyes	Ez.1:18	Eyes	Rev. 4:6
11. Firmanent (Crystal)	Ez.1:22	Sea (Crystal)	Rev. 4:6
12. Throne	Ez.1:26	Throne Rev.4:10;Rev. 5:6	

1. Jerusalem fell Aug.10 or 24 just before the Tishri year therefore the "report" of the fall (not prophecy) dates from Sept. in year twelve.

Now, that we have harmonized the "first year" of Ezekiel's prophecy with the "fifth year of Jehoiachin's Captivity", and confirmed that both events fell exactly on our "Foundation Date", 592/1 B.C., in our Biblical Chronology, let us consider in detail the remaining years before the destruction of Jerusalem.

The next year, which was the sixth year of Zedekiah (591/0 B.C. - Ezekiel's second year of prophecy), was also the tenth year since Nebuchadnezzar battled with Egypt (601/0 B.C.). Nebuchadnezzar discovers that Zedekiah has broken his covenant (the sixth year of "Jehoiachin's Captivity" - Ezekiel 8:1;17:13,15). Therefore, "he (Zedekiah) rebelled against him (Nebuchadnezzar) by sending his ambassadors into Egypt, that they might give him horses and much people" Ezekiel 17:15. This sixth year of Zedekiah (591/0 B.C.) was the fifth year of Psammetichus II (596 - 589 B.C.). Nebuchadnezzar (14/15 year - see chart page 47) had sent forces south to Jerusalem while continuing the isolation of Tyrus, but Psammetichus II responds to Zedekiah's request and moves north into Palestine causing Nebuchadnezzar to withdraw (Ezekiel 17:17). For those who question Psammetichus' involvement, we quote the following statement: "An equally problematic event of Psammetichus II's reign, is an expedition to Phoenicia mentioned in a late demotic papyrus".[1] An attack at this time against Phoenicia was in opposition to Nebuchadnezzar's isolation of Tyrus, and also gave help to Judah at Jerusalem. These army maneuvers encouraged Zedekiah's rebellion. Jeremiah, the prophet, continues the story: "Now Jeremiah came in and went out among the people: for they had not put him into prison. Then Pharaoh's (Psammetichus II) army was come forth out of Egypt (591/0 B.C.): and when the Chaldeans that besieged Jerusalem heard tidings of them, they departed from Jerusalem. - - - And it came to pass that when the army of the Chaldeans was broken up from Jerusalem for fear of Pharaoh's army, then Jeremiah went forth out of Jerusalem to go into the land of Benjamin, to separate himself thence in the midst of the people" Jer.37:4,5,11,12.

As a result of Jeremiah's action, his continual prophecy that they should surrender to Babylon, and that Nebuchadnezzar would return, Jeremiah was put in a dungeon for "many days" Jer. 37:16. In the following seventh year of "Jehoiachin's Captivity" (590/9 B.C.), Ezekiel prophesied the message (his 3rd year of prophecy) found in the 20th to 23rd chapters of his book. He rebukes them for rebellion (Ezek. 20:8,13,21) and warns the elders that he will not be inquired of by them (Ezekiel 20:31). He tells them that they will be put under the rod (Ezekiel 20:27) until they loathe themselves (Ezekiel 20:43). Ezekiel admonishes them that even though Nebuchadnezzar has withdrawn, he stands at Rabbath and is now making a choice of two ways to attack through information given by divination, by consulting images, and looking in the liver (Ezekiel 21:19-21). He tells them that they have become as dross and like brass, tin, iron, and lead in the midst of a furnace (Ezekiel 22:18). He states, "And I sought for a man among them, that should make up the hedge, and stand in the gap before me for the land,

1. Gardner, Sir Alan, "Egypt of the Pharaohs", page 360, 1961;Publisher: Oxford University Press.

that I should not destroy it: but I found none" Ezekiel 22:30. As though to emphasize his point he lapses into silence during his fourth year of prophecy. He finds no one to help and he prophesied nothing during the eighth year of "Jehoiachin's Captivity" (589/8 B.C.). Jeremiah is in the dungeon so there is no voice heard for God until the ninth year of "Jehoiachin's Captivity" (Ezekiel 24) when Ezekiel prophesies during his fifth year.

During the silent year in Ezekiel's prophecies (8th year of "Jehoiachin-'s Captivity" - 589/8 B.C.), Pharaoh Hophra takes the throne in Egypt. The following year (588/7 B.C., Jan. 10th), Egypt threatens Nebuchadnezzar, as Hophra moves towards Jerusalem with a large armed force, but Hophra withdraws in face of the overwhelming odds. Nebuchadnezzar's final attack against Jerusalem begins (Nisan year 588/7 B.C., Jan. 10th or 24th) which brings Jerusalem's final doom three years later in 586/5 B.C. (Inclusive - see Jer. 39:1,2; 52:4,5,12; Ezek. 24:1,2; 26:1,2). II Kings 25:1-4a,8-9 gives a sad resume of the events: "And it came to pass in the ninth year of his (Zedekiah's) reign (Nisan year 588/7 B.C.) in the tenth month in the tenth day of the month (Jan. 10, or 24, 587 B.C.[1]), that Nebuchadnezzar King of Babylon came, he, and all his host, against Jerusalem, and pitched against it; and they built forts against it round about. And the city was besieged unto the eleventh year of King Zedekiah (586/5 B.C.). And on the ninth day of the fourth month (July 9, or 23, 586/5 B.C.) the famine prevailed in the city, and there was no bread for the people of the land. And the city was broken up - - And in the fifth month (August), on the seventh day of the month (Aug. 7 or 21, 586/5 B.C.), which is the 'nineteenth year' of King Nebuchadnezzar of Babylon (Therefore his 18th year would have been Zedekiah's 10th - Jer. 32:1), came Nebuzuradan, captain of the guard, a servant of the King of Babylon, unto Jerusalem: and burnt the house of the Lord, and the kings house and all the houses of Jerusalem, and every great man's house burnt he with fire".

In the tenth year of Zedekiah (587/6 B.C.), during the second year of the siege of Jerusalem, Jeremiah was under guard in the "Court of the prison" (Jer. 38:13,28) after a time in the dungeon (Jer. 38:6-11). This event in Zedekiah's tenth year correlates perfectly with our chronology by synchronizing exactly with the eighteenth year of Nebuchadnezzar (Jer. 32:1,2 - see Chart - Therefore the first attack against Jerusalem was in Nebuchadnezzar's seventeenth year - Nisan year 588/7 B.C., Jan. 24th).

Gedaliah was selected by Nebuchadnezzar as governor of the land of Israel, but was killed by Ishmael, the Ammonite, on the first day of the seventh month (October 1, or 14 - II Kings 25:22-26; Jer. 41:1-18). Ten days later (Jer. 42:7 - 43:7, Oct. 10, or 24, 586 B.C.), after rejecting Jeremiah's message to stay in the land, the remnant of the Jewish people went into the land of Egypt to the city of Tahpanhes (Jer. 43:8) where

1. The 10th of a Nisan year would be the 24th (1st day of a Nisan year is the 14th - Passover)

Jeremiah prophesied the coming destruction of Egypt (Jer. 43:8-13).

It should be repeated here that the eleventh year of King Zedekiah (597/6 - 586/5 B.C.) was also the eleventh year of the "Captivity of Jehoiachin" (Not the eleventh year since Jehoiachin began to reign - 598/7 B.C.). Ezekiel 26:1,2 clearly confirms this as follows: "And it came to pass in the eleventh year (the eleventh year of "Jehoiachin's Captivity") in the first day of the month (the first day of the new Nisan year) the gates of Jerusalem were broken (March 1 or 14, 586/5 B.C.) that the word of the Lord came unto me saying, Son of man, because that Tyrus hath said against Jerusalem, Aha, she is broken that was the gates of the people (exactly eleven years since Zedekiah began his reign - Mar. 16 - II Kings 25:2; Ezek. 26:1,2 [1]): she is turned unto me: I shall be replenished, now she is laid waste" (This statement identifies the exact day that Nebuchadnezzar broke the gates of Jerusalem introducing the last phase of the attack - Nisan year March 14, 586/5 B.C. - the first day of the new Nisan year).

We will consider the information involving the "Sabbatic Year" later, but we can say that each of the first three of Nebuchadnezzar's attacks against Jerusalem all fell on a "Sabbatic Year". This was shrewd strategy on the part of King Nebuchadnezzar. The three attacks took place on the three consecutive "Sabbatic Years" as follows: 605/4 B.C.; 598/7 B.C.; 591/0 B.C.. The latter attack is when Nebuchadnezzar withdrew in the face of the Egyptian attack (see chart page 47). He did not wait for the fourth Sabbatic Year for his final attack on Jerusalem in 586/5 B.C., as in the situation when Zedekiah, the King of Judah, had rebelled against him causing his retreat before the Egyptians (see Ezek. 17:15).

One hypothesis we must clarify. If we pin-point the first attack on Jerusalem by Nebuchadnezzar in January, 588 B.C. (Nisan year 589/8 B.C.), as some historians have suggested, applying the universally accepted Nisan year, we would then be forced to locate the first attack in the "fourth" Nisan year before the final fall of Jerusalem (the 8th Nisan year of Zedekiah's reign - 589/8 B.C.). This is absolutely impossible. The attack continued only during the 9th (588/7 B.C.), 10th (587/6 B.C.), and the 11th (586/5 B.C.) Nisan years of Zedekiah's reign. In the 10th Nisan month (Jan., 587 B.C.) of the "ninth" Nisan year (588/7 B.C.), Nebuchadnezzar first attacked Jerusalem, and in the 4th and 5th Nisan year months (II Kings 25:1-10) of tne "eleventh" Nisan year (July and August, 586/5 B.C.) Jerusalem fell:

Nisan	Nisan	Nisan		
: 9th yr	: 10th yr	: 11th yr	:	
: 588 BC	: 587 BC	: 586 BC	: 585 BC	:
:Julian C.	:Jan	: Jul-Aug.:		:
: :	:/ -:	: :/	: :	:

(588/7 BC: 587/6 BC : 586/5 BC)
Nisan yr - 10 mo:9th 4th,5th mo: 11th - Nisan year
First attack: Final attack: 1. When no day mentioned - is 14th.

(Confirmation see II Kings 25:1-9;Jer.39:1,2;52:4-12;Ezek.24:1,2;26:1,2).

The wonderful part about the above correlation of events, they continue to synchronize with the exact details given: Ezekiel 40:1 and II Kings 25:27 (see chart page 47). "In the five and twentieth year of our Captivity (Jehoiachin's), in the beginning of the year, in the tenth day (24th) of the month, in the fourteenth year after that the city was smitten, in the self same day the hand of the Lord was upon me, and brought me thither" Ezekiel 40:1. "And it came to pass in the seven and thirtieth year of the "Captivity of Jehoiachin", King of Judah, in the twelfth month, on the seven and twentieth day of the month, that Evil-Merodach (Amel Marduk) King of Babylon in the year that he began to reign did lift up the head of Jehoiachin King of Judah out of prison" II Kings 25:27.

Since Jer. 52:5,12 confirms that the Nisan year 586/5 B.C. was the eleventh year of Zedekiah and the 19th of Nebuchadnezzar, we also know that the eleventh year of Zedekiah was the eleventh year of "Jehoiachin-'s Captivity". Jerusalem fell that year. So the first complete year since "Jerusalem was smitten" was year 585/4 B.C. (Ezekiel 40:1). Therefore, her fourteenth year since "Jerusalem was smitten" fell on the year 572/1 B.C., on the 33rd year of Nebuchadnezzar (see chart - 33/34 year), and also on the 25th year of "Jehoiachin's Captivity" (25/26 year - Ezekiel 40:1 - see chart page 47).

Thus, the year Nebuchadnezzar died (43/44 year) was 562/1 B.C.. This was also the Jewish people's "Jubilee" year, and truly introduced the "year of release" for the captive Jehoiachin. If on Oct. 8, 562 B.C. Nebuchadnezzar died making year 562/1 B.C. the accession year (Tishri yr - Sept.) of Amel-Marduk , therefore, his first regnal year was 561/0 B.C., during which according to II Kings 25:27, in the twelfth month, on the twenty-seventh day, Amel-Marduk released Jehoiachin from prison on Aug. 27,560 B.C.. This meant that Jehoiachin was freed in August only a few days before the new Tishri year 560/59 B.C.. Therefore, the first month (Sept.) of 560/59 B.C. was the beginning of his first year of complete release, and was also the 37th regnal year of "Jehoiachin's Captivity" (II Kings 25:27 - see chart).

As for the three references in the twelveth year of "Jehoiachin's Captivity" (Ezekiel 32:1; 32:17; 33:21), we purpose that these "three" references must be dated according to the "Tishri Year" system 586/5 B.C., and starting in the month of "September" immediately following the fall of Jerusalem. Otherwise it postpones the following scriptural event too long. This would actually postpone the final "report" of the fall of Jerusalem (Ezekiel 33:21) in reaching Babylon to a year and five months later. This does seem a extremely long time. Therefore, the only answer might be that since the final fall of Jerusalem and their complete "Captivity" took place in August just before the beginning of the new "Tishri Year" (Sept. 586 B.C.), Ezekiel may have dated the "report" (not prophecy) of "our (complete) captivity" from that point (Ezek. 33:21). Therefore, only the 10th and 12th months of the 12th year in the above "three references" would date from the final fall of Jerusalem (Tishri Year). The two months would fall, respectively, in June and August of the year 585 B.C. rather than January and March 584 B.C.. This might

24

be a possible solution, unless Nebuchadnezzar's battle with Tyrus in his twentieth year postponed all moves or messages back to Babylon for some time. This would seem illogical. In other words, "Jehoiachin's Captivity" dates from the first or second month of the "Nisan year" (Religious year), while the "Jerusalem Captivity" dates from the first month of the "Tishri year" (Civil year).

Since the coastal city of Tyrus did not fall until 571 B.C., the Tyrus enclave must have been like a thorn in the side of Nebuchadnezzar from his first to his 34th year.[1] Even then the "island" city stronghold continued to thrive until the time of Alexander the Great.

1. Saggs, H.W.F., "Greatness that was Babylon", page 143, Sidgwick and Jackson Publishers

Now that we have located and dated the "Foundation Stone" of our Chronological Blue Print as Nisan year July 5 or 19, 592/1 B.C., we can now consider the second chronological yard-stick and the two references which point back to the beginning of Israel's sin. Some have suggested that this scripture has a prophetic reference to Israel's future sin. Yet, we see no indication for such an interpretation. The quotation spans perfectly the time, in the past, since Israel had become a real entity as follows: "Lie thou (Ezekiel) also upon thy left side, and lay the iniquity of the house of Israel upon it: according to the number of the days that thou shalt lie upon it thou shalt bear their iniquity. For I have laid upon thee THE YEARS of their (past) iniquity, according to the number of the days, THREE HUNDRED AND NINETY DAYS: so shalt thou bear the iniquity (the past iniquity) of the house of Israel" (390 years spans the years of Israel's sin) Ezekiel 4:4,5- chart page 20.

The above prophecy of Ezekiel was recorded in the "fifth" year of King "Jehoiachin's Captivity" (Ezekiel 1:2), or in the very year of our "Foundation Date" of our chronology July 5, or 19, 592/1 B.C.. Therefore, we add 390 years to the year of Ezekiel's prophecy (592/1 B.C.), and we arrive at the exact date of the beginning of Israel's sin (982 B.C.).This was also the very year the "Kingdom was divided" between Judah and Israel in the year 982 B.C. (I Kings 11:31-38). Then Ezekiel continues his prophecy and extends the 390 years another 40 years to pin-point the beginning and added total of Judah's sins (1022 B.C.). This dates the end of David's righteous reign (I Kings 14:8) but spans Solomon's sinful (I Kings 11:9) 40 year reign (1022 - 982 B.C.) as follows: "And when thou (Ezekiel) hast accomplished them (390 days representing years), lie again on thy right side, and thou shalt bear the iniquity of the HOUSE OF JUDAH FORTY DAYS: I have appointed thee EACH DAY FOR A YEAR" (40 years added to the 390 years designated the beginning of Judah's sin - which includes Solomon's reign) Ezekiel 4:6. David followed after God, during his 40 year reign, with all his heart except in the matter of Uriah the Hittite, therefore, David's reign was a pinnacle of righteousness (Typified the Messiah) falling "exactly central" between the two captivities (Egyptian and Babylonian - see chart page 27).

Some may glance at the following "Chart of the Kings" and immediately see that certain dates disagree with historical known dates by one or two years. This is true. For example, we know that Jehoiakim began to rule in 609 B.C. rather than 610 B.C.. Therefore, in correlating the dates of the period of the kings, dates will vary by one or two years when comparing the two columns in the following chart (corrected in the far left column). The chronological structure, according to Ezekiel's prophecy, proves that the two extra 9 months periods, which are added by

"CHART OF THE KINGS"

(Note: Dates below are inaccurate in some places by one or two years. This is corrected as the extra months allowed for Jehoahaz and Jehoiachin are prorated over the whole 390 years in the chart, "Israel's and Judah's Chronology Correlated", pages, 35 - 47. Left column corrected below.*

B.C. 1102 Saul begins his 40 year reign

"Turned back from following me" I Samuel 15:11

1063 David's accession - Jerusalem falls 477 years later (Josephus[1]) - - - - - - - :

1062 David began his 40 year reign - <u>436 yrs to EgyptianCaptivity - 1498 BC</u> :

     I Kings 2:11   :

1042 "Followed me with all his heart" - <u>Central year between captivities</u> (456 yrs) :

     I Kings 14:8   :

1022 Solomon's reign 40 years - - - - - - <u>436 yrs to Babylon Captivity - 586 BC</u> :

1018 Temple - 480 years to Egypt Captivity   :40 years - see Ezekiel 4:6 '

Corrected "Lord angry with Solomon" II Kings 11:9   :

982 - 982 Rehoboam	17	I Kings 14:21;II Chr.12:30: plus - - - - - - - -982 :	:
965 - 965 Abijam	3	I Kings 15: 2;II Chr.13: 2	B.C. :	:
962*- 962 Asa (9 mos - 42)	41	I Kings 15:10;II Chr.16:13		:
920 - 921 Jehoshaphat	25	I Kings 22:42;II Chr.20:31	:	:
895 - 896 Jehoram	8	II Kings 8:17;II Chr.21: 5	equals :	
887 - 888 Ahaziah	1	II Kings 8:26;II Chr.22: 2	:	:
886 - 887 Athaliah	6	II Kings 11: 3;II Chr.22:12	"For I have laid	:477
880 - 881 Joash	40	II Kings 12: 1;II Chr.24: 1	upon thee the YEARS	:yrs
840 - 841 Amaziah	29	II Kings 14: 2;II Chr.25: 1	of their iniquity,	: :
811 - 812 Uzziah	52	II Kings 15: 2;II Chr.26: 3	according to the	: :
759 - 760 Jotham	16	II Kings 15:33;II Chr.27: 8	number of the days,	: :
743 - 744 Ahaz (9 months	16	II Kings 16: 2;II Chr.28: 1	THREE HUNDRED NINETY-390:	
727*- 728 Hezekiah 30)	29	II Kings 18: 2;II Chr.29: 1	days: so shalt thou	: :
697 - 699 Manasseh	55	II Kings 21: 1;II Chr.33: 1	bear the iniquity of :	:
642 - 644 Amon	2	II Kings 21:19;II Chr.33:21	the house of ISRAEL" :	:
640 - 642 Josiah	31	II Kings 22: 1;II Chr.34: 1	Ezekiel 4:5	: :
609 - 611 Jehoahaz (3 mos) * 1		II Kings 23:31;II Chr.36: 2		: :
609 - 610 Jehoiakim	11	II Kings 23:36;II Chr.36: 5	plus :	
598/7 599 Jehoiachin(3 mos)* 1		II Kings 24: 8;II Chr.36:9,10		:
: 598 Captivity	1	Babylon attack Dec.6,598; wins March 16, 597 B.C.		:
: 597 To Babylon	1	Jehoiachin taken; Zedekiah Acc.; Apr. 22,597 B.C.		:
6 596 Zedekiah's	1st year	; Jehoiachin's 1st year captivity		:
yrs 595 Zedekiah's	1 - 2nd yr	; Jehoiachin's 2nd year captivity		:
: 594 Zedekiah's	1 - 3rd yr	; Jehoiachin's 3rd year captivity		:
: 593 Zedekiah's	1 - 4th yr	; Jehoiachin's 4th year captivity		: :

592/1 B.C. TOTAL 390 years and also Jehoiachin's 5th year - Ezek.1:2;4:5 - 592 BC	:
591 B.C.	6th year - Ezek. 8:2	:
590 B.C.	7th year - Ezek.20:1	:
589 B.C.	8th year - No prophecy	:
588 B.C.	9th year - Ezek.24:1	:
587 B.C.	10th year - Ezek.29:1	:
586/5 B.C. Jerusalem falls II Kings 25:2-8	11th year - Ezek.26:1; 586/5 B.C. -:
	Jer. 29:2; 52:5-12	:

* 9 months added for 3 months reign. Extra months in Asa's and Hezekiah's reign. This puts both reigns over into the next year for the one year allowed for Jehoahaz & Jehoiachin.

1. Josephus Dissertation V, p. 955, Loeb Classic Library, Harvard University Press

allowing, historically, a full year for both "Jehoahaz" and "Jehoiachin" (each ruled three months), must be correlated into the chronology at two other points, rather than where they are in the right column. For example, we add the one extra 9 months period during the reign of Asa. This extends the reign of Asa over into the next year, thus adding one full year to the chronological plan, rather than just 3 months. This permits Asa's 15th year to fall rightly and exactly "36 years" after the division of the"Kingdom" fulfilling the scripture perfectly according to the Hebrew translation of II Chronicles 16:1, "In the six and thirtieth year of the"Kingdom"(Malkut - not "reign") of Asa, Baasha king of Israel came up against Judah - -" (see chart p. 36 - year 947 B.C. inclusive). We add 36 years to 947 inclusive and we have the year 982 B.C., which is perfect evidence of the inspiration of scripture.

Again, we needed the other 9 month period during the reign of Hezekiah which allows his reign to begin in the historically known date of 727 B.C.. This allows Hezekiah's 14th year (II Kings 18:13) to fall rightly on a Sabbatical year, 713 B.C., just before a Jubilee year, 712 B.C., according to Isaiah's prophecy (Isa. 37:30), and this absolutely proves that the extra 9 months are needed during Hezekiah's reign. Therefore, Manasseh's reign rightly begins in 697 B.C. rather than in 698 B.C. (Another evidence of the inspiration of the scriptures). As we stated above, "For the two three months reigns of Jehoahaz and Jehoiachin, we add two nine months periods which are prorated (in two places) over the 390 year period" (corrected in left column in the chart of the Judean Kings page 27).

The following reasons are listed as proof that only a long chronology of the kings, as we have set forth, fits perfectly and dramatically in our historical and chronological plan:

1. The Biblical total of the sum that the Hebrew kings reigned demands a long chronology (Judean Kings).

2. Ezekiel's 390 year prophecy demands a long chronology.

3. Josephus' 477 years measurement from David's accession year to the Babylonian Captivity correlates only with a long chronology (see chart page 27).

4. The 500th year anniversary of the temple of Bast celebrated in the 21st year Zerah (Osorkon I) extends back to the fourth year of Amenophis II (1445 - 1421 B.C.) and synchronizes only with a long chronology of the kings (1441 - 941 B.C. - see chart of the Egyptian Kings at the back of the book).

5. The short chronological measurement of 910 years from the second "Call of Abraham" (1928 B.C.) to the fourth year of Solomon (1018 B.C.) correlates with a long chronology of the kings.

6. Only an early date for Shalmaneser III (904 - 868 B.C.) and a long chronology of the kings allows Shalmaneser III to be contemporary with King Badezarus (Baalazar II - 909 - 903 B.C.) of Tyre.

7. The long chronology of the kings correlates Ashurdan's 8th year with the eclipse which is known, astronomically, to have appeared on June 13, 809 B.C., rather than in 763 B.C..

8. The long chronology of the kings permits David's 40 years to fall "central" between the two captivities (see "Chart of the Kings"). As a result we have these following computations:

   a. It was exactly 396 years from the beginning of Saul's reign (1102 B.C.) back to the end of the Egyptian Captivity (1498 B.C.).

   b. It was exactly 396 years from the end of Solomon's reign (982 B.C) forward to the beginning of the Babylonian Captivity (586 B.C.).

   c. It was exactly 436 years from the beginning of David's reign (1062 B.C.) back to the end of the Egyptian Captivity (1498 B.C.).

   d. It was also exactly 436 years from the end of David's reign (1022 B.C.) forward to the beginning of the Babylonian Captivity (586 B.C.).

   e. Finally, it was exactly 456 years from the "central" year of David's reign (1042 B.C.) back and forward to both Captivities (David's life and reign typifies the pinnacle and future glory of Christ Jesus the Messiah's reign).

It is generally accepted that the Egyptian King Siamun obtained the throne around 1000 B.C. and ruled seventeen years. Therefore, Shishak (Sheshank I) reigned from around 984/3 B.C.. Sheshank I's accession was only a year or two before the division of the Kingdom and just before the death of Solomon. Therefore, during the last few years of Solomon's reign (1022 - 982 B.C.), Jeroboam fled to Shishak I (in Egypt - I Kings 11:40) for fear of the king (Solomon) after Ahijah's prophecy concerning the division of the kingdom. The division came when Rehoboam (Judah) and Jeroboam (ten northern tribes) began their separate reigns in 982 B.C., thus, the correlation of these facts demands a "Long Chronology" of the kings (I Kings 11:31-38).

CHAPTER III
TWO KINGDOMS CORRELATED
(Please follow carefully on charts pages 35 - 47)

In the fifth year of Rehoboam (II Chron. 12:2; I Kings 14:25) Shishak I
threatened Jerusalem; therefore, as tribute, he carried away much gold
and treasure of the house of the Lord and of the house of King Solomon
(II Chron. 12:9 - 978/7 B.C.). Rehoboam ruled seventeen years (I Kings
14:21), and Abijam(h) succeeded him on the throne of Judah, in Jeroboam's
eighteenth year (I Kings 15:1,2; II Chron. 13:1,2), reigning three years.
During the early years of the division there was continual war between
the two kingdoms (II Chron. 13:3-22; I Kings 15:6). The following Judean
King, Asa, ascended the throne during their Jubilee year (962 B.C.). He
was their first "good" king since David and thereby established
peace for the "first ten years" of his reign (962 - 952 B.C. - II Chron.
14:1-6; I Kings 15:11-15). Jeroboam (Israel) died in the 22nd year of
his own reign which was the 2nd year of Asa (I Kings 15:25). His reign
was succeeded by Nabat's two year reign. Asa's third year was Baasha's
accession year and he reigned 24 years (I Kings 15:28-34, see chart).

By prorating the extra "nine months" allowed for "Jehoahaz's" (three
months reign) into our chronology at this point (956 B.C.), we extend
our chronology over into the next year allowing "ten full years" of
peace at the beginning of Asa's reign according to II Chron. 14:1ﬤ, plus
"five full years" without peace (II Chron. 15:5; 952 -947 B.C.) until
the "fifteenth year" of Asa (II Chron. 15:10,19). This <u>"15th year"</u> of
his reign was also the <u>36th year</u> of Asa's <u>Kingdom</u> (not "reign" - II
Chron. 16:1) when Baasha, in his 13th year, came against Judah at Ramah.
In the face of this threat, Asa made a league with Benhadad of Syria,
and Syria attacked Ijon, Dan, Abel-Maim, and all the store cities of
Naphtali (II Chron. 16:4), thereby causing Israel to withdraw from
Ramah. Asa's success with the withdrawal of Israel in his "fifteenth
year", which was also the "thirty-sixth year" of the "Kingdom", introduced revival (II Chron. 15:10-15), and a period of "twenty year peace"
until Asa's "thirty-fifth year" of his reign (II Chron. 15:10,19; 947 -
927 B.C.). Thereafter, Asa sinned against God and had war and was dis-
eased in his feet (II Chron. 16:9-12).

Since Shishak I ruled twenty-one years (984/3 - 963/2 B.C.), this meant
that Zerah (Osorkon I) began to reign in Egypt the same year that Asa
mounted the throne in Judah (962 B.C.). Asa, at the beginning of
his reign, had no war for ten years, then Asa smote Zerah (Osorkon I) to
Gerar (II Chron. 14:9-15; 952/1 B.C.). In the "twenty-first year" of
Asa, King of Judah (941 B.C. - also the 21st of Zerah), King Zerah of
Egypt celebrated the 500th year anniversary of the temple of Bast. The
Temple of Bast dated back 500 years to the fourth year (1441 B.C.) of
Amenophis II (1445 - 1421 B.C.) which dramatically verifies the stabil-
ity of our chronological frame work. Amenophis' fourth year naturally
falls in the year 1441 B.C. when we date Hatshepsut's coup d'e tat in

30

the well known historical year 1501 B.C., but her reign alone (1498 - 1477 B.C.) began after the death of her husband, Thutmose II, during Israel's Exodus, three years after the coup d'e tat, in the year 1498 B.C.. Then Thutmose III reigned from 1477 to 1445 B.C., and Amenophis II's reign followed (1445 - 1421 B.C.). This makes 1498 B.C. correlate exactly with the year of the "Biblical Exodus" according to our seven step chronological plan. Thus, the 500th anniversary of the Temple of Bast, dating from the fourth year of Amenophis II, correlates perfectly only into our long chronology of the kings (See Apendix G).

I Kings 16:6-11 states that when Baasha died in his 24th year (I Kings 15:33), then Elah began to reign for two years in Asa's twenty-sixth year, and was succeeded the following year (Asa's 27th - I Kings 16:8-10,15) when Zimra reigned only seven days. He was succeeded by the co-regency of Omri and Tibni in 935 B.C. (I Kings 16:21). The two reigned for four years in Israel (I Kings 16:21 - from the 27th to 31st years of Asa), then in Asa's 31st year (I Kings 16:23) Omri continued reigning alone during his fifth and sixth years (I Kings 16:15,23). His reign discontinued in the city of Tirzah at the end of his sixth year, after-which the capital was moved from Tirzah to Samaria in 929 B.C.. (Asa's 33rd year was the first year of the last six years of Omri's twelve year reign   I Kings 16:23). Therefore, Omri's capital at Samaria was known from that year as "Beth-Omri", "Bit-Humri", or the "House of Omri" for the last six years of Omri's reign and for years afterwards. The name continued to be famous in history and among the heathen for years. For example Ahaziah was twenty-two years old when he reigned one year (II Kings 8:25,26; 9:29), and he was called "the son of the 42nd year of Jehoram" his father; therefore, chronologically, he was also "the King Son" of the 42nd year of the "Kingdom of Omri" (II Chron. 22:1,2). Thus, as Ahaziah reigned the one year (887 B.C.), it was exactly the 42nd year since Omri's capital had been moved to Samaria (929 B.C.). The best translation of II Chron. 22:1b,2a could be, "So King Ahaziah, the son of the 42nd year of Jehoram, King of Judah, reigned, and when he began to reign, he reigned one year in Jerusalem - - ". That year of Ahaziah's reign (887 B.C.) was also the 40th year of Moab's oppression by the "House of Omri" (926 B.C. was the 36th year of Asa and Omri's 3rd year in his capital at Samaria). Now, let us correlate perfectly the inform-ation within the 40 year period (926 - 887 B.C. inclusive):

In the 38th year of Asa, Omri died in his 12th regnal year (three years after he smote Moab) which was Ahab's accession year (II Kings 16:29). Asa reigned 41 years (I Kings 15:9,10; II Chron. 16:13); therefore, Jehoshaphat's first year was Ahab's fourth year as quoted below: "And Jehoshaphat, the son of Asa, began to reign over Judah in the fourth year of Ahab King of Israel (920 B.C.). Jehoshaphat was thirty and five years old when he began to reign; and he reigned twenty and five years in Jerusalem - - - " I Kings 22:41,42 (see II Chron. 20:31).

Since the above is true then the seventeenth year of Jehoshaphat would be Ahab's twentieth year (904 B.C.). According to the following scripture Ahaziah began to co-reign with Ahab in that year: "Ahaziah the son of Ahab began to reign (co-reign) over Israel in Samaria the seven-

31

teenth year of Jehoshaphat, King of Judah, and (later) <u>reigned two years over Israel</u>" I Kings 22:51 (After the death of Ahab). This, two year period, could refer to either of two periods during Ahaziah's (Israel) reigning years; two years of which he co-reigned (with good health - inclusive) with Ahab, or to the time in which he reigned two full years after Ahab's death, while Joram co-reigned with him because of his sickness. The latter view is proven true by the following scripture: "So Ahab (King of Israel) slept with his fathers; and Ahaziah, his son, reigned in his stead" I Kings 22:40 (He still reigned after the death of Ahab two years even though he was still sick). Therefore, the more likely interpretation of I Kings 22:51 is that Ahaziah began to reign as a co-regent with Ahab in the seventeenth year of Jehoshaphat and three years later after the death of Ahab, even though he was still sick, he ruled two full years. This interpretation is verified by II Kings 3:1, "Now Jehoram (Joram), the son of Ahab, began to reign (pro-reign with the sick co-regent, Ahaziah) over Israel in Samaria the eighteenth year of Jehoshaphat, King of Judah (therefore, Joram pro-reigned and co-reigned for a total of five years with both Ahab and Ahaziah) and reigned (alone later) twelve years". In both I Kings 22:51 and II Kings 3:1, the Hebrew language uses the Kal future third person masculine singular which means "He shall reign" denoting a reign in the future. It was not yet an established fact. This clearly gives sufficient proof for the above interpretation, and also is clearly portrayed on the Chart (page 37) with scriptural proofs.

Thus far, we have the following information. Ahaziah began to co-reign with Ahab in Jehoshaphat's 17th year. In the 18th or second year (903 B.C.), Ahaziah fell through the lattice in his upper chamber in Samaria, and was sick (II Kings 1:2). He did not die at this time since he ruled two full years after Ahab's death (during the Battle of Karkar in 902 B.C. - I Kings 22:29-40) one year later. Since Ahaziah took sick in his second year of co-reign, Ahab decided to make Joram (Jehoram) pro-regent with sick co-regent Ahaziah (II Kings 3:1) because of the "Battle of Karkar" in 902 B.C. and during Jehoshaphat's 19th year, which was Ahab's twenty-second and last year. Therefore, Joram was pro-regent with sick co-regent Ahaziah (Israel) the last two years of Ahab's reign (903 - 902 B.C. inclusive). Then Ahaziah, still sick after Ahab's death at Karkar in 902 B.C., began his Kingly reign in Jehoshaphat's 20th year (I Kings 22:40) with Joram as his co-regent (901 B.C.). He reigned as King two full years, though sick after the death of Ahab (902 B.C.) ,and early in his third year, which was Joram's 5th year of consecutive co-regency with Ahab and Ahaziah, Ahaziah dies (899 B.C. - II Kings 8:16). This event happened to be in the beginning of the first year of Jehoram's (Judah) co-regency with Jehoshaphat and in Jehoshaphat's 22nd year. These assumptions are confirmed by the following scripture: "And in the fifth year of Joram (the fifth year of his co-regency, since his name had not been changed to Jehoram yet) the son of Ahab King of Israel, Jehoshaphat being then King of Judah (22nd year), Jehoram the son of Jehoshaphat King of Judah began to reign (co-reign

with Jehoshaphat during his last four years)" II Kings 8:16 (there are
two Jehorams, and two Ahaziahs mentioned in our chronology in the next
few years).

Therefore, Jehoram, of Judah, co-reigned with Jehoshaphat in Jehoshaphat-
's 22nd year. This same year, Ahaziah (Israel) died, which was Joram's
5th year of co-regency and also his accession year to Israel's throne.
Thus, Jehoshaphat's 23rd year, and Jehoram's (Judah) 2nd year of co-
reign with Jehoshaphat, both fell on Jehoram's (Joram - Israel) first
year of a twelve year reign (898 - 887 B.C., inclusive). These scriptures are quoted below as confirmation: "So he (Ahaziah - after
5 years of sickness - 903 to 899 B.C., inclusive) died according to the
Word of the Lord which Elijah had spoken. And Jehoram (Israel) reigned
in his stead in the second (co-reign) of Jehoram the son of Jehoshaphat,
King of Judah, because he had no son" II Kings 1:17 (please see
chart on page 37). "Now Jehoram the son of Ahab began to reign (co-
reign) over Israel in Samaria the eighteenth year of Jehoshaphat King of
Judah and reigned (he shall reign) twelve years" (later, not including
his five year co-regency) II Kings 3:1.

Therefore, Jehoshaphat's 23rd year was Jehoram's (Israel) first year,
and Jehoshaphat's 25th, and last year, was Jehoram's (Israel) 3rd year.
Judah's succeeding King's (Jehoram) first year was Jehoram's (Israel)
4th year. Thus, Jehoram's (Judah) 8th, and last year, was Jehoram's
(Israel) 11th year (II Kings 9:29), which was Ahaziah's (Judah) accession year. Ahaziah reigned one year until the twelfth year of Jehoram
(Joram - Israel), and they were both killed by Jehu that year (II Kings
8:25 - 9:29). That same year, Moab won her freedom after 40 years of
bondage to Israel, which year was known, historically, as Shalmaneser's
18th year (887 B.C.). Therefore, Shalmaneser's first year of reign was
904 B.C., which was two years before the Battle of Karkar (902 B.C.).
When we also include Shalmaneser's years of co-regency, according to the
Royal Eponym List starting in 908 B.C., then Shalmaneser's sixth year of
co-regency fell rightly, historically, in the very year of the "Karkar
Battle" (902 B.C.). Other events within the "40 year span" between Asa's
35th year (927 B.C. - II Chron. 15:19) until the year of the death of
both Ahaziah (Judah) and Jehoram (Israel) in 887 B.C. is the final proof
that the intervening chronology is exact (If we figure years inclusive
the period would date from 926 B.C.).

For added historical confirmation concerning this period, we must go to
the famous "Moabite Stone" or "Stele of Mesha", King of Moab, now in the
Louvre in Paris, which gives us the following historical information.
This Stele may have been erected around the years 887 to 885 B.C.. The
inscription reads in part (a free translation of the Hebrew): "I am
Mesha, son of Kemosh (Chemosh) - - King of Moab, the Dibonite - - - - My
father reigned (ruled) over Moab thirty years. I reigned after my
father. And I built (made) this high place ("Moabite Stone"), to my god
Kemosh, because of my victory (the deliverance of Mesha), and because
Kemosh (he) saved me from all the kings and caused me to see my aspirat-

ions (desire) upon all who hated me. Omri, King of Israel, oppressed Moab (by yearly tribute) many days because Kemosh was angry with his land. And Omri's sons (son) succeeded him, and they (he) also said we (I) will (also) oppress Moab. In my days Jehoram (he) spake (thus, according to his words), but I saw my aspirations (have triumphed over) come upon him and upon his house, and Israel (has) perished (forever), with an everlasting loss (Both Ahaziah and Jehoram were killed that year - 887 B.C.). Now Omri had posssessed all the land of Medeba (Mehedeba - Moab) and controlled (dwelt in) it in his days, and half the days of his sons' days, FORTY YEARS, but my god, Kemosh, restored (dwell or reigned there) it in my days - - - - ".

Therefore, Omri subdued Moab in his ninth year which was the 35th year of Asa (927 B.C.) and he ruled them for four years; Ahab succeeded Omri ruling over Moab for another twenty-two years plus Ahaziah's (Israel) two years, and then followed Jehoram's twelve years which adds to exactly FORTY YEARS OF BONDAGE UNDER THE FOUR KINGS OF ISRAEL and verifying the above 40 years, and also correlating perfectly with the chronology of Judah according to our following chronological chart of the kings of the divided kingdom (982 to 560 B.C.).

Before we continue on with our chronology, another word concerning the seeming contradiction between the two texts II Kings 8:26 and II Chron. 22:2. II Kings 8:26 states that Ahaziah was twenty-two years old when he began to reign, while II Chron. 22:2 seems to imply that he was forty-two years old (his father was).

Dr. Anstey seems to have the correct answer found in Philip Mauro's book.[1] He suggests that the Hebrew reads in II Chron. 22:2 as follows: "A son of forty-two years was Ahaziah when he began to reign". Literally, Ahaziah became a "Kingly son" in Jehoram's 42nd year. Also, Ahaziah was a "Kingly son" of the "Kingdom of Beth Omri" (Bit Humri), or the "House or Capital of Omri" in its forty-second year. Omri and Tibni ruled in the Capital of Tirzah four years (I Kings 16:21), and Omri ruled alone two more years (I Kings 16:23,24). He then moved his Capital to Samaria after his 6th year in 929 B.C.. Thus, Ahaziah was twenty-two years old when he began his one year reign, and that year was also the forty-second year of his father Jehoram, and also the forty-second year since the Capital had been moved to Samaria. This is clearly portrayed on the chart (page 37), which dates the forty-two years from 929 to 887 B.C.. Thus, this measurement inter-weaves with our chronological scheme perfectly. Therefore, we find that Ahaziah's accession year (II Kings 9:29) was Jehoram's (Israel) 11th year and also the 42nd year of the life of Jehoram his father (II Chron. 22:2). Jehoram ruled 8 years at the age of 35, but co-reigned 3 years before this with Jehoshaphat at the age of 32 years (II Kings 8:16,17). Therefore, Jehoram (his father) was born 42 years before Ahaziah, his son, became king, and also born the very year Samaria was made the Capital by Omri in 929 B.C..

Since Jehu killed Ahaziah (King of Judah), and Jehoram (King of Israel) in the year 887 B.C., therefore Athaliah's (Judah) and Jehu's

1. Mauro, Philip, "The Chronology of the Bible", page 67

ISRAEL'S AND JUDAH'S CHRONOLOGY CORRELATED

B.C.	(JUDAH'S KINGS)	I Kings 13:18;11:23,25	Rezon (Hezion) at Damascus
	Sabbatic year	(Shishak's accession)I Kings 12:2,25	Hadad (Edom) over Syria
984			
983	Shishak reigns in Egypt - I Kings 11:40	(ISRAEL'S KINGS)	Tabrimon - I Kings 15:1-19
982	REHOBOAM (JUDAH'S)	1 (41) Kings 14:21 (Tirzah) JEROBOAM	1 Built Shechem, Penuel IK.12:25
981		2 Prophet Ahijah - I Kings 11:29	2 Sister of Queen Tahpenes wed
980	II Chron. 11:17	3 Prophet Shemaiah - I Kings 12:22	3 Hadad - I Kings 11:19,40
979	II Chron. 12:1,5 -	4 - forsook the law of the Lord	4
978/7	II Chron. 12:2 -	5th-Shishak with Lubims,Sukkiims take-	5 - treasures and gold shields
977	Sabbatic year	6 Ethiopians attack - I Kings 14:25	6 ASSHUR-DAN II
976		7 I Kings 11:40	7
975		8 II Kings 12:2,3	8
974		9 Iddo (Seer) & Shemaiah - Prophet	9 Genealogy - II Chron. 12:7,15
973		10th year Rehoboam (Judah)	10th year of Jeroboam (Israel)
972		11 Gold calf-Dan & Bethel IKings 12:28	11
971		12th year of Rehoboam (Judah)	12th year of Jeroboam (Israel)
970	Sabbatic year	13	13
969		14 "War" - II Chron. 12:15	14 Golden calf - II Chron. 13:8
968		15	15
967		16	16 Tabrimon - Syria
966	II Chron. 12:13	17 Rehoboam dies at age 57 years	17
965	ABIJAM - - - - -	1 a league w/Tabrimon in Jeroboam's -	18th year - I Kings 15:1,2,18,19
964	"War"	2 Abijam took Bethel, Jeshamah, and -	19 - Ephriam-500,000 slain-IIC.13
963	Sabbatic year	3 Asa's accession year in Jeroboam's-	20th year - I Kings 15:9
962	Jubilee ASA	1 I K.15:9,10 - 10 YR PEACE(OSORKON)	21 II Chron. 14:1,6
961		Asa's 2nd year I K.15:25 began reign-NABAT	22 (1) NABAT
960		Asa's 3rd year I K.15:28 began reign Baasha	(2) Gibbethon seige-IK. 15:27
959		BAASHA	1
958			2
957			3
956	9 months added - -(6)-prorated for 3 mos rule-Jehoahaz		4
955	Sabbatic year		5
954		8 Jehoshaphat born, son of Asa	6 ADAD-NIRARI II
953		9th year Asa (Judah)	7th year of Baasha (Israel)

```
952 B.C.        Until     10th year Asa had no war - II Chron. 14:1    8th year of Baasha
951 Ethiopians - - -      11 Asa smote Zerah to Gerar-Osorkon I      9 II Chr.14:9-15;16:8
950                        12 Oded and son Azariah prophets - - - -   10 II Chron. 15:1,8
949 For five years -      13 "There was no peace" II Chron. 15:5      11 Benhadad I - Syria
948 Sabbatic year         14 I Kings 15:17;II Chr.16:1-6 Baasha -     12 - built Ramah
947 II Chron.15:9-12      15th REVIVAL part Simeon, Ephraim,Manasseh13 36th of Kingdom-IIC.16:1-Asa
946                       16 Sought God with hearts - IIChr.15:10-    14 gave Benhadad treasures to
945                       17 (No War from 947 to 927 B.C.)            15 smite Israel - I Kings 15:18
944 "PEACE FOR 20 YRS"18 "No War until Asa's 35th year"              16 "Asa destroys Ramah & built
943                       19      II Chron. 15:19                     17 Geba & Mizpah" I K. 15:22
942                                                                   18
941 Sabbatic year         21st & Osorkon's 21st-500th yr celebration19 - of the Temple of Bast
940                       20
939                       21
938                       22 - I Kings 16:7,12
937                       23
936                       26th year of Asa began the reign of Elah   24 (1) 2 years-I Kings 16:8-9
935 I K.16:10,11,15 -      27th year of Asa OMRI & TIBNI co-regency - 1 (2)Zimra 7 days-destroyed
934 Sabbatic year         28      Omri & Tibni co-regency            2 house of Baasha(Gibbethron)
933                       29      Omri & Tibni co-regency            3
932                       30      Omri co-regency until yr 4th-IK.16:21 TUKULTI-NINURTAII
931            Asa's 31st year Omri reigns alone-IK.16:23 OMRI 5 - accession year alone
930                       32      Omri at Tirzah until his - - -      6th year - I Kings 16:23
929 Jehoram 1 yr old      33rd year BETH OMRI CAPITAL AT SAMARIA 1ST(7)of 6;IK.16:24;IIK.8:26;IIC22
928 Jehoram 2nd           34                                         ASHUR-NASIR-PAL II-929 B.C.
927 Sabbatic yr     (To 35th of Asa no War IIC.15:19)(TAKELOTH I) 8 OMRI RULED MOAB FOR 40 YEARS
926 See 947 B.C. - -     -36th or 36th year of "Kingdom" IIChr.16:1 10 Moabite bondage - 1st year
925 Jehoram 5th          37      Omri wrought evil - I Kings 16:25   11        2nd year
924 I Kings 16:29 -      38th year of Asa began reign of King AHAB 12 Ethbaal, Jezebel-IKings 16
923 Jehoram 7th          39th of Asa diseased - II Chr.16:12 AHAB 1 Jezebel I Kings 16:29,31
922 Jehoram 8th          40      Elijah - I Kings 17:1               2        5th year
921 Jehoram 9th          41 - II Chron. 16:13 - Asa dies             3        6th year
920 Sabbatic year        1st JEHOSHAPHAT WAS THE 4TH YEAR OF Ahab- 4th IK.22:41;IIC.20(OSORKON II)
919 Jehoram 11th         2       Jahaziel prophet                   5        8th year
```

918 II Chron.17:7 - - 3rd Obadiah, Zechariah taught law of God 6 Moabite bondage - 9th year
917 Jehoram 13th 4 Attempted to build Jericho - - - - - 7 I Kings 16:34 10th year
916 Jehoram 14th 5 8 11th year
915 Jehoram 15th & of(6)the Samarian Kingdom (Beth Omri) 9 12th year
914 Jehoram 16th 7 10 13th year
913 Sabbatic year 8 Elijah 11 14th year
912 JUBILEE 9 Obadiah saved 150 prophets I Kings 18:4 12 15th year
911 Jehoram 19th 10 Elijah's famine for three years 13 16th year
910 Birth - 20th of 11 IK.18:1 famine-Elijah at Cherith - - - 14 - IK.17:5Zarepheth 17th year
909 Ahaziah-21st yr 12 IK.18:19 famine Elijah's alter & 450 15 - prophets of Baal 18th year
908 Jehoram 22nd 13 Anoint Hazael, Jehu, & Elisha-IK.19:15- 16 Shalmaneser co-reg.-access.
907 Jehoram 23rd 14 Israel fights Benhadad I-IK.20:1(32Kings17 127,000 killed)1st yr co-reg
906 Jehoram 24th 15 Benhadad Aphek IK.20:26(No War 3 years) 18 Shalmaneser's 2nd 21st year
905 Sabbatic year 16 IK.21:1 Naboth's vineyard(Peace-IK.22:1)19 Royal Eponym- 3rd 22nd year
904 Ahaziah co-reg.- 17th IK.22:51 Ahab's cov. w/Syria-IK.20:34 20 SHALMANESER REIGNS 23rd year
903 IK. 22:51 - - - 18th IIK.3:1 Ahaziah sick & Ahab old-Joram-21-co-reg.(BATTLE OF 24th year
902 Jehoshaphat's - -19th w/Ahab to Ramath G.(Ahaziah sick-Joram22pro-reg.-KARKAR)IIK.22:29,33.
901 Jehoram 29th 20 IK.22:40 "AHAZIAH KING"but sick - Joram- 0-co-reg. ELIJAH RAPTURED
900 II Chr. 20:1- 21 IIK.3:5 Moab's revolt fails(Ahaziah sick)1-4th co-reg.(Elisha-IIK.3:11)
899 JEHORAM JUDAH AS 22 CO-REG. ACC.IIK.8:16(Ahaziah dies in the 2-5th year of Joram's co-reg.)
898 Sabbatic year(32)23 yrs old-1st yr-IIK.8:32;IIK.3:1 JERORAM 1 IIK.5:1 Naaman.Famine Elisha
897 Jehoram 33rd 24 (TAKELOTH co-reg.)Shunem woman-II Kings(2)4:2,8,38 Famine 30th year
896 Jehoram 34th and 25 4th & last year co-reg. Shalmaneser's (3) 9th year Famine 31st year
895 JEHORAM KING 1 IIChr.21:5-7,20 married Athaliah-Ahab 4 II Chr.22:2 Famine 32nd year
894 Jehoram 36th 2 Shalmaneser's (5) 11th year Famine 33rd year
893 Jehoram 37th 3 Famine - ate son - II Kings 6:26,29 6 Famine 34th year
892 Samaria is - - - 4 attack by Benhadad & killed by Hazael 7 II K.6:24 Famine ends
891 Sabbatic year 5 (TAKELOTH II) Shunem's land-IIK.8:5 8 Shalmaneser's 14th year
890 Jehoram smote - - 6 Jair,Edom-Libnah revolts;Philistines & (9) Arab's attack-IIK.8;IIChr.21
889 Jehoram 41st 7 IIChr.22:2;IIK.8:26-Jehoram sick-II C.(10) 21:19;Edom IIC.21 38th year
888 22 yrs- 42nd year(8) Ahaziah accession(IIK.8:29;9:29) in - 11th. MOAB FREE 40TH 39th year
887 AHAZIAH ONE YR 1 Ahaziah & Jehoram to Ramah Gilead-die- 12th.Shalmaneser's 18th-40TH YR.
886 ATHALIAH 1 Joash's 1st year IIK.11:21;IIK.10:36 1 JEHU IIK.8:25;II Chron.22:5
885 2 Joash's 2nd year 2 885 B.C.-Moabite Stone

Year		Jehu / Israel
884 Sabbatic year	3 Joash's 3rd year Jehonadab son of	3
883 B.C.	4 Joash's 4th year Rachab whose children	4
882	5 Joash's 5th year would not drink wine	5
881	6 Joash's 6th year & helped Jehu-IIK.10	6
880 JOASH	1 reigns - 7th year of age. It was in the	7th year Jehu-IIK.12:1;IIChr.24
879	2 Jehoiada prophet - II Kings 11:9	8
878	3 II Chron. 23:8	9 Jehu destroyed all the sons
877 Sabbatic year	4	10 of Ahaziah, Ahab's sons, Jez-
876	5	11 ebel, and Baal, but not the
875	6	12 gold calves in Dan and Bethel
874	7	13 II Kings 10:29.
873	8	14
872	9	15 "Four generations of Jehu
871	10 "And Jehoiada made a covenant between	16 shall sit on the throne"
870 Sabbatic year	11 the Lord and the king and the people,	17 II Kings 10:30
869	12 that they should be the Lord's people	18
868	13 between the King also and the people"	19 SHAMSHI-ADAD V
867	14 II Kings 11:17	20 "Jehu took no heed to walk
866	15 II Kings 12:2	21 in the law of the Lord God
865	16	22 with all his heart" II Kings
864 Amaziah born	(17) a son of Joash in his 23rd year.	23 10:31.
863 Sabbatic year	18	24 "Hazael began to cut Israel
862 JUBILEE	19 Amos the prophet - Amos 1:3,4	25 short from the Jordon east-
861	20	26 ward" II Kings 10:32,33
860	21	27
859	22	28
858	23rd year of Joash's reign began JEHOAHAZ	1 II Kings 13:1;18:1
857	24	2
856	25 Repaired temple - II Kings 12:6	3
855 Sabbatic year	26	4 ADAD-NIRARI III-Q.Semaramers
854	27	"Hazael (5) oppressed Israel all the
853	28	6 days of Jehoahaz" IIK.13:21-

```
852 B.C.                                              7
851                          29                       8
850                          30 Jehoiado died at 130 years – IIC.24:24    9 Zechariah, son of Jehoiada
849                          31                       10 the priest, killed –
848 Sabbatic year            32                       11    II Chron. 24:20ff.
847                          33                       12    II Kings  13:10
846                          34                       13
845                          35 Hazael took Gath & attacks Jerusalem      14
844            Joash's       36 but left being paid – II Kings 12:18      -15th year – II Kings 13:10
843                          37th year Jehoash co-regent with Jehoahaz    16th of Jehoahaz
842                          38 Joash diseased – II Chron. 24:25          17th. Hazael dies–IIK.13:24
841 Sabbatic year            39        Jehoash's accession in             1 Benhadad II – IIK.13:1-25
840 AMAZIAH(Age 25)          40–dies age 46. IIC.24:23;IIK.13:9 JEHOASH   2nd year–IIK.14:1,2;12:21
839                          1st year of Amaziah was also Jehoash's –     3
838                          2                                            4
837                          3 Destroy Syrians at Aphek – II K.13:17      5
836                          4 Elisha dies – II Kings 13:20               6 Three times did Jehoash
835                          5 Moab invaded the land – II K. 13:20        7 recover the cities from
834 Sabbatic year            6                                            8 Benhadad – II Kings 13:25
833                          7                                            9
832                          8              (PAMAY – Egypt)               10 Israel envied Judah and
831                          9 Smote Seir (Joktheel – Selah)II C.25:11    11 smote 3000 and takes Amaziah
830                          10             II Kings 14:7                 12 and destroys 400 cubit of
829                          11                                           13 Jerusalem's wall II Kings 14;
828                          12                                           14 II Chron. 25.
827 Sabbatic year            13                                          15        SHALMANESER IV
826 Amaziah's                14 Uzziah born     (SHESHANK IV)            16 – II Kings 14:23
825 II Kings 14:17           15th year was also Jehoboam's accession –    1
824                          16 – 2 – reigned 15 years more    JEHOROAM   2
823                          17 – 3 after death of Jehoash                3
822                          18 – 4 in the year 826 B.C. – IIC.25:25      4
821                          19 – 5 Jonah – II Kings 14:25                5
820 Sabbatic year            20 – 6                                       6
819                          21 – 7                                       7
                             22 – 8
```

818	23 - 9		Shalmaneser 8 subjected Hamath area.	
817	24 - 10th year since Jehoash died		9	1st year of ASHUR-DAN III
816	25 - 11		10	2nd year
815	26 - 12		11	3rd year accession
814	27 - 13		12	4th year
813 Sabbatic year	28 - 14		13	5th year
812 JUBILEE	29 - 15th Amaziah dies at Lachish-age 53 -		14	6th year - II Kings 14:15-19
811 UZZIAH (Age 16)	1 (52)Built Elath-IIK.14:21;II Chr.26:2		15	7th year - trouble in Hamath
810	2 Nineteen districts revolt to Uzziah in-(16)		8th year of Ashur-Dan III	
809,13th of June - -(3)	- ECLIPSE-month Simanu.Revolt-Assyria-(17)		-9th year or 8th regnal year	
808	4		18	Revolt in Assyria
807	5		19	Revolt in Assyria
806	6		20	Revolt in Assyria
805 Sabbatic year	7 Warred against Philistines and destroy-		21	Benhadad II with coalition
804 II Chr. 26:5,6	8 ed Gath, Jabneh, and Ashdod.		22	attacked Hadrach-Zech. 9:1.
803	9 Zechariah the prophet		23rd year of Jehoboam	
802	10 II Kings 15:1 - "In the twenty and		24	Ashurdan's 17th & 18th years
801	11 seventh year of Jehoboam, King of		25	recovers Hamath - - - - -
800	12 Israel began Azariah son of Amaziah King		26 Ashur-Dan to Hadrach-dies	
799	13 of Judah to reign(over Israel) - - - -		27th year.ASHUR-NIRARI V	
798 Sabbatic year	14	Jehoboam's	28th year-co-regent with Uzziah	
797	15	Jehoboam's	29th year-co-regent with Uzziah	
796	16	Jehoboam's	30th year-co-regent with Uzziah	
795	17 Amos the prophet	Jehoboam's	31st year-co-regent with Uzziah	
794	18 Amos 1:1,2	Jehoboam's	32nd year-co-regent with Uzziah	
793	19	Jehoboam's	33rd year-co-regent with Uzziah	
792	Armenia attack-(20)-Revolt in Calneh(Nimrud)		Ashur-Nirari (34) and family killed by Urartu.	
791 Sabbatic(PSDIBULAT)	Jehoboam recovers Damascus & Hamath to		35 ASSYRIA RULED BY URARTU	
790	22 to Judah(to Uzziah) - II Kings 14:25,28.		36th w/Uzziah. URARTIAN KINGDOM	
789	23	Jehoboam's	37th w/Uzziah. URARTIAN KINGDOM	
788 IIChr.26:10 - -	24 "Uzziah had vine dressers in Carmel"		38th w/Uzziah. URARTIAN KINGDOM	
787 Amos 1:1 - - -	24 Jehoboam alive 2 yrs before earthquake.		39th w/Uzziah. URARTAIN KINGDOM	

B.C.					URARTIAN KINGDOM
786	EARTHQUAKE	26	Perhaps Jehoboam dies in the earthquake. 40th w/Uzziah.		URARTIAN KINGDOM
785	OF UZZIAH - - -(27)-		during 27th year of his reign,and in - 41st JEHOBOAM DIES.		"
784	Sabbatic year	28	UZZIAH IS RULER OVER BOTH KINGDOMS FOR 10 YEARS ALONE.		URARTIAN KINGDOM
783	Jotham one year(29)		old, son of Uzziah aged 43. ASSYRIA RULED BY		URARTIAN KINGDOM
782		30	Uzziah ruling over Israel - II Chron. 26:10		"
781		31			"
780	II Chron.26:16-(32)-		Uzziah had 2600 mighty men - 307,500 soldiers - engines.		"
779	The Fertile - -(33)-		Crescent divided between Uzziah's Kingdom and the big		URARTIAN KINGDOM
778		34	Jotham six years old.		
777	Sabbatic year	35	Assyria still ruled by		URARTIAN KINGDOM
776	B.C.- - - - -		36th year of Uzziah - DATE OF THE FIRST OLYMPIADS.		URARTIAN KINGDOM
775	II Kings 15:5 -(37)-		Jotham nine (co-regent) Uzziah's heart lifted up - Leprosy-IIChro.26.		
774	In Uzziah's		38th year Israel revolts - - - ZACHARIAH 1		4th generation-Jehu-IIK.15:8f
773	In Uzziah's		39th year(IIK.15:13,14) the reign of SHALUM 1		PUL REVOLTS FROM URARTIAN K.
772		40	Jotham co-regent MENAHEM 1		AND CONTROLS HAMATH AREA.
771	Uzziah's	41st.	Pul control's Hamath area and 2		URARTIAN KINGDOM
770	Sabbatic year	42	receives tribute from kings who fell 3		Pul receives tribute from
769		43	to Uzziah in 810 B.C. 4		Menahem - II Kings 15:19,20.
768		44	Jotham co-regent (OSORKON III) 5		URARTIAN KINGDOM
767		45	Jotham co-regent For the next 20 yrs Pul6		URARTIAN KINGDOM
766		46	Jotham co-regent gradually gains land 7		"
765		47	" from Urartian Kingdom. 8		"
764		48	" 9		"
763	Sabbatic year	49	Menahem's last year 10th		"
762	JUBILEE		50th year of Uzziah there reigned PEKAHIAH 1	II Kings 15:23	"
761			51st Jotham's last year co-regency 2		"
760	Uzziah dies		52nd-Jotham's accession (PSAMMUS) PEKAH 1	II Kings 15:27	"
759	JOTHAM (Age 25)	1	(16 yrs) - - - Jotham began reign the 2nd	II Kings 15:32	"
758		2	: - Smote Ammon - II Chron. 27:3 3		"
757		3	: - 4		"
756		4	5		"
755	Sabbatic year	5	Built wall of Ophel - II Chron. 27:3 6		URARTIAN KINGDOM

	6th of Jotham (Judah)	Pekah's 7th (Israel)	URARTIAN KINGDOM

754 — 6th of Jotham (Judah) — Pekah's 7th (Israel) — URARTIAN KINGDOM

753 — 7 — 8 — "

752 — Ahaz 12 yrs old-(8)-when Hezekiah, his son, was born. — 9 — "

751 — Ahaz 13 yrs old-(9)-when Hezekiah 1 yr. Pul takes tribute(10) & conquered Nabu-shum-ukin I

750 — 10 (PIANKHY) — 11 — . URARTIAN KINGDOM

749 — 11 — 12 Urartian Kingdom weakens and

748 Sabbatic year — 12 — 13 Pul (Tiglath Pileser III)took

747 — 13 Oded prophet - II Chron. 28:9 — 14 Ijon, Kedesh, Hazor, Gilead,

746 — 14 — 15 Galilee, Naphtali & carried

745 B.C. - PUL OR -(15)-TIGLATH PILESER III RULER OF ASSYRIA — 16 them to Assyria - IIK. 15:29.

744 II Kings 16:1 — 16 Jotham alive.Ahaz accession in Pekah's — 17th year.(Urartian K. weakened)

743 AHAZ (Age 21) — 1 (16 yrs) 17th of Jotham in prison. — 18 PUL-MASSACRE'S LAND-URARTIANS

742 Ahaz wounded by-(2)- Syria & Israel.Jotham in prison. — 19 IIC. 28:5 - Resin - II K. 16

741 Sabbatic year — 3 Ahaz attacked by Philistines and Edom-(20)- II Chron. 28:17

740 II Kings 15:30 -(4)-Jotham's 20th-prison. Hoshea conspires(1) HOSHEA against Pekah

739 — 5 "They overthrew Pekah their King and

738 Urijah priest — 6 I (T. Pileser) made Hoshea to be King

737 II Kings 16:11 — 7 over them"(Gaza, Damscus campaign). (Hoshea a puppet under Pul)

736 — 8 (Hoshea subjected to Tiglath Pileser III)

735 — 9 Ahaz was subject to Assyria who smote Syria - II Kings 16:7-10

734 Sabbatic year — 10 (BOCHORIS) Tiglath P. at Damascus - - - Ahaz compromised-IIC.28:20

733 — 11

732 Ahaz's 12th Hoshea gains independence from Tiglath Pileser - II Kings 17:1

731 — 13 (SHABALAKA - SO - SIBAKE) — (1 yr) — (2 yrs)

730 — 14 Hoshea's 1st year as King of Israel-HOSHEA 1 - II Kings 17:1

729 B.C. - - - -(15)- - - - - - - Pul or Tiglath Pileser-(2)- ruled over Babylon

728 — 16 Hezekiah's accession & begins in Hoshea's 3rd year - II Kings 18:1

727 Sabbatic year — 1 HEZEKIAH(Age 25)Repair temple-IIC.29Hoshea 4th year SHALMANESER V

726 — 2 Smote Philistines - II Kings 18:8 — 5 II Kings 17:3 - gifts

725 — 3 Conspiracy - II Kings17:4 — 6

724 II Kings 17:5 — 4th year of Hezekiah which was also Hoshea's 7th - II Kings 18:9 Prison

723 — 5th year of Hezekiah (Judah). — 8th year Hoshea (Israel)

722 BC IIK.17:6;18:1 6th year of Hezekiah Samaria taken in Hoshea's 9th (MERADACH-BALADEN)
721 BC   7 - 1st year of SARGON II (Sharrukin) - ISRAEL'S CAPTIVITY TO ASSYRIA
720 Sabbatic year   8 - 2nd year attack Egypt & Israel: "The cities Gil(ead), Abel(Beth-
719   9 - 3rd   maacha?) which is in boundary of the land of Beth-Omri I turned
718(SIBAHKI-SHEBITKU)10- 4th   its entire area into the territory of Assyria. I set officers &
717   11 - 5th   viceroys over it" (IIK.15:29;16:9-16;I Chr.5:6-25;Isa.7:1-9).
716 Prorate 9 mos.- 11 - 6th - over 390 years for 3 mos rule of Jehoiachin (gain one year)
715 Campaign - 12 - 7th of Sargon against Tirhakak Ethiopian General - II Kings 19:9, Isa.37:9
714   13 - 8th   Sennacherib (Gen.) to Lachish & Jerusalem 713 BC-IIK.19:9
713 Sabbatic(IIK18:)14 - 9th - Hezekiah sick prays for Jerusalem - IIKings 19:1-8 1st year-Isa.37:30
712 JUBILEE Is37:30 15 -10th-Meradach visits Hezekiah-IIK.20:1,12;19:29-ate not - -2nd year
711 Sennacherib's - 16 -11th-campaign to Ashdod - Isa. 20:1-6 (IIK.19:29 - ate of -3rd year)TIRHAKAK
710 Isa.39:5-8 - - 17 -12th of Sargon attacked Babylon (Meradach's rule ends the following yr)
709 Manasseh born -(18)-when Hezekiah was age 48. Sargon accession year rule of Babylon-Meradach
708   19   1st interregnum year-5 yrs
707   20th year of Hezekiah. (He asked for 15 years   "
706   21   of extended life, therefore Manasseh, most   "
705 Sabbatic year   22   evil king, was born to him during that time)."   "
704   23rd year Sargon dies & SENNACHERIB reigns   "
703   24 Sennacherib rebuilds Nineveh. MERADACH-BALADEN reigns in Babylon
702 5th campaign - (25) - Sennacherib puts Bel-ibni on Babylon throne - BEL-IBNI
701   26   Meradach and Bel-ibni rebell against Sennacherib
700   27 Son of Sennacherib rules Babylon - - - - - ASHUR-NADIN-SUM
699   28
698 Sabbatic year   29th Hezekiah dies - II Kings 18:2
697 MANASSEH(Age 12) 1 (rules 55 years)
696   2 Manasseh - "hath done wickedly above all that Amorites did" - IIK.21:11
695   3
694   4   "Worshipped all the hosts of heaven - observed times and use enchant-
693   5   ments, and used witchcraft, and dealt with a familiar spirits and with
692   6   wizards" II Chron. 33:3-6.
691 Sabbatic year   7
690   8

689 B.C.	9th yr Manasseh. Sennacherib destroys Babylon-city in ruins 8 years
688	10 "Behold I am bringing evil upon Jerusalem and Judah 1st year
687	11 that whosoever heareth of it, both his ears shall tingle. 2nd year
686	12 3rd year
685	13 - - And I will wipe Jerusalem as a man wipeth a dish, 4th year
684 Sabbatic year	14 wiping it, and turning it upside down. And I will for- 5th year
683	15 sake the remnant of mine inheritance - " I Kings 21:12-14.6th year
682	16 7th year
681 B.C.	17 ESARHADDON reigns and rebuilds Babylon in 1st year - - -8th year
680	18
679	19 Esarhaddon "bound Manasseh with fetters & carried him to Babylon"
678	20 II Chron. 21:33:11
677 Sabbatic year	21 Also carried to Babylon: Baala King of Tyre
676	22 Haushgabri King of Edom
675	23 Musurri King of Moab
674	24 Milki-ashapa King of Byblos
673 B.C.	25 Esarhaddon defeats Taharka - Isa. 19:2-4; Ezra 4:2 (Tirhakah-Isa.37:9)
672/1 B.C.	26 (TANUTAMON)
671	27
670 Sabbatic year	28
669/8	29 (NECHO I) ASHURBANIPAL reigns
668	30
667	31 Library of Ashurbanipal
666	32
665	33
664/3 B.C.	34 (PSAMMETICHES I)
663 Sabbatic year	35th year of Manasseh
662 B.C. JUBILEE	36 Amon, son of Manasseh was two years old.
661	37 II Chron. 33:12ff
660	38 "And when Manasseh was in affliction, he besought the Lord his God, &
659	39 humbled himself greatly before the God of his fathers, & prayed unto
658	40 Him, & He was intreated of him, & heard his supplication, & brought him to
657	41 Jerusalem into his Kingdom, then Manasseh knew that the Lord He was God".

B.C.		Event
656 B.C.	42nd year of Manasseh	
655 Sabbatic year	43	
654	44	
653	45	"Manasseh built a wall without the city of David, on the West side of
652	46	Gihon in the valley, even to the entering in at the fish gate; and
651	47	compassed about Ophel, and raised it up a very great height – –"
650	48	II Chron. 33:14.
649	49	
648 Sabbatic year	50	Josiah born son of Amon when he was 16 years old.
647	51	
646 B.C.	52	KANDALANU (Babylon)
645	53	
644	54	
643	55th year Manasseh died at 66 years of age.	
642/1 AMON (Age 22)	1 (2 yrs)	
641 Sabbatic year	2	His servants slew him (23 years old) – II Kings 21:23.
640/9 JOSIAH(Age 8)	1 (31 yrs)	
639/8	2	
638/7	3	
637/6	4	
636/5	5	
635/4	6	
634/3 Sabbatic year	7	
633/2 II Chron. 34	8th year Josiah began to seek the God of David his father.	
632/1	9	Josiah 16 years old when Jehoahaz was born. ASHURBALLIT
631/0	10	
630/9	11	
629/8 II Chron.34:3	12th year of Josiah he purged Judah. ASHUR-STILLU-ILI	
628/7 Jer.25:1 0/1	13th year was the 1st year of 23 years of Jeremiah's prophecies.	
627/6 Sabbatic 1/2	14	
626/5 2/3	15	NABOPOLASSER
625/4 3/4	16	
624/3 4/5	17th year of Josiah (Judah)	

B.C.			
623/2 B.C.	5/6	18th	year of Josiah they repaired temple (Passover) – II Kings 22:3
622/1 Ezek.1:1	6/7	19th	year Ezekiel born. "Hilkiah, priest, and Huldah, prophetess, found
621/0	7/8	20	a book of the law of the Lord given by Moses" – II Chron. 34:14.
620/9 Sabbatic	8/9	21	
619/8	9/10	22	
618/7	10/11	23	
617/6	11/12	24	
616/5	12/13	25	
615/4	13/14	26	
614/3	14/15	27	
613/2 Sabbatic	15/16	28	
612/1 JUBILEE	16/17	29	Fall of Nineveh and the fall of Assyria – ASHUR-UBALLIT II
611/0	17/18	30	
610/9	18/19	31st	year Josiah dies after attack on NECHO II – IIK.23:29;IIC.35:20
609/8 JEHOAHAZ (3mos)	0/1(July-Sept)		JEHOIAKIM(Sept) Age 25 ruled 11 yrs – IIC.26:1,5f.
608/7	20/21	1/2nd	year of Jehoiakim. Jehoiachin co-reigns age 8 – II C.36:9
607/6	21/22	2/3rd	accession year.
606/5 Dan. 1:1	22/23	3/4th	year – Daniel's captivity. Nebuchadnezzar's attack – CAPTIVITY
605/4 Sabbatic	23/24	4/5th	Jer.46:2;25:1;36:1 NEBUCHADNEZZAR'S 0/1st & 23rd yr prophecy(Father dies)
604/3 Jer.36:9,22,23	5/6th		Tribute 3 years Nebuchadnezzar's 1/2nd accession year
603/2(Dream 2nd year)	6/7th		Jehoiakim subject Nebuchadnezzar's 2/3rd Capt.Dan.1:5;2:1
602/1	7/8th		Jehoiakim subject Nebuchadnezzar's 3/4th attack Hatti land
601/0 II Kings 24:1	8/9th		Jehoiakim rebells Nebuchadnezzar's 4/5th attack Egypt
600/9	9/10th yr of Jehoiakim.		Nebuchadnezzar's 5/6th year at home
599/8	10/11th yr of Jehoiakim.		Nebuchadnezzar's 6/7th Hatti Arab land
598/7 Sabbatic	11: Dec.6 Mar.16 JEHOIAKIM		Nebuchadnezzar's 7/8th Jer.52:28;II Chron.36:9
597/6 ZEDEKIAH Apr.22	0/1st accession – Jehoiachin's Captivity–		8/9th IIKings24:8,12,18
596/5 Captivity year	1/2nd accession (PSAMMETICHUS II)		9/10th Nebuchadnezzar
595/4	2/3rd year of Jehoiachin's Captivity		10/11th "
594/3	3/4th year of Jehoiachin's Captivity		11/12th "
593/2 Jer.51:59	4/5th Zedekiah's cov. w/Babylon Ez.17:13		12/13th accession year
592/1 Ezek.1:2;4:5	5/6th(June-July) 390 yrs add to 592 BC={13/14th)years Israel's sin		

591/0 Sabbatic	6/7th Babylon withdraws before Egypt-Jer.37:5,11	14/15 Zedekiah rebells
590/9 Ez.20:1	7/8th year Zedekiah & Jehoiachin's Captivity	15/16 Ezek.8:1;17:15
589/8	8/9th accession year (HOPHRA)　No prophecy	16/17th of Babylonian King
588/7 Jan. 10th	9/10th attack Jerusalem-IIK.25:1;Jer.52:29;Ez.24	17/18th " "
587/6 Ez.29:1	10/11th accession year of Zedekiah　Jer.32:1	18/19th " "
586/5 CAPTIVITY	11th Jerusalem fell (Aug.11) IIK.25:2,8;Ez.26:1,2	19/20 Ez.30:20;31:1 Tishri yr
585/4	12/13th accession Jehoiachin Capt. 1st yr fr fall	20/21 Ez.32:1,17;33:21
584/3 Sabbatic	13/14th " 2nd yr fr fall	21/22 access. Nebuchadnezzar
583/2	14/15th " 3rd yr fr fall	22/23th "
582/1 Regnal yr	15/16th " Jer.52:30 4th yr fr fall	23/24th "
581/0	16/17th " 5th yr fr fall	24/25th "
580/9	17/18th " 6th yr fr fall	25/26th "
579/8	18/19th " 7th yr fr fall	26/27th "
578/7	19/20th " 8th yr fr fall	27/28th "
577/6 Sabbatic	20/21st " 9th yr fr fall	28/29th "
576/5	21/22nd " 10th yr fr fall	29/30th "
575/4	22/23rd " 11th yr fr fall	30/31st "
574/3	23/24th " 12th yr fr fall	31/32nd "
573/2 Sabbatic	24/25th " 13th yr fr fall	32/33rd "
572/1 Ez.40:1	25/26th accession Jehoiachin Capt.14th yr fr fall	33/34th — Vision of the temple
571/0	26/27th " The fall of the shore city of Tyrus —	34/35th access.Nebuchadnezzar
570/9 Sabbatic	27/28th " Captivity.(AMASIS)	35/36th "
569/8	28/29th "	36/37th "
568/7	29/30th "	37/38th "
567/6	30/31st "	38/39th "
566/5	31/32nd "	39/40th "
565/4	32/33rd "	40/41st "
564/3	33/34th "	41/42nd "
563/2 Sabbatic	34/35th "	42/43rd "
562/1 JUBILEE	35/36th Nebuchadnezzar dies Oct.8,562/1 B.C.	43/44th "
561/0 IIK.25:27	36/37th AMEL MARDUK 1st yr(Sept)Jehoiachin released 12th mo. Aug.560 B.C.	
560/9	37th Jehoiachin's 1st year of release(Sept.560/59 B.C. in 37th year-IIK.25:27)	

(Israel) first years (II Kings 11:21). Athaliah ruled 6 years followed by Joash who began his rule in the 7th year of Jehu (II Kings 12:1). Joash's 22nd year was Jehu's last or 28th year. Jehoahaz reigned instead of Jehu in Joash's 23rd year (II Kings 13:1). If the 23rd year of Joash was the first year of Jehoahaz then Joash's 37th year was Jehoahaz's 15th year at which time Jehoash began to co-reign with Jehoahaz during his last three years of reign (II Kings 13:10). If the above is true then Joash's 39th year was Jehoahaz's 17th and last year (842 B.C.). Jehoash (Israel) ruled the following year (841 B.C.) which was the 40th and last year of Joash (II Chron. 24:23; II Kings 13:9,10). Amaziah ruled in Joash's stead in Jehoash's 2nd year (II Kings 14:1,2; II Kings 12:21 - 840 B.C.); therefore, Amaziah's 15th year was Jehoash's 16th and last year (826 B.C.). That year was also Jeroboam's accession year (II Kings 14:23) which was followed by his first year, and in turn coincided with Amaziah's 16th year (825 B.C.). Amaziah's last and 29th year fell on Jeroboam's 14th year (812 B.C.). Uzziah reigned in Amaziah-'s stead the following year (811 B.C.), therefore Uzziah's 13th year fell on Jeroboam's 27th year (799 B.C.). At which time Uzziah began to reign OVER ISRAEL with Jeroboam as his CO-REGENT.[1] This is confirmed by the following scripture: "In the 27th year of Jeroboam King of Israel (which would be Uzziah's 13th year) began Uzziah son of Amaziah King of Judah to reign (over Israel)" II Kings 15:1. Therefore, Jeroboam was a co-regent with Uzziah from his 27th year until his last or 41st year (for 14 years - 799 to 785 B.C. - Jeroboam probably died shortly before or during the year of Uzziah's earthquake), during which time, he recovered Hamath to Judah. This interpretation is confirmed by the following two scriptures: "He (Jeroboam) restored the coast of Israel from entering of Hamath unto the sea of the plain, according to the Word of the Lord God of Israel, which he spake by the hand of his servant Jonah, the son of Amittai, the prophet, which was of Gath-hepher. - - - Now the rest of the acts of Jeroboam, and all that he did, and his might, how he warred, and how he recovered Damascus, and Hamath to Judah (Uzziah) for Israel (co-regency), are they not written in the book of the chronicles of the Kings of Israel?" II Kings 14:28; "Uzziah (Judah) also had vine dressers in Carmel (Israel)" II Chron. 26:10. The above two scriptures contain clear proof that Uzziah began to reign over Israel in Jeroboam's 27th year which was Uzziah's 13th year of reign. Therefore, Jeroboam helped recover Hamath again to Judah.

History confirms that Hamath and other cities belonging to Assyria went over to Uzziah in Ashurdan III's 8th year which would have been Uzziah's 3rd year (809 B.C.). The Assyrian Eponym List (allowing 46 years gap while the Urartian Kingdom ruled Assyria - 792 to 745 B.C.) also tells us that in Ashurdan's 17th and 18th years he attacked Hatarika (Hamath and area - 802 to 801 B.C.) and no doubt recovers Hamath. But two years later (799 B.C.) Jeroboam begins his co-regency with Uzziah (II Kings 15:1) and soon after recovers Hamath to Judah (II Kings 14:28). Therefore, Uzziah (Judah) and Jeroboam (Israel) united their kingdoms for 14 years in the face of the Urartian Kingdom's advance south which flourished during the 46 year eclipse of the Assyrian Kingdom.

1. Aharoni, Y., "The Land of the Bible", page 315 ,Burns & Gales Ltd 1968

(Historians have coordinated the Assyrian Eponym List with the wrong eclipse June 15, 763 B.C.. It should be correlated with the eclipse June 13, 809 B.C.. Shalmaneser III began to reign in 904 B.C. as contemporary of Baalazar, 909 - 903 B.C., of the city of Tyre)

Therefore, the following had taken place. Shalmaneser IV (827- 817 B.C.) began to have trouble with Armenia (Urartu) during his reign, therefore, King Zakir of Hamath and Luash began to oppose and seek independence from Assyrian rule in 819 B.C.. During Shalmaneser's last two years he was forced to intervene and to subject Damascus and Hatarika (Hamath area) once again under his suzerainty. The succeeding Assyrian monarch, Ashurdan III (817 - 799 B.C.), also had trouble with Hamath area.

Since Israel had been in a weakened state up to this time (II Kings 14:26), Damascus (Syria) provided a suitable prey as Hamath gained freedom from Assyria; therefore, King Zakir came out against Damascus. "Benhadad II, King of Syria, responded with a coalition of 12 to 16 kings against Zakir (II Kings 14:28). Zakir mentioned that only seven of the kings took part in the attack on Hazred (Hadrach - Zech. 9:1) the capital city of the northern Syrian principality on the Orontes".[1]

This southern coalition was greatly weakened by Zahir's victory which no doubt prepared the way for Uzziah (Azariah - Judah) to come into preeminence. Uzziah began his reign in the 6th year of Ashurdan and that same year Ashurdan did battle with Hatarika (Eponym list - Hamath - 811 B.C.). In his eighth year there was a terrible revolt in the city of Assur which lasted two years and spread later to the cities of Arrapha and Cozan. This internal upheaval in Assyria was so drastic that the whole Hamath area went over to Uzziah in Uzziah's third year (809 B.C.) King Ashurdan III gradually regained control and re-established peace in his country. He was once again able to attack Hatarika (Hamath - Eponym List) in his 16th year and no doubt recovered the lost territory. Ashur-Nirari V (799 - 792 B.C.) did not fare so well. The Assyrian kingdom was fast entering its death throes brought on by the lost territory in the south,and by the Urartian forces from the north, who continued their tireless attack southward. Ashur-Nirari's final year was climaxed with a revolt in Calah when the situation went beyond control, and King Ashur-Nirari V and his whole family were murdered by the Urartian forces. King Sarduris II had mastered the great Assyrian Empire in 791 B.C.. This is verified by Georges Roux's statement in his book "Ancient Iraq" where he says concerning the conditions in Assyria at this very time, "About 790 B.C., for several years 'there was no king in the country' confessed a chronicle".[2] This date exactly correlates with our 46 year gap theory in the Royal Assyrian Eponym List page 232 of this book. Urartu had finally gained full control, which she kept for the next forty-six years. The

1. Unger,M.F., "Archaeology & the Old Testament",p.251,Zondervan Pub.Hse
2. Roux, Georges, "Ancient Iraq" p. 251; Publisher Allen and Unwin
   Press, England

Assyrian bondage is also verified by another statement by Georges Roux as to the latter part of this forty-six year period when he says, "Thus, for 36 years (781 - 745 B.C. - added), Assyria was practically paralysed".[1] These victories brought the shroud down on the Assyrian Empire for the next 46 years, which they, so carefully, have been able to hid from the eyes of historians.

Uzziah, after a series of battles (II Chron. 26:1-15), now controlled all the territory from Egypt to Hamath. Uzziah's co-regent, King Jeroboam, died (perhaps just before or during the year of Uzziah's earthquake - Amos 1:1,2) in Uzziah's 27th year, and Uzziah ruled both kingdoms ten years alone (II Chron. 26:10; II Kings 15:1 - "over Israel"; II Kings 14:25,28 - 784 to 774 B.C.), during which time he had 2600 mighty men,307,500 soldiers, and engines of war (II Chron. 26:16). Thus, the great fertile crescent was providentially placed under the complete dominion of two great new powers, Urartu (King Sarduris II, Armenia), and Judah (Uzziah), until Pul, the descendant of King Adad-Nirari III, a soldier of fortune, began his persistent strikes at the foundation of the Urartian Empire in the Hamath area from 772 B.C.. He then pushed south to bring Menahem (Israel) and Uzziah (Judah) under tribute, after which, he gradually moved eastward putting Nab-shum-ukin I, King of Babylon, under tribute in 751 B.C.. Then, finally after 27 years of persistent war, Pul re-established the Great Assyrian Empire in 745 B.C. under his new name and as the great King, Tiglath Pileser III.

ASSYRIAN CHRONOLOGY
CORRELATED WITH THE EPONYM LIST

(Co-reigned from 908 B.C.) - - - Shalmaneser III		904-855 B.C.
Adad-Nirari III		955-827 B.C.
Shalmaneser IV		927-817 B.C.
Ashur-dan III		817-799' B.C.
Eclipse in the eighth regnal year - Ashur-dan's reign- - June 13,809 B.C.		
King and family killed by Urartu - Ashur-Nirari V		799-791 B.C.
(Armenian) Urartian Kingdom: Sarduris II		791-745 B.C.
Assyrian Empire re-established - - Tiglath Pileser III		745-727 B.C.
Shalmaneser V		727-722 B.C.
Sargon II		722-704 B.C.

1. Roux, Georges, "Ancient Iraq" page 251; Publisher Allen & Unwin Press England

The two kingdoms 887 - 760 B.C.
(Correlated Events - 127 years)

Years Reign	Year B.C.		Years Reign
ATHALIAH 6 years	886 1st	year was the 1st year of Jehu II Kings 10:36; 11:21	JEHU 28 years
JOASH 40 years	880 1st	year was the 7th year of Jehu II Kings 12:1	
	858 23rd	year was the 1st year of Jehoahaz II Kings 13:1	JEHOAHAZ 17 years
	844 37th	year Jehoash co-regent with Jehoahaz II Kings 13:10	
	841 40th	year was the 1st year of Jehoash II Kings 13:9,10; II Chron. 24:23	JEHOASH 16 years
AMAZIAH 29 years	840 1st	year was the 2nd year of Jehoash II Kings 14:1,2; 12:21	
	826 15th	year was accession year of Jeroboam II Kings 14:17-21	
	825 16th	year was the 1st year of Jeroboam Chart - correlation of Kings	JEROBOAM 41 years
UZZIAH 52 years	811 1st	year was the 15th year of Jeroboam II Kings 14:21; II Chron. 26:1	
	799 13th	year was the 27th year of Jeroboam (co-regency) II Kings 15:1 (reigned over Israel - Uzziah)	
	785 27th	year was 41st and last year of Jeroboam(Earthquake) Amos 1:1,2 - Earthquake of Uzziah	
	784 28th	year Uzziah reigns over both Kingdoms II Kings 14:25,28; II Chron. 26:10	UZZIAH* 10 YEARS
	776 36th	year was date of the first Olympiads Historical known date	ALONE
	775 37th	year Jotham co-regent at age of nine (Uzziah-Leprosy) II Kings 15:5; II Chron. 26:10-21	
	774 38th	year was the 1st year of Zechariah II Kings 15:8,12 (6 months)	ZECHARIAH 1 year**
	773 39th	year accession of Shalum & Menahem II Kings 15:13,14 (1 month)	SHALUM 1 year**
	772 40th	year was the 1st year of Menahem II Kings 15:17	MENAHEM 10 years
	762 50th	year was the 1st year of Pekahiah II Kings 15:23	PEKAHIAH 2 years
	760 52nd	year was the 1st year of Pekah II Kings 15:27 (Uzziah dies)	PEKAH 1st year

127 years (inclusive) - - - - - - - TOTAL - - - - - - - - 127 years

* Uzziah reigned over both kingdoms - II Chron. 26:10; II Kings 15:1
  (over Israel); II Kings 14:25,28; ** Historic chronology allows 1 year

During the reigns of Adad-Nirari III (855 - 827 B.C.) and Shalmaneser IV (827 - 817 B.C.) there was a growing movement south by armed forces from Urartu gradually gained power over the Medes to their East around Lake Urmia by destroying Sangibutu and Namri. Then they turned south and west and began to whittle away at the Assyrian Empire during the reigns of Ashurdan III (817 - 799 B.C.) and Ashur-Nirari V (799 - 792 B.C.). These attacks from the north by Urartu in coalition with Uzziah's (Azariah) attack from the south encouraged the defection from Assyria of other north Syrian states. The following quotation gives us an interesting historical resume of what transpired:

"Urartian advances along almost the whole of the northern frontier of Assyria continued during the years succeeding the reign of Adad-Nirari III, and, in the absence of a ruler of his calibre to minimize the consequences of this trend, Assyria suffered severely. Urartu gained a firm grasp on the regions immediately south of Lake Urmia and so controlled the trade routes from northern Iran. More serious still was the situation in the west, where the Urartian thrust dispossessed Assyria of almost the whole region north and west of Carchemish, thereby taking from Assyria control of the metal trade of Asia Minor. Besides the economic consequences, this must have had a direct effect upon the military efficiency of Assyria, since almost the whole of the area upon which Assyria depended for the supply of horses was now in Urartian hands. The economic effects of the cutting of the routes into Asia Minor led to disturbances in Syria and a number of campaigns were undertaken against Hatarika (Biblical Hadrach - probably coordinated with Azariah), Arpad and Damascus. It is during this period of Assyrian weakness that the reign of Jeroboam II of Israel is to be placed: the lack of any strong central control enabled him to extend his (co-regency with Azariah of Judah - added) borders at the expense of Hamath and Damascus (II Kings 15:25-28). There was also trouble from the districts along the Tigris south of Assyria proper, whilst within Assyria itself the economic distress arising from the cutting of trade led to revolts in a number of cities. There was a revolt in the capital itself, Calah (792 B.C. - added), and Ashur-Nirari V, the last of the three kings following Adad-Nirari III, was murdered with the whole of the royal family".[1]

The death of Ashur-Nirari V and the collapse of the Assyrian Empire took place in the 20th year of Azariah (Uzziah) and in the 34th year of Jeroboam, just seven years after Jeroboam became subjugated to Uzziah and just seven years before Jeroboam's death (785 B.C.). In the face of their common enemy, Urartu, Jeroboam was willing to be subjugated to and fight for Uzziah, which in turn, brought Uzziah (Azariah) renown and fame before all nations as a powerful force and influence in the Middle East as the following scriptures confirm: " - - And his (Uzziah's) name

1. Saggs, H.W.F., "The Greatness that was Babylon", page 104, Sidgwick & Jackson, Publishers

spread abroad even to the entering in of Egypt for he strengthened himself exceedingly. - - - And his name spread far abroad; for he was marvellously helped (by Jeroboam), till he was strong" II Chron. 26:8,15b.

Then Jeroboam's death left Uzziah in complete control of the northern and southern kingdoms (for 10 years) with their combined military strength under his hand: "Moreover Uzziah had an host of fighting men, that went out to war by bands, according to number of their account by the hand of Jeiel the scribe and Maaseiah the ruler, under the hand of Hananiah, one of the king's captains. The whole number of the chief of the fathers of the mighty men of valour were two thousand and six hundred. And under their hand was an army three hundred thousand and seven thousand and five hundred, that made war with mighty power, to help the king against the enemy" II Chron. 26:11-13. That enemy mentioned above was no doubt Urartu, and if it had not been for Uzziah's great strength at this time, Israel and Judah would also have been prey for the Armenian Kingdom (Urartu).

II Chron. 26:10 states that Uzziah had "- - husbandmen also, and vine dressers in the mountains, and in Carmel: for he loved husbandry". The mention of Mount Carmel above is the final proof that Uzziah was in full control of the "Northern Kingdom" where the mount is located which acts as a shroud for the present day city of Haifa, Israel.

His fame, success, and power no doubt went to his head and he grew proud, arrogant, and forgot God who had been the very means of his rapid rise to such a pinnacle (see Dan. 2:21). Therefore, God smote him with leprosy when he trangressed in the temple of his God (II Chron. 26:16-21). Because of his sickness, Jotham (about nine years old; 775 B.C. - II Kings 15:33; II Chron. 26:21) became co-regent with Uzziah during, at least, the last sixteen years of his 52 years reign. Ten years had passed since he had begun his rule alone over the Northern Kingdom (Israel). Because of Jotham's youth and Uzziah's sickness, Zachariah (Israel) saw his chance to revolt and to re-establish the kingly rule, after 10 years, in the Northern Kingdom. This he accomplished in the 38th year of Uzziah (II Kings 15:8-12). Uzziah's sickness and his weakening power was also a sign to a great soldier of fortune in Assyria that it was time for him to move to regain a portion of the former Assyrian Empire and gain a foothold against the Urartian Kingdom which now had ruled Assyria for twenty years. This man was Pul, a descendent of Adad-Nirari III, a rightful heir to the Assyrian throne. He heard rumors of Uzziah's sickness, of young Jotham's co-regency, and of Zachariah's grab to recover the "Northern Kingdom", which all happened within several years (775 - 773 B.C.). This was the proper time to move and reclaim Hamath, Assyria's rightful territory, which was under Uzziah's rule, and regain his first foothold on the border of the Urartian Kingdom. This Pul began to do in 772 B.C. with success and with little opposition as extant history declares: "Of Azariah (Uzziah) my hand mightily captured - - nineteen districts of the town Hamath, together with the towns in their circuit, which are situated on the sea of the setting sun, which in their faithlessness made revolt to Azrijahu

53

(Azariah - Uzziah), I turned into the territory of Assyria. My officers my governors I placed over them".[1]

In the light of this information we must be careful in separating the historical events available,since some may have reference to the period of twenty-seven years during which Pul was gradually regaining the kingdom, while other events would have reference to a later period beginning with the year 745 B.C. when Pul ruled as Tiglath Pileser III. This will take a great deal of further study on the part of interested historians. Perhaps this information is already in the notes of some historians and they are waiting for just this information to correlate their finds.

Therefore, Pul's reference to "the tribute of Kustaspi, of Kummuha, Resin of Syria, Menahem of Samaria, (here followed the names of 14 other kings and one queen), I received".[2] This, no doubt, took place during the first few years after he took Hamath in 772 B.C.. Even though Resin, of Syria, was later killed by Tiglath Pileser in 742 B.C. (II Kings 16:9), Resin's reign extended back through Pekah's, Pekahiah's and Menahem's reigns; therefore, he could easily have paid tribute with Menahem as the quotation states. Uzziah also paid tribute to Pul at the same time as the quotation from the Assyrian Eponym Canon states:

"In the course of my campaign - - tribute of the Kings - -Azrijahu of Judah (Azariah - Uzziah)".[3]

Uzziah died in 760 B.C. in the first year of Pekah (II Kings 15:27,32) just two years after the death of Menahem; therefore,Uzziah's tribute had to be during the days of Menahem.

Pul no doubt conquered Hamath in Uzziah's 40th year, which was Menahem's first year (772 B.C.), or at least by 770 B.C.. By his third year 770 B.C. he had consolidated his position where he felt that he could move south into Syria and Palestine. This Pul did by putting Syria (Resin), Israel (Menahem), and Judah (Uzziah - Jotham) under tribute. This move reinforced his position and prestige and brought in needed financial support to sustain his first moves north against the Urartain Kingdom. This was a slow process and continued for the long period of twenty-seven years during the remainder of Menahem's, Pekahiah's, and until the 16th year of Pekah's reign (745 B.C.). At which time Pul overthrew the Urartian Kingdom and re-established Assyrian's monarchal rule under his new name Tiglath Pileser III. Assyria had been wrestled from them by King Sarduris II of Urartu just forty-six years before. Pul ruled over only a portion of Assyria for twenty-seven years (772-745 B.C.) and later ruled an additional eighteen years (745-727 B.C.), as Tiglath Pileser III, over the reunited Assyria.

Amos the prophet prophesied in 787 B.C., during the days of Jeroboam just two years before Uzziah's earthquake, concerning the revived Assyrian power under Pul and stated: "But, behold, I will raise up (one

1,2,3, Anstey,M., "Romance of Bible Chronology", P. 201

who rises - Pul = Tiglath Pileser III) against you a nation, O house of
Israel, saith the Lord the God of hosts; and they shall afflict you from
the entering in of Hamath unto the river of the wilderness" Amos 6:14.

The early importance of the Urartian Kingdom which faced Pul is well
verified by history and proves that their move at this time into
the middle east was not a sudden invasion, but a gradual growth of power
which threatened Asssyria for many years, as the following quotation
confirms: "The new kingdom of Urartu, first mentioned by Ashur-nasir-pal
II (928 - 904 B.C. - added), became a formidable power in the time of
Shalmaneser III (904 - 868 B.C.) and completely altered the character of
the opposition to be met in the north. There now commenced a struggle
which last over a hundred years - - ".[1]

As to the deleted forty-six years in the Eponym List during which the
Urartians ruled over Assyria, M. Anstey suggests that "Assyria was over-
taken by some disaster", or that the "names were either lost by
accident, or destroyed by design".[2] Syncellus states that the records
for this period were tampered with, and he assigns this as the reason
why Ptolemy's Canon went back no further than 747 B.C..[3]

Edwin R. Thiele says, "The historical records of Tiglath-Pileser were
mainly engraved upon stone slabs which originally lined the walls of his
palace at Calah (Nimrud), but which were later removed by Esarhaddon to
be used in his palace in the same city. Here they were found by Layard
in his excavation of what he termed 'the Southwest Palace' of Nimrud:
Layard gives a vivid discription of the mutilated and disordered
conditions in which these slabs were found,[4] some sawn in two with only
a portion of the original slabs remaining, many with the original carv-
ings completely chiseled away to be replaced by new inscriptions, and
yet others that had been exposed to fire with the stone nearly reduced
to lime and too cracked and fragile to permit removal".[5]

In the light of the above and the following quotations, it would only
seem logical for Esarhaddon, son of King Sennacherib, to have destroyed
the records of the forty-six years of Urartian rule, which included records of the Urartian attack, and Pul's first twenty-seven years assault
against the Urartian Kingdom before gaining his full and final power:
"So Sennacherib king (only a General at this time - Rabshakeh under
Sargon - II Kings 19:8) of Assyria departed (from Jerusalem - 713 B.C.),
and went and returned, and dwelt at Nineveh (for another 32 years - ruling 704 - 681 B.C.). And it came to pass, as he was worshipping in the
house of Nisroch his god, that Adram-Melech and Sharezer, his sons,
smote him with the sword (681 B.C.), and they escaped into the land of
Armenia (Urartu). And Esarhaddon, his son, reigned in his stead" II Kings
19:36,37.

1. The Cambridge Ancient History, Vol. II, p.20, Cambridge Univer. Press
2. Anstey, M., "Romance of Bible Chronology", Vol. I, page 40
3. Ibid, page 220
4. Layard, A.H., "Nineveh and its Remains", Vol. II, p.26, Eyre & Methuen
5. Thiele, E.R., "The Mysterious Numbers of the Hebrew Kings" p.94, Eerdmans

It is interesting to note that history confirms that Esarhaddon took immediate revenge for his father's death, as confirmed by broken clay cylinder (III, Rawlinson 15, Col. 1, 18ff.), when he attacked Urartu and overthrew them: "The terror of the great gods, my lords, overthrew them. They saw and dreaded the meeting. Istar the mistress of conflict and battle, who loved my priesthood, raised my hands, broke their bow, cleft through their battle array; in their assembly resounded the cry 'This is our King'".[1]

Thus, sixty-five years had passed and only four Assyrian kings had sat on the throne since the dark forty-six year period had overshadowed Assyria. And here, once again, the Armenians(Urartu) had either engineered the plot, or, at least, promised refuge to Sennacherib's two sons, Adram-melech and Sharezer, who killed their father with the sword (II Kings 19:37). This was another indirect incursion by Assyrian's former enemy the Urartians (681 B.C.). Therefore, in his pursuit to get every possible revenge for the death of Sennacherib, his father, Esarhaddon, after destroying them in battle, attempted also to erase every record of the Urartian power during those tragic forty-six years while Assyria was subject to their oppression. As a result of the mutilation of those records he has, literally, confused the issue from that day until now.

Later, those same Urartian tribes were united with Minni, and Ashchenaz and were designated as the kings of the Medes (Jeremiah 51:27,28). They, with Persia destroyed Babylon in 539 B.C. accomplishing final revenge over the Middle East by bringing the Babylonian Empire, which included Assyria, to its end. They helped to enthrone King Cyrus, who ruled and instigated the Persian World Empire, which lasted for the next two hundred and eight years (539 - 331 B.C.).

1. Anstey, M., "Romance of Bible Chronology", Vol. 1, page 215

Since the 52nd and the last year of Uzziah (760 B.C.) was the first year of Pekah (II Kings 15:27); therefore, Jotham's first year was Pekah's 2nd year (II Kings 15:32 - 759 B.C.).Therefore, the 15th year of Jotham (745 B.C) was Pekah's 16th year. This was the year that Tiglath-Pileser III, as King, gained control of most of Assyria from Urartu. Ahaz succeeds Jotham in Jotham's 16th year (accession year - II Kings 16:1 - 744 B.C.). The following year (743 B.C.) Tiglath-Pileser III is in Arpadda and completed the "Massacre in the land of Urartu". During that year Ahaz began his first year of reign in the 18th year of Pekah, and Jotham, no doubt, is either put in retirement or imprisonment at the orders of Assyria.

The tables are now completely turned to what they were fifty years before when Urartu and Judah had gained full control of the fertile crescent. Now, Tiglath-Pilesers III had reconquered his own country and had brought areas of Urartu under his subjugation. Palestine states also, under the reigns of Ahaz and Hoshea, were bowing completely to Assyria.

In the mean time Resin of Syria and Pekah of Israel attacked Ahaz (Judah) during his second year "but could not overcome him" II Kings 16:5. Later, "they smote him and carried away a great multitude of them captive, and brought them to Damascus" II Chron. 28:5. At the same time Pekah smote many of Judah and carried 200,000 away to Samaria. They were later returned after the rebuke from the prophet Oded (II Chron. 28:9-11). Because of these attacks Ahaz was desperate and "sends unto the kings of Assyria to help him" II Chron. 29:16; II Kings 16:7-9. He also took the silver and gold which was found in the house of the Lord and paid Tiglath-Pileser III to come against Damascus, at which time Resin, King of Syria, was slain. Ahaz responded by going to Damascus personally and later introducted heathen customs into the temple worship.

In the fourth year of Ahaz and the 20th year of Jotham (he had been imprisoned four years), Hoshea, an all out collaborator with Assyria, slew Pekah (II Kings 15:30), and was made governor (King) of Israel under the hand of Tiglath-Pileser III as information of his Gaza and Damascus campaign confirms: "They overthrew Pekah their King and I (Tiglath-Pileser III) made Hoshea to be king (governor) over them". "The cities Gil(ead), Abel (Beth-maacha?) - - which is the boundary of the land of Beth-Omri (Samaria) I turned in its entire extent into the territory of Assyria. I set my officers and viceroys over it". "The land of Beth-Omri (Samaria) - - the goods of its people and their furniture I sent to Assyria. Pekaha (Pekah) their King (I had killed?) - - Asiri (Hoshea) I appointed over them - - - their tribute of them I received".[1]

Hoshea began his rule as an Assyrian puppet in Ahaz's 4th year (II Kings 15:30) which was 740 B.C.. Hoshea continued in this position until the

1. Anstey, M., "Romance of Bible Chronology", Vol. 1, page 203

12th year of Ahaz (II Kings 17:1 - 732 B.C.) at which time he gained freedom from Assyria. These eight years of puppet rulership under Assyria are not mentioned in the Bible except for the first year (II Kings 15:30) and the final year of deliverance (II Kings 17:1). During these eight years Ahaz (Judah) was also completely subject to Assyria (II Kings 16:7-10) and compromising in many ways (II Chron. 28:20). In the sixth year of this eight year period Ahaz was attacked by the Philistines and Edom (734 B.C.), so Ahaz again called on Assyria for help (II Chron. 28:16-20). The Eponym List confirms that Assyria heeded his call and came against Philistia in that very year. The following two years (733 -732 B.C.) Assyria attacks Damascus, and during the second year (732 B.C. - the 12th of Ahaz), Hoshea obtains his freedom or independence from Assyria. This is stated as Hoshea "began - - to reign" (or gained freedom from Assyria), yet there was no doubt a transition period of two years before Hoshea was crowned as King in Israel. Chronological study proves this clearly. Hoshea's enthronement falls in Ahaz's 14th year (730 B.C.). In that year, in which Hoshea was crowned, there was no danger from Assyria, according to the Eponym List (pages 233-234), since Tiglath-Pileser III stayed in his land. Also, during the following two years he was involved with Babylon where he made himself (Pulu - Pul) King of Babylon in 729 B.C.. His interest in the east eased the pressure on Hoshea. Pul's second year in Babylon (728 B.C.) ended his reign over both kingdoms by means of his death.

Shalmaneser V succeeded Tiglath-Pileser III in 727 B.C., which was the the same year that Hezekiah began to rule Judah. In the meantime, Shalmaneser V had heard of Hoshea's conspiracy, while his father was involved in Babylon, and he immediately deployed his troops toward Damascus during the first year of his reign (727 B.C. - Eponym List). He put Hoshea under tribute allowing him to continue to reign as King of Israel. These tributes and presents demanded by Assyria were too much of a burden to Hoshea and resulted in another conspiracy (II Kings 17:4). Hoshea sought help from the Egyptian King So (Shabalaka). This brought an immediate reaction from Shalmaneser V, who, "in the fourth year of King Hezekiah (724 B.C. -inclusive), which was also the seventh year of Hoshea son of Elah King of Israel - -", besieged Samaria (II Kings 17:5; 18:9). He captured Samaria after three years in the 6th year of Hezekiah (722 B.C.), which was the 9th and last year of Hoshea (II Kings 17:6; 18:10). Samaria went into Captivity to Assyria that year.

In our exposition thus far, we have been logical and honest in our approach. We have used only the Judah chronology as the true, authentic, and uninterrupted measurement. The three hundred and ninety year span of time has been our year guide in the chronology of Judah, thereby giving appropriate and simple answers to the correlation with Israel's side of the chronology. There is no logical reason to doubt the integrity of these answers, nor deny the unalterable proof of them. If we do, as some have tried, we only create for ourselves a dozen more problems even more difficult to solve. We feel that the seeming impossible problems that the inspired Hand of God has providentially allowed as stumbling blocks for the unbeliever (I Peter 2:8; II Peter

58

3:36) are far more easy to solve, than the multitude of contradictions which result in an attempt to overlap reigns in a short chronology.

Therefore, we have arrived at the date 722 B.C. which is the 6th year of Hezekiah and the ninth and last year of Hoshea. The Northern Kingdom of Israel has fallen, and many of its inhabitants have been carried away into Captivity in Assyria. Sargon, the new King of Assyria, had been enthroned during the last year of the battle for Samaria. If by chance Sargon ruled any part of 722 B.C., then his first year would definitely be dated 721 B.C.. There is a problem of historical reference to Sargon taking Samaria rather than Shalmaneser V? Sargon II (Sharrukin) was, no doubt, a co-regent of Shalmaneser V, or perhaps a general during the three year attack on Samaria; therefore, later as King, he could claim rightful honor for the capture of Samaria. Again, there is a possibility that Shalmaneser V may have died or was killed during the last phase of the attack on Samaria, thereby, both kings could rightfully claim honor for taking the city. Personally, we do not feel that this creates such a problem as some make out. Should not an army general, who later becomes king, claim that he captured the city when he actually led the attack? We admit that it hedges on plagiarism, used for self glory, yet, it is undeniably a half truth, and adds self esteem when a king is writing his own autobiography.

The use of such plagiarism is a simple answer to the so called problems involved with Sennecherib's first four campaigns as general (Rabshakeh - II Kings 19:8), all of which took place under Sargon's rulership. These victories he (later as king) claimed as his own since he was the general who led the actual attack. To beg the question and make an insurmountable problem out of it is intellectual foolishness. To call this a discrepancy simply proves that some are more interested in disproving the Holy Word of God for self-justification, rather than vindicating the same.

George Smith states, "Sennacherib held some official rank during his father's reign, and it is quite possible that he commanded the expedition in 711 B.C. as his father's deputy. In the Tablet K 2169, Sennacherib is called 'Rabsaki' (Rabshakeh) or General, and 'great royal son', that is, heir to the throne; and he is said to possess his own scribe. The passage reads: "Tablet of Aia-suzubu-ilih the Scribe of the Rabshakeh, of Sennacherib, the great royal son of Sargon, King of Assyria".[1] (It could even be a co-regency).

There is another inscription on tablet No. 105 on Table-case D in the Nineveh Gallery of the British Museum that Sennacherib sent to his father extracts from despatches which he had received concerning imperial affairs. In the account of his fourth campaign, Sennacherib says: "Merodach-baladan, on whom I, in my first military expedition (720 B.C.) inflicted a defeat, and whose force I had broken in pieces, dreaded the onset of my powerful weapons and the shock of my mighty battle (during his 4th campaign in 711 B.C. - added)"[2] (Merodach - 722-709 B.C.).

1,2. Anstey, M., "Romance of Bible Chronology", Vol. 1, page 213

Therefore, in the light of the above facts we can read II Kings 19:8 as follows: "So Rabshakeh (Sennecherib as General) returned, and found the king of Assyria (Sargon) warring against Libnah: for he had heard that he was departed from Lachish" (See also Isa. 20:1; 37:4).

Now, let us reconstruct Sargon's battles under the generalship of Sennacherib. It is generally accepted that Sargon did not go to battle during his first reigning year (721 B.C.). During his second reigning year, Sargon sent Sennacherib on his "first campaign" against Merodach-baladan. Afterwards, he sent him on against the Egyptian King So (Shabalaka in 720 B.C., whose army was under the generalship of Egypt's Sibahki or Shubitku, who was later King - 718 to 715 B.C.). On the way, Sennacherib attacked Damascus, Samaria, captured Hanun, King of Gaza, sending him and his family to Ashur, and defeated So in his attack at Raphia.

It might be interesting to note that Assyria also subtly chose two sabbatical years for their attacks through Palestine at this time. One attack in the Sabbatic year 720 B.C. during Sennacherib's <u>first campaign</u>, under the Kingship of Sargon, and the next attack in the Sabbatic year 713 B.C., during Sargon's 9th year, and Hezekiah's 14th year (II Kings 18:13). Of course during a Sabbatic year they left their land to lie idle and food could possibly be at a shortage (Lev. 25:1-17). A very appropriate time to take advantage of their enemy. Later, just before the fall of Jerusalem in 586 B.C., Babylon used the same tactics. Three of Babylon's attacks on Jerusalem also fell on the Sabbatical years 605 B.C., 598 B.C., and 591 B.C.. The latter attack ended in a forced withdrawal by Babylon in the face of a threat from Egypt (Jeremiah 34).

Sennacherib's "second campaign" was in Sargon's 7th year (715 B.C.) when "Sargon(was) conqueror of the Tamudu (an Arabian tribe), Ibadidi, Marsimani, and Hayapa, who the rest of them enslaved, and caused them to be placed in the land of Beth-Omri (Samaria)". The "third campaign" is definitely identified as the attack on Hezekiah in Sargon's ninth year (713 B.C. - Sabbatic)which was the 14th year of Hezekiah (II Kings 18: 13). The Jubilee year followed (712 B.C.), and the third year (711 B.C.) they ate of the land (Isa. 37:30). In this first year (713 B.C.), Isaiah truly prophesied the destruction of Egypt three years later as he dramatically speaks forth the Words of God found in Isa. 20:1-6: "In the year that Tartan (Sennacherib) came unto Ash-dod, and took it; at the same time spake the Lord by Isaiah the son of Amoz, saying, Go and loose the sackcloth from off thy loins, and put off thy shoe from thy foot. And he did so, walking naked and barefoot. And the Lord said, Like as my servant Isaiah hath walked naked and barefoot <u>three years for a sign</u> and wonder upon Egypt and upon Ethiopia; so shall the King of Assyria lead away the Egyptians prisoners, and the Ethiopians captives, young and old, naked and barefoot, even with their buttocks uncovered, to the shame of Egypt. And they shall be afraid and ashamed of Ethiopia their expectation, and of Egypt their glory. And the inhabitant of this isle shall say in that day, Behold, such is our expectation, whither we flee for help to be delivered from the king of Assyria: and how shall we

1. Anstey, M., "Romance of Bible Chronology", Vol. 1, page 207

escape?". Thus, the mighty power of Assyria was manifest strongly through out the Middle East.

During Sennacherib's "third campaign" (713 B.C.), Jerusalem was threatened (Hezekiah), and Haran, Rezeph, Telasar, Hamath, Arphad, Sepharvaim, Hena, Iva, Samaria (Isa. 37:12,13), and many of the cities of Judah (Isa. 36:1) had been taken. Sennacherib was now in the process of taking Lachish (Isa. 36:2), Libnah (II Kings 19:8), Ashdod (Isa. 20:1), Askelon, Ekron, Timnath, and Eltakeh where he later defeated Egypt (Taylor Cylinder). Even though there were warnings of Egypt's interference (Isa. 37:9; 36:6) at that time, Egypt's defeat took place three years later in 711 B.C. fulfilling Isaiah's prophecy (Isa. 20:1-6 quoted above). That year of attack on Egypt was also exactly the third year after Hezekiah's sickness (Isa. 37:30), during which time God had promised Hezekiah that Judah would eat freely of the land. Therefore, the defeat of Egypt and the withdrawal of Assyria's armies left Judah in comparative peace for sometime after 711 B.C.. The reason being that Assyria was directly involved with Babylon during the following years.

Therefore, during Sennacherib's "fourth campaign" in Sargon's 11th year, Sennacherib repeated his attack on Ashdod because of a coup d'etat which had taken place there. During the former battle (9th year of Sargon) Sennacherib replaced King Azuri of Ashdod and appointed Ahimite (Achimite), his brother, over the kingdom. The people soon revolted after Sennacherib had left through Hittite influence, drove away Ahimite and appointed Yavan as King. This rebellion at Ashdod was the introductory disturbance which engineered Sennacherib's "fourth campaign" which included Egypt's final defeat according to God's Word. When Yavan heard of Sennacherib's approach, he fled to Egypt to the "boundaries of Meroe". Sennacherib took Ashdod and appointed Metinti as King of Ashdod. He continued his attack and reached Eltaketh where he was met by the Egyptian army. Egypt was defeated and forced to return Yavan to Assyria.

Many historians have sought to correlate Sennacherib's "eight campaigns" completely separate from Sargon's. But when we synchronize all facts, details, dates, and known reigns of contemporary kings, we come to one conclusion. The "first four" of Sennacherib's "eight campaigns" are identical to "four of the campaigns" of King Sargon during which Sennacherib was Commander-in-chief (Tartan) of Sargon's forces (II Kings 17:6). Please notice that these first four campaigns had to be during Merodach's first reign (722 -710/9 B.C.), which in turn is absolutely confirmed by scripture and statements quoted about Sennacherib on page 60. (Also see correlating details on chart page 43).

Since we prorate the extra 9 months during the year 716 B.C., this appropriately defers Hezekiah's 14th year and his sickness to the 9th year of Sargon, which was the year which he approached Jerusalem. This also postponed the threat to Jerusalem correctly to the Sabbatic year 713 B.C.. That Sabbatic year suitably pinpoints the first of the "three consecutive years", and during the last of the three, they would eat of the land according to Isaiah 37:30. Therefore, this correctly proves, and establishes our chronology, and provides an appropriate place to prorate the extra 9 months for the prophetically allowed "year" for Jehoiachin's 3 months reign.

SENNACHERIB'S EIGHT CAMPAIGNS

721/0 B.C. - New Years Day - Sargon's army against Babylon

720 B.C. Sennacherib's 1st Campaign (General) Sargon's 2nd-to Babylon

715 B.C. Sennacherib's 2nd Campaign (General) Sargon's 7th-to Egypt

713 B.C. Sennacherib's 3rd Campaign (General) Sargon's 9th-to Ashdod

711 B.C. Sennacherib's 4th Campaign (General) Sargon's 11th-to Egypt

   709 B.C. Sargon took Babylon for five years

   704 B.C. Sennacherib becomes King of Assyria

702/1 BC Sennacherib's 5th Campaign (King's 3/4th year) to Babylon

694 B.C. Sennacherib's 6th Campaign (King's 10th year) to Elamites

691 B.C. Sennacherib's 7th Campaign (King's 13th year) to Khalute

689 B.C. Sennacherib's 8th Campaign (King's 15th year) Destroys Babylon

Hezekiah's extra 15 years of life would now date from 713 B.C. rather than 714 B.C., whose 15 years includes the years 712 B.C. to 698 B.C. inclusive. The first year of Manasseh now falls on the correct year 697 B.C.. Judah's chronology follows until the fall of Jerusalem in 586 B.C.:

Manasseh	55 years	697 - 642 B.C.
Amon	2 years	642 - 640 B.C.
Josiah	31 years	640 - 609 B.C.
Jehoahaz	1 year(3 mos)	609 B.C. (July to September)
Jehoiakim	11 years	609 - 598 B.C. (Dec. 6th)
Jehoiachin	1 year(3 mos)	598 - 597 B.C. (Mar. 16)
Zedekiah	11 years	597 - 586 B.C. (Aug. 10 or 24)

Now, we can see, even though prophetically Ezekiel's prophecy allows a full year for Jehoahaz and Jehoiachin's reigns, why we do not immediately add the full year during their actual reigns, since it would throw the chronology out of balance. We prorate the extra months at two appropriate places in the chronology where they are needed. This clearly balances the chronology rightly and solves all scriptural problems.

This third phase or step in our chronological framework is comparatively short and has no serious problems to consider, and very few events to correlate. This phase simply extends back from the division of the king- dom (982 B.C.) for exactly forty years. These forty years are identical to Solomon's forty year reign (I Kings 11:42; II Chron. 9:30). His reign, even though he built the temple prepared by his father, David, and experienced the Lord's appearance twice, ended in evil as the fol- lowing scripture confirms:

"And the Lord was angry with Solomon, because his heart was turned from the Lord God of Israel, which had appeared unto him twice, and had com- manded him concerning this thing, that he should not go after other gods: but he kept not that which the Lord commanded. Wherefore, the Lord said unto Solomon, 'Forasmuch as this is done of thee, and thou hast not kept my covenant and my statutes, which I have com- manded thee, I will surely rend the kingdom from thee, and will give it to thy servant. Not withstanding in thy days I will not do it for David thy father's sake: but I will rend it out of the hand of thy son. Howbeit I will not rend away all the Kingdom; but will give one tribe to thy son for David my servant's sake, and for Jerusalem's sake which I have chosen'" I Kings 11:9-13.

Therefore, Ezekiel's additional prophecy extends and adds to the 390 year period of Israel's sin. This extra span of another forty years were added making, as the following scripture verifies, a total of 430 years of sin for the Southern Kingdom of Judah, and dating from the fifth year of "Jehoiachin's Captivity" (390 + 40 = 430 years, or from 592 to 1022 B.C.): "And when thou (Ezekiel) hast accomplished them (390 days equal- ing 390 years) lie again on thy right side, and thou shalt bear the iniquity (past sins) of the House of Judah forty days: I have appointed thee each day for a year" (40 days equal 40 years) Ezekiel 4:6.

Thus, it was 430 years from the beginning of Solomon's reign (1022 B.C.) to the first year of Ezekiel's prophecy (592/1 B.C.), which was also the fifth year of "Jehoiachin's Captivity" (Ezekiel 1:1,2). This clearly identifies the total years of Judah's sins at this time through Ezekiel's two prophecies, and limites the total period of his personal prophecies to six years (seven years inclusive) during which Ezekiel prophesied before the Babylonian Captivity of Jerusalem (592/1 B.C. - 586/5 B.C.). This also makes a total of 436 years from the end of David- 's reign, or the beginning of Solomon's reign, to the Babylonian Captivity. Then providentially, it was also exactly 436 years from the beginning of King David's reign (1062 B.C.) back to the Egyptian Cap- tivity (1498 B.C.). Therefore, we see a miraculous scriptural pattern unfolding before our eyes.

Thus, these last forty years, which were added to 982 B.C., brings us to the year 1022 B.C., the first year of Solomon's reign. Therefore, his fourth year was 1018 B.C., which was the year that Solomon's temple was begun according to I Kings 6:1. (Later, this point, the 4th year of Solomon, 1018 B.C., will be the beginning year of our measurement for our next or fourth phase of our chronology).

Now, let us consider the scriptural confirmation concerning the date of the beginning of Solomon's temple, the years of building, and several strange coincidence involved around these dates: "In the fourth year (of Solomon's reign - see I Kings 6:1) was the foundation of the house of the Lord laid, in the month Zif (May). And in the eleventh year, in the month of Bul (November), which is the eighth month, was the house finished throughout all the parts thereof, and according to all the fashion of it. So was he seven years in building it" I Kings 6:37,38 (Nisan year - March 14 to March 14).

If we used the accession year theory, then the date of the beginning of the temple would be Nisan year 1019/18 B.C., while the regnal year would be Nisan 1018/17 B.C.. In the light of other scriptural considerations the regnal year is more acceptable.

According to the above quoted scripture the total time taken in building and fashioning the temple was exactly seven years and six months. Since the latter part of the verse states it took seven years to build, we conclude that the last six months were used to fashion and decorate the temple in all its beauty. Thus, using the regnal year system the years would fall as follows:

SOLOMON'S TEMPLE
(Nisan year regnal system)

Solomon's	4th year 1018/17	(May to May)	1st year of building
Solomon's	5th year 1017/16		2nd year of building
Solomon's	6th year 1016/15		3rd year of building
Solomon's	7th year 1015/14		4th year of building
Solomon's	8th year 1014/13		5th year of building
Solomon's	9th year 1013/12	(Sabbatic year)	6th year of building
Solomon's	10th year 1012/11	(Jubilee year)	7th(Temple completed)*
Solomon's	11th year 1011 B.C.	(May to Nov.)	Temple decorated

* I Kings 6:38 1. Josephus, Book 15, chapter 11, section 1

Since the first Adam's expulsion from Eden was by God on the first Jubilee disaster year 4012 B.C. (all Jubilee years fall on the 12th and on the 62nd year of every century B.C.), then we discover that the temporal building of Solomon's Temple was completed exactly 60 Jubilees later (on the Jubilee year - 1012/11 B.C.), or after 420 Sabbatic years, and also exactly 3000 years since the expulsion of Adam. It is also rather a strange coincidence that while Solomon's Temple was begun in 1018 B.C., Herod's temple was begun exactly 1000 years later on the very anniversary (18 B.C.)[1] (Herod ruled from 37 B.C. and the temple was first begun in his 18th year, which was 19 B.C., but it took some time to gather materials for the building). Is this the reason that

Herod began to build the temple during the year 18 B.C. to celebrate the 1000th anniversary of the beginning of Solomon's temple? Solomon completed his temple during the Jubilee year seven years later in 1012 B.C. (inclusive - see the chart above). Could it be possible that seven years after Herod began to build the temporal tabernacle in 18 B.C., Christ's physical temple was completely formed seven years later by His birth during the 80th Jubilee year 12 B.C. (inclusive - see Appendix A)? This date, 12 B.C., was the 80th Jubilee since the expulsion of Adam, and was preceded by 560 Sabbatic years. Also, 12 B.C. was exactly 4000 years since the expulsion of Adam from the Garden of Eden. Christ Jesus of Nazareth would have been conceived nine months before or during the Sabbatic year 13 B.C. (Since a Sabbatic year immediately precedes every Jubilee year). Therefore, according to historical data, Christ Jesus absolutely had to be born sometime during the years between 12 and 6 B.C.(A complete discusion on the possible date of the birth of Christ follows at the end of the book).

While we are on this subject, it is interesting to note that Moses tried to deliver his people out of Egypt during the 49th Jubilee period and failed, but finally, after 40 years in the wilderness and after the 50th Jubilee (which was the Jubilee of Jubilees since Adam's expulsion, or 2500 years later in 1512 B.C.), on the exact second Sabbatic year after that, he succeeded in bringing the children of Israel out of Egypt in the Sabbatic year 1498 B.C..

It is also interesting to note that David, a type of Christ, took the kingly throne on a Jubilee year (1062 B.C.). This should make an interesting study for any one interested after one establishes a true chronology, and correlates it with the true Sabbatic and Jubilee years (see Appendix B).

As a closing confirmation it might be interesting to many to mention that Ussher's chronology agree exactly, at this point, with our date of Solomon's Temple.[1]

1. I Kings 6:1, Scofield Reference Bible, 1945, Oxford University, N.Y.

We have now completed the 390 year period, and have considered the third phase revealed in Ezekiel 4:6, which extends our mensuration another forty years. Thus, pinpointing the beginning of Solomon's reign of 40 years, which was considered evil (I Kings 11:9-11 - 1022 to 982 B.C.). Now, we must retrogress four years to the 4th year of Solomon (1018 B.C.) to find our launching pad for our fourth phase or stage in our chronological plan: "And it came to pass in the <u>four hundred and eightieth year</u> after the children of Israel were come out of the land of Egypt, in the·<u>fourth year of Solomon's reign</u> (1018 B.C.) over Israel, in the month of Zif (May), which is the second month (Nisan year), that he (Solomon) began to build the house of the Lord ("Solomon's Temple" I Kings 6:1).

Since the temple took seven years to build (1018 - 1012 B.C. inclusive), and the fourth year of Solomon was the first year of the building of the temple (1018 B.C.), this point is the place of our departure for our fourth phase. The above scripture confirms that it was exactly 480 years from 1018 B.C. back to the point of Israel's release from the Egyptian Captivity and the Exodus of Israel from the land of Egypt. This Exodus date (1498 B.C.) is known to be the time of the death of Thutmose II (Egypt) and the beginning of Hatshepsut's reign. There has been some confusion concerning the details of these events, but it is now possible, with information at hand, to correlate the events in detail. With the present knowledge of the exact date of the Exodus, we can come to some definite conclusions.

Thutmose II, the fourth king of the Eighteenth Dynasty, reigned over Egypt the last years of the sixteen century (1510 - 1498 B.C.). He, as a half-brother, was married to his sister Hatshepsut (Common among the 18th Dynasty kings), the true heir to the throne. On the earlier monuments of their joint reign Hatshepsut is depicted as occupying the ordinary position of a Egyptian queen and high-priestess in temple liturgy, and otherwise playing a subordinate role.[1] But this did not long prevail. It is known that she staged a coup d'etat in the year 1501 B.C.. Thus, the reliefs began to depict her as on an equal footing with her brother-husband for the last three years of his reign (1501 to 1498 B.C.). Thutmose II was a weakling, physically, and perhaps also intellectually; therefore, Hatshepsut, no doubt, took advantage of this and took full power in 1501 B.C., while Thutmose II continued on as a figure-head with Hatshepsut pulling the strings of authority behind the throne. This continued, according to our chronological measurement, for a period of three years during which time Moses appeared once again on the scene after an absence of forty years (1538 - 1498 B.C.) spent amongst the Midianites in the Sinai Peninsula. He returned with the

1. Blackman, A.M., "Luxor and its Temples", page 126, **A & C Bloch Ltd**

direct command from (El or Jah) God to Pharaoh to "Let my people go". The following quotation justifies such a view, and confirms that Moses was dealing personally with these two monarchs: "On her beautiful monument at Thebes she (Hatshepsut) is represented with masculine attire and beard, and boasting of the Ammon's favour and of her own gracious manners. Each fit of terror which each fresh plague excited in the monarch soon gave way to renewed hardening of <u>his heart under her influence</u> until the door of repentance was forever shut against him" (compare II Cor. 7:10 with Prov. 21:1)[1].

It was probably Hatshepsut's "change of mind" and orders from behind the throne which led Thutmose II and his charioteers to follow Moses and the children of Israel to the Red Sea. There, very appropriately, he was drowned with the rest of the Egyptian forces (Compare Exodus 14:4,10,17, 18,23,28;15:19 with Psa. 106:11). After these catastrophes, Hatshepsut still remained to pick up the pieces which is so vividly pictured by her description of the situation (Thutmose's body could have easily been recovered from the waters and embalmed - see Exodus 14:30b): "She (Hatshepsut) recorded her good work upon a rock temple of Pakht at Beni-Hasan, saying, 'I have restored that which was ruins; I have raised up that which was unfinished since the Asiatics (Hyksos - 1580 B.C. - added) were in the midst of Avaris (Biblical name Zoan in the day of Joseph but later it was called Per-Re'emases and then Tanis[2] - see Exodus 1:11) of the northland, and the barbarians (vagabonds and Hebrews) in the midst of them, overthrowing that which had been made while they ruled in ignorance of (the Egyptian god) Re'"[3] (The Asiatics could instead be the Hebrews and the vagabonds may have been the mixed multitude and Lepers among them, but this is very doubtful).

Also, it might be well to quote the following statement concerning some of Hatshepsut's reconstruction work which further verifies Thutmose II as King, and identifies them as the rulers at the time of the Exodus: "The eighth plague, the locusts eating every tree, attacked what the Egyptians so prized that Egypt was among other titles called 'the land of the sycamore'. The destruction at the Red Sea took place probably under Thutmose II, and it is remarkable that his widow later imported many trees from Arabia Felix"[4] (see Exodus 9:25;10:5,15).

Historically, we know that Thutmose III (7 years old) co-reigned with Hatshepsut from 1498 B.C. for twenty-two years, therefore, her death took place in the year 1477 B.C. (See chart at the back of the book). Thutmose III mounted the throne in his 28th year and immediately attacked Megiddo the following year (1476 B.C. - exact date according to history). During the next 19 years he had 14 campaigns into Palestine to providentially destroy and weaken the Canaanite enemies of Israel before their entrance into the land in 1458 B.C.. In fact, Israel waited in the wilderness until Thutmose III finished the job (If we do not walk

1. Faussets Bible Dictionary, page 187 ,1954 by Merrill F. Unger
2. Unger, M.F., "Archeology and the Old Testament", p.149,Zondervan, Co.
3. "The Cambridge Ancient History", Vol. II, page 65, Cambride Univ. P.
4. "Fausset Bible Dictionary", page 187 , by Merrill F. Unger, Copyright
   1954 by Zondervan Publishing House. Used by permission

with God and go up by faith, God providentially completes the work by a long road of suffering and by some other means). In his 19th year, he finally conquered Syria and Palestine which was in the year 1458 B.C. (See chart). This was the very year (40 years since 1498 B.C.) that Joshua and Israel entered Canaan Land. Dramatically, Thutmose III was the first Egyptian 18th Dynasty King who used the symbol of the "Hornet" on his military uniform. Thus, he fulfills Exodus 23:28 and Deut. 7:20, as the mysterious power, designated as "the hornet", which was to help the Israelites in their conquest of Canaan.

If the Exodus took place in 1498 B.C., and Moses was 80 years old at that time, then Moses was born in 1578 B.C. two years after the establishing of the 18th Dynasty in Egypt under Ahmose I when he expelled the Hyksos from Egypt. Therefore, Moses' birth would have fallen in the reign of the first King (Ahmose I - 1580 - 1557 B.C.) of the Eighteenth Dynasty. This "New Dynasty" fulfills the Scripture concerning the "new kings" who "knew not Joseph" (Exodus 1:8). Joseph had died just 64 years before this under the Hyksos in 1642 B.C.. Secondly, Moses' name would compare readily with Ahmose's under whose reign he was named. Please listen to this interesting and convincing confirmation: "That Moses was born in Egypt and reared under strong Egyptian influence is independently attested by his clearly Egyptian name supported by the Egyptian names current among his Aaronid kinsmen for two centuries. The name itself is apparently nothing more than Egyptian 'Mase', pronounced 'Mose' after the twelfth century B.C., 'the child', a word preserved in composite like Ah-mose ('son of Ah', the god of light).

It is, in fact, quite probable that Pharaoh's daughter did not give a special name to this unknown infant, a child of an alien race, and that she contented herself simply to name him 'the child'. The interpretation given by the sacred writer; on the other hand, by a peculiar coincidence of sound and a circumstance in the story is connected with the Hebrew root masha,'to draw out', because Pharaoh's daughter drew the infant out of the water (Exodus 2:10)".[1]

Thirdly, Nefretiri, the wife of Ahmose, is often pictured with a dark or almost black face. Since Moses grew up in the palace of the Pharaoh he no doubt had a close relationship with the Queen mother, who was malatto. Therefore, Moses' affinity, in later selecting an Ethiopian wife, irratated Miriam in the wilderness (Numbers 12:1). Also, "The name of Moses's brother Aaron's grandson, 'Phinehas', is also Egyptian, meaning 'the Nubian', and is interesting as providing an independent (and absolutely reliable) confirmation of this circumstance".[2]

Fourthly, Ahmose I had three daughters whose names were Ahotep, Merit-Amon, and Sat-Kamose. One of them must have discovered Moses in the bulrushes (Exodus 2:5). Later historians have suggested such names

1. Unger, M.F., "Archeology and the Old Testament", p.135,136, op. cit., Gardiner, A.H. in Journal of Egyptian Archeology, V(1918)p.221, Mallon, op. cit.,p.133,Cf. Albright,"From the Stone Age to Christianity",p. 193
2. Unger, M.F., "Archeology and the Old Testament",p. 136, op. cit.,p. 193f, Cf. Albright, "From the Stone Age to Christianity"; by Merrill F. Unger, Copyright 1954 by Zondervan Publishing House. Used by permission

as Tethmose or Thermuthis which might compare closely to the name of the above daughter "Sat-Kamose" as the princess who rescued Moses from the waters of the Nile River, and spared his life for his God given mission.

For historical confirmation, we quote the following: "Julius Africanus, in his canon condensing the list of Dynasties of Manetho, added this remark about the list of kings of the Eighteenth Dynasty, 'The first of these (Kings) was Amos (Ahmose I), in whose reign (Dynasty) Moses went forth from Egypt, as I have declared, but according to the convincing evidence of the present calculation it follows that in this reign (Ahmose I), Moses was still young'"[1]

Julius Africanus' statement above confirms our chronological conclusion perfectly. There are too many events falling in to place to doubt the integrity of this God given plan from the Holy Word of God.

The next Eighteenth Dynasty King, Amenophis I (1557 - 1536 B.C.), reigned during the last 20 years of Moses' first 40 years while in Egypt. This makes the second King of the Eighteenth Dynasty, Amenophis I, the author of the first Israel Egyptian bondage and slavery, since he ruled when Moses attempted to protect one of his brethren by slaying an Egyptian before fleeing into Sinai (Exodus 2:11-15 - 1538 B.C.). Please notice how the following quotation verifies this conclusion: "The story is that the King, whose name is given as Menophis or Amenophis I (second ruler under the 18th Dynasty during Moses' first 40 years - added), - - resolved to propitiate the gods by purging the land of Egypt of all lepers (Moses later had a sign of leprosy) and unclean persons. These, to the number of 80,000 (these lepers may have been forced to live with the Jewish people in Avaris Egypt shortly after the Hyksos were expelled in 1580 B.C. - added), were banished to the city of Avaris (Pelusium). There they chose for their leader Osarsiph, whose name was changed to Moses. He gave[2] them new laws, bidding the people to sacrifice the sacred animals".

With such a close association with the outcast lepers in Avaris, Israel certainly needed the laws concerning cleansing in their relationship with these leperous outcasts. Many of these people, no doubt, represented the mixed multitude that went out with Israel from Egypt, and "all", those "over" twenty years at the time of the Exodus, died in the wilderness (Numbers 32:11 - with the exception of Caleb and Joshua). Not only does the above quotation corroborate our theory, but it documents and clarifies other conditions and situations which the nation of Israel had to bear and suffer. The scriptures refers to the mixed multitude that went out of Egypt with Israel in Exodus 12:38 and Neh. 13:3. If only those under twenty, at the Exodus, went into the land of Israel, then only those under 60 years entered Caanan, with the exception of Caleb and Joshua, who gave a good report and were older.

One Egyptian official relates his obscure origin in Egypt under the Eighteenth Dynasty which compares closely with Moses' story. Could these

1. Manetho (trans. W.G. Waddell; Loet Class. Libr.) Cambr.Mass.1941,p111
2. Anstey, M., "The Romance of Bible Chronology", page 161

be the words of Moses?

"Ye shall talk of it, one to another, and the old man shall teach it to the youth. I was one whose family was poor and whose town was small, but the Lord of the Two Lands (Pharaoh) recognized me, I was accounted great in his heart, the king - - - in the splendor of his palace saw me. He exalted me more than the courtiers, introduced me among the princes of the palace - - ".[1]

Now, let us reconsider a few facts and conclusions concerning the reigns of the early 18th Dynasty Kings. Historians have known that Hatshepsut's coup d'etat definitely took place in 1501 B.C., but they were uncertain as to the date of the death of her husband, Thutmose II. Since his death correlates with the Exodus three years later (1498 B.C.), this solves that problem. Also, history tells us that he, Thutmose II, was thirty years old and Hatshepsut was 37 at the time of his death. Thutmose II began to reign in his eighteenth year making his wife, Hatshepsut, 25 years old when she gained her queenly position (1510 - 1498 B.C.). She was, in turn, the daughter of the previous King Thutmose I. If so, then it would seem logical for us to claim that she was born during his reign. For this to happen, we must allow Thutmose I 26 years which makes Hatshepsut born in the second year of his reign (1536 - 1510 B.C.). Historians are uncertain concerning the length of his reign. The King who reigned before Thutmose I was the second King of the Eighteenth Dynasty, Amenophis I, who introduced the Israeli oppression. His reign is known to be 21 years which takes us to 1557 B.C. (1557 - 1536 B.C.). This leaves us a total of either 22 or 23 years for the first King of the Eighteenth Dynasty, Ahmose I, and we have chosen the latter number of years which takes us to exactly 1580 B.C. (1580 - 1557 B.C.) for the introduction of the Eighteenth Dynasty. This date, 1580 B.C., the early archaeologists and historians have always advocated. It allows Moses to have been born the third year (1578 B.C.) from the beginning of the Eighteenth Dynasty, and fulfilling the precious Word of God (Exodus 1:8-11). The Hyksos Dynasties had known Joseph, but they were expelled from Egypt by Ahmose I around 1580 B.C..

It does not seem logical to think that Ahmose I began immediately to persecute the Jewish people since Moses was reared in his palace, but 23 years later under the reign of his son, Amenophis I, it could be quite feasible. His reign also included Moses' 40th year when he escaped to Sinai, while his people were suffering greatly (1538 B.C.). Thus, Amenophis I falls heir as the first instigator of the heavy bondage and slavery.

Eusebius, one of the early church fathers, further attests that the Eighteenth Dynasty was involved with the Exodus, when he refers to the writings of Georgius Syncellus: "Eusebius, another Father of the church, in his canon wrote a gloss to the name Ceneheres of one of the late Kings of the Eighteenth Dynasty (his identity is not known): 'About this time Moses led the Jews in their march out

1. Cambridge Ancient History, Vol. II, page 48, Cambridge University P.

of Egypt"[1] (Thutmose II was the 4th, Hatshepsut the 5th and Thutmose III the 6th ruler of the Eighteenth.Dynasty and all three were involved during the time of the Exodus).

For our final and best attestation that Thutmose II was the King then, we quote from that well known book, Faussets Bible Dictionary, which says:

"In an inscription of the 22nd year of Aahmes I (Ahmose I), Fenchu are described as transporting limestone blocks from the quarries of Rufu to Memphis and other cities (the real bondage just beginning); the name 'Fenchu' means 'bearers of the shepherd's staff', an appropriate designation of the nomad tribes of Semitic origin near Egypt, including the Israelites, who are designated by no proper name in the 18th Dynasty . . . Thutmose II was probably the Pharaoh who perished in the Red Sea . . .". ". . At his (Thutmose II) death the confederate nations north of Palestine revolted, and no attempt to recover them was made till the 22nd year of Thutmose III. . . .Moses returned from Midian at the close of the reign of Thutmose II found him at Zoan (i.e. Tanis or Avaris), the city taken by Aahmes I in lower Egypt".[2]

Thus, we see that the stones continually cry out to help correlate the historical pattern hid away on rocks of ages past to consummate the last day revelation of the treasures also buried in the pages of the Holy Word of God.

Now let us recollect and reiterate some of the information which we have gathered. Since we now know the date of the Exodus (1498 B.C.), we also know from Biblical and extant History that Thutmose II died that year at the age of thirty; his son, Thutmose III, by a concubine, Isis, was exactly seven years old, and Hatshepsut, who began to reign at this time with the seven year old Thutmose III, was exactly thirty-seven years of age. Therefore, this gives us the exact following information: Hatshepsut was born in the year 1535 B.C. while Thutmose II was born seven years later in 1528 B.C.. The two married and came to power in 1510 B.C. at the ages of 25 and 18 respectively. Thutmose III, who was seven years of age at the deaths of his father's first born half-brother, and of his father, Thutmose II (1498 B.C.), could not rule because of his young age. Hatshepsut was in direct royal line and was legally rightful heir to the throne both by birth and by authority of her god's Amun's prophecy. To clinch her hold on the throne she married the child Thutmose III and reigned the following 22 years with an iron hand and died at the age of 58 in Thutmose III's 28th year and in his 22nd year of co-regency. Since she held the throne until Thutmose III was 28 years of age, she no doubt held it by simple force and power of her dominant character which greatly antagonized the young man. Therefore, after either her natural or unnatural death, he attempted to erase her name from as many monuments as possible to obliterate every possible effect of her memory (The "Fausset's Bible Dictionary; Cambridge Ancient History; M.F. Unger's book "Archeology & the Old Testament; along with later authors David Livingston and John J. Bimson, lecturer in O.T. at Trinity College, Bristol, England; & Hans Goedicke confirm a early date of the Exodus).

1,2. "Fausset's Bible Dictionary", page 190 , by Merrill F. Unger, Copyright 1954 by Zondervan Publishing House. Used by permission

Hatshepsut was born in the year 1535 B.C., seven
years before Thutmose II. She married the child
Thutmose III to secure her hold on the throne,
ruling for 22 years. She died at the age of 58
years.

    Statue of Hatshepsut, represented as king, kneeling and
    offering jars of wine to Amun.

Therefore, Thutmose II reigned exactly twelve years (1510 - 1498 B.C.), Hatshepsut reigned 22 years with Thutmose III as her co-regent (1498 - 1477 B.C.), and then the energetic Thutmose III reigned 32 years alone totaling 54 years, which includes his co-regency with Hatshepsut. He died at 60 years of age in 1445 B.C. (1477 -1445 B.C.). The enclosed chart at the back of the book sets forth these facts clearly.

Now, how do these facts correlate with the Biblical information? The Israelite Egyptian bondage began to take the form of real persecution under the second ruler of the 18th Dynasty, Amenophis I (1557 - 1536 B.C.), during whose reign Moses escaped to Sinai and lived with the Midianites (1538 B.C). During this time Israelite captives were being sent to the mines near Serabit el Khadim. Serabit el Khadim itself had an Egyptian temple which honored one of their many gods (Hathor), the mistress of turquoise, and was a three day journey into the wilderness. Moses' demand from Pharaoh equalled the demands of many Egyptians who attended worship at the Egyptian temple at Serabit el Khadim. Moses said to Pharaoh, "The Lord God of the Hebrews hath met with us: and now let us go, we beseech thee, three days journey into the wilderness, that we may sacrifice to the Lord our God" Exodus 3:18 (see Exodus 8:27;15:22). "Sir Flinders Petrie has suggested that 'the three days journey' into the wilderness was an expression used to denote the route to the temple of Serabit in the centre of the Sinai Peninsula, where the then existing ceremonies and ritual of the Hebrews were observed".[1] As to the latter statement, we would suggest that after the Red Sea deliverance , Moses led them to the wilderness of Shur (close to Serabit), which is known as Marah and also is designated as "three days in the wilderness" (Exodus 15:22). Israel's true worship first began a short time later at Sinai (at least as a nation). After completing their three days journey they found themselves close to Serabit el Khadim and the mines of Sinai. So they approached the Egyptian guards and delivered the Israelite slaves (These Hebrew slave workers left evidence of Israelite worship in writings there) along with many other captives from Palestine and surrounding countries. Israel destroyed the temple at Serabit el Khadim and then continued on to Sinai where they received the Law and many Biblical instructions. All of the other Sinai captives slaves were also released and many became a part of the mixed multitude and moved with the Israelites, while Israel received the Law. Therefore, these other released captives were greatly impressed by the new God given ceremonies, and by seeing and hearing of the miracles of the God of Abraham, Isaac, and Jacob. They were intreged by God's Shekinah Glory, the fire by night and the cloud by day, and the miracle of the Manna each morning. Gathering a great knowledge of the tabernacle procedure and of the God of Glory, these released slaves gradually infiltrated back through the surrounding countries to their own lands.

Among these Sinai captive slaves released were also a number from Ras Shamra north of Sidon near Antioch, a Phoenician city, and they introduced the Jewish laws, ceremonies, and customs into their own worship.[2]

1. Marston, Sir Charles, "The Bible Comes Alive" p.46, Eyre & Spotteswoode
2. Halley, H.H., "Pocket Bible Handbook", 1948, page 54

The Temple Library at Ras Shamra (Ugarit) which was excavated by a French Expedition from 1929 A.D. reveals this information, and they date these added ceremonies back to 1496 B.C..

Israel finally, after two years in Sinai, arrived at Kadesh Barnea in 1496 B.C.. It is known, historically, that Hatshepsut went to Serabit el Khadim in her third year of reign (inclusive) and rebuilt the Serabit Temple there in 1496 B.C.; the very year that Israel arrived at Kadesh Barnea.[1] (What a wonderful correlation of facts vindicating perfectly our God given Biblical chronology!). There was also much reconstruction work to be done further afield than just in Egypt proper, and Hatshepsut also needed the continued resources of the mineral wealth of Sinai to re-establish her kingdom. (It should be stated that the Hebrew writings found at Serabit el Khadim are dated around 1500 B.C., correlating with and confirming perfectly our chronology. Israel's heavy slavery would date from 1538 to 1498 B.C.).

The next 38 years of the Hebrews' sojourn in the wilderness includes the remainder (20 years) of the 22 years reign of Hatshepsut. Then in 1477 B.C., Hatshepsut died and Thutmose III, the young energetic King, at 28 years of age, immediately gathered his armed forces and headed up the Palestian coast for Megiddo. Early in the next year, 1476 B.C., he attacked and destroyed Megiddo, which was the first of 14 campaigns into Palestine and Syria. During this time Moses and the Hebrew children continued their "wandering in the wilderness" at a settlement called Kadesh Barnea. Thutmose's 6th campaign took place in 1469 B.C., and his eighth in 1466 B.C., at which time, he destroyed Carchemish and Aleppo. That year, 1466 B.C., was the first year of "ten years" (see chart page 221). The last year of which was to climax his battles in the north country, after which, he remained in Egypt. During the "eighth year" after the Carchemish battle, Moses led Israel to Mt. Nebo. The "ninth year" was the 19th year of Thutmose's reign alone. He was 46 years of age when he finally took Syria and Palestine in 1458 B.C.. That very same year, 1458 B.C., was also the 40th year since the Exodus (1498 B.C.), and the year Joshua led the Hebrew children into the land of Canaan following the completion of Thutmose III's defeat of Palestine and Syria. Does scripture correlate with these facts? Sir Charles Marston quotes Prof. Garstang and states as follows: "Archaeological discoveries in Palestine, taken as a whole, favour the hypothesis, that if the Israelites did not actually conquer and occupt Canaan with the aid of the Egyptians, they did so with their connivance. There are two passages in the Pentateuch (Exodus 23:28 and Deut. 7:20) which refer to a mysterious power, designated as 'the hornet', which was to help the Israelites in their conquest of Canaan. Now the 'Hornet' was the uniform badge of Thutmose III and his successors. Professor Garstang has suggested that it is a symbolical allusion to Egyptian power".[2]

God's cognisance of the "Hornet's" help is also significant. Speaking through Joshua God tells the Israelites, after they had conquered

1. Marston, Sir Charles, "The Bible Comes Alive" page 41
2. Ibid, page 65, Eyre & Spotteswoode Printers Ltd

Canaan: "I sent the 'Hornet' (Thutmose III) before you, which drave them out (troubled and weakened them) from before you, even the two kings of the Amorites, not with thy (Israel's) sword, nor with thy bow" Joshua 24:12.

These two kings, who had been specially "troubled and defeated" by Thutmose III's attacks on Palestine, are specifically referred to as Sihon and Og (Deut. 2:24; 3:4-6). It is known that Thutmose III had military outposts located at Bethshan and Jericho, so this is a good indication that he had truly troubled and subjugated the Amorite and Moabite forces beyond Jordan to his authority.

Later, the Amorites defeat by the children of Israel made the Israelites masters of the whole country east of the Jordan. "If that were with Egyptian help or connivance, it would account for the fact that Mut Baal was referring the Egyptian official to Joshua, and others, for information about Pella, a city east of the Jordan in Israelite occupation. Further light is cast on the course of political events by the religious history of this period".[1]

Thus, the "Hornet", Thutmose III, was providentially sent before the Israelites to subjugate Palestine and establish military posts. During these campaigns, Thutmose III so weakened Israel's enemies so that they had very little power to resist the forces of Joshua. You will notice on the enclosed chart(page 222) that Thutmose III followed up the defeat of Palestine and Syria the next year (1457 B.C.) with only one other battle which took place at Kadesh on the Orontes River. He, no doubt, stayed on the Shalfala, the coastal plain of Palestine, during this last northern campaign while Joshua was in the process of fighting his first year's battle on the hills of Judea (1457 B.C.). The following year (1456 B.C.) was the first of 12 years during which Thutmose III stayed in Egypt from the age of forty-nine until his death at sixty years of age.

Joshua finally captured the land after seven years (Joshua 14:7-10) in the year 1451 B.C. and extant history confirms this clearly when it states that Egypt, a year later, regarrisoned Bethshan in 1450 B.C.[2] (Please see the Biblical confirmation in Judges 1:27,28; Joshua 17: 12-16). Since the Israelites did not control the plains, then Egypt immediately sent small forces back to hold the Shafala coastal plains and the valley of Megiddo in whose eastern end Bethshan is located.

For information concerning some of the conditions in Palestine and details concerning these Egyptian outposts in Palestine at the time of Israel's entrance into the land let us quote W.F. Albright's statement: "The local princes of Palestine and Syria were permitted to govern their subjects as before, and were not molested as long as they paid their

1. Marston, Sir Charles, "The Bible Comes Alive" p.91,Eyre & Spotteswoode
2. Albright, W.F., "From the Stone Age to Christianity", page 207, The Johns Hopkins University Press, Baltimore/London, 1957

tribute properly and performed their part of the compulsory labor for the crown (Corvee). However, as time went on repeated uprisings brought reprisals and fortresses were built by the Egyptians in various parts of the country in order to exercise a more effective control. The Egyptian garrison of <u>Bethshan</u> was first <u>established about 1450 B.C.</u>, (re-established - added) to judge from the most recent excavations there, but it was rebuilt again and again before the final decline of the Egyptian Empire in Asia. Gaza and Joppa remained for a long time provincial capitals, and they were probably not the only ones. There was a curious double administration of government, which may have worked well under strong Pharaohs like Thutmose III and Harmais, but which undoubtedly worked badly under a weak one like Akhenaten, as we learn from the Amarna letters. - - This double system consisted of the local feudal princes, each of whom was called 'governor' (Khazidnu in Accadian), on the one hand, and the Egyptian commissioners, called 'inspectors' (rabisu), on the other. The native governors were supported by local forces of patrician chariotry and plebeian footmen (khupshu), with which they were even permitted to wage internecine warfare, provided it did not interfere with tribute or reach dangerous dimensions. The Egyptian commissioner raised tribute, saw to the execution of work on the roads and in the royal grain-lands of Jezreel or the forest-reserves of Lebanon; they were supported by contingents of Egyptian, Nubian, Bedouin, or Mediterranean slave-troops and mercenaries, usually armed as bowmen. Since the Egyptian bureaucracy was notoriously corrupt, as we know from innumerable documentary allusions, the troops often failed to receive their wages or maintenance, whereupon they plundered the unlucky provincials. The native princes were allowed to correspond directly with the court, and employed for this purpose Cuneiform tablets written in Accadian, which had become the lingua franca of Western Asia".[1]

Since Thutmose III died in 1445 B.C., the first year of Amenhotep II was 1444 B.C. (Amenophis II). History confirms that during his third year 1442 B.C., he had a Syrian campaign. In his 7th year (1438 B.C.), Amenophis II attacks Shemash-Edom and carried away 3600 Apiru (Hebrews) captive.[2] This would have been the 20th year since the Israelites had entered the land. During his 9th year (1436 B.C.), he advanced up the coastal plains and in through the valley of Megiddo and his campaign reached as far as the Sea of Galilee.

Joshua was still alive at that time although very aged. Joshua died just three years later in 1434/33 B.C. just twenty-five years after entering Canaan land.[3] The year 1434/33 B.C. (see Joshua 24:1,25) was the last year that Israel was able to celebrate the feast of tabernacles to such an extent. It was not celebrated in such a way until exactly 1000 years later as Nehemiah states (Neh. 8:17; 13:6 - 434/33 B.C.) at the close of his ministry in Jerusalem before he went back to Babylon (He had promised the king to return).

1. Ibid, page 207,208
2. Wilson, J.A., "The Culture of Ancient Egypt", page 201; Publisher: University of Chicago Press
3. Josephus, "Antiquities of the Jews", Book 2, Ch. 2,S. 29, page 150

Since Thutmose III died in 1445 B.C., the first year of Amenophis II was 1444 B.C., and his fourth (inclusive), in which he built the temple of Bast, was 1441 B.C..[1] Five hundred years later in the 21st year of Zerah (Osorkon I, 962 - 927 B.C.), which was 941 B.C., they celebrated the 500th anniversary of Osiris in the Temple of Bast. A perfect test of our God given chronological plan from the Holy Word of God. Only a long chronology with the division of the kingdoms in 982 B.C. would correlate with these facts.

1. Gardiner, Sir Alan, "Egypt of the Pharaoh's", page 199, 1961, Publisher: Oxford University Press

Judges

2:6	Joshua possesses the land in - 7 years		1458 - 1451 B.C.
	(See Joshua 14:6-10)		
2:8	Joshua dies - 25 yrs in land	18 years	1451 - 1333 B.C.
	(Josephus, Antiq.,V,1:29) :*	1:	
2:10	New Generation(100 yrs-1518 BC:15 years[1]:		1433 - 1418 B.C.)
	(Caleb & Joshua & 20 yr old at Exodus died-Num. 32:11)		
3:8	Mesopotamia - Cushan-Nishathaim 8 years		1418 - 1410 B.C.
3:11	Othniel delivers Israel	40 years	1410 - 1370 B.C.
3:14	Eglon (Moabites overrun land)	18 years	1370 - 1352 B.C.
	⇕	:	
3:30	Ehud delivers Israel : The	20 years :	1352 - 1332 B.C.
	: 80 :		
4:3	Jabin's incursions : year : 20 years		1332 - 1312 B.C.
	:period :		
5:31	Barak delivers Israel: rest : 40 years		1312 - 1272 B.C.
6:1	Midianites(overrun land to Gaza)7 years		1272 - 1265 B.C.
8:28	Gideon delivers Israel	40 years	1265 - 1225 B.C.
9:22	Abimelech (Egypt petty King)	3 years	1225 - 1222 B.C.
	(See 12th Dynasty - Amenemhet = Ammenemes = Abimelech)		
10:2	Tola judges Israel	23 years	1222 - 1199 B.C.
10:3	Jair judges Israel	22 years	1199 - 1177 B.C.
	"that year"*	:* :	
10:6	Israel sins-Philistine-Israel-: 1 year :		1177 - 1176 B.C.
	(Westbank) (Eastbank)		
10:8	Philistine(20)Ammonite(18)	18 years	1176 - 1158 B.C.
Judges	: - (1176 - 1156 B.C.)		
11:26	Jephthah rules 6 yrs after -	300 years	1158 - 1152 B.C.

*   Adjustable years

The above chart should total 300 years. To do this we must solve one of
God's scriptural puzzles. The chart covers the three hundred years from
the year Joshua entered the land (1458 B.C. - the year Israel began to
live in Heshbon and Aroer) according to Judges 11:26: "While Israel
dwelt in Heshbon and her towns, and in Aroer and her towns, and in all

1. Marston, Sir Charles, "The Bible Comes Alive"p.117;Eyre & Spotteswood

the cities that be along by the coasts of Arnon, three hundred years, why therefore did ye (Ammon) not recover them?".

There are "five historical notations" which we must keep foremost in our minds as we correlate History and the Bible in the above chart. First, the eight year rule over Israel by Mesopotemia; second, the Moabite reign which agrees with the El-Amarna period; third, the famous Egyptian and Hittite Treaty (1273/2 B.C.) which introduces the Midianite rule over Israel; fourth, Abimelech's reign, as the Egyptian petty ephemeral king (Ammenemes), must fall during the reign of the Egyptian King Merneptah (1228 - 1218 B.C.); and the fifth, as the first year of the Philistine reign over Israel correlates with the known historical date of 1176 B.C.. All these above "five events" synchronize perfectly at the right point in time in our above chronological schedule.

In correlating these "five events" in the above chart, we must first consider Judges 2:6. Joshua had just conquered the land of Israel and he allowed every man to go unto his inheritance to possess it. How long did it take? According to Caleb's following statement, we know that it took "seven years" for the Israelites to conquer Canaan land after coming over the Jordan under the leadership of Joshua in 1458 to 1451 B.C.: "- - Caleb the son of Jephunneh the Kenezite said unto him (Joshua), Thou knowest the thing that the Lord said unto Moses the man of God concerning me and thee in Kadesh-Barnea (1496 B.C. - Deut. 2:14). Forty years old was I when Moses the servant of the Lord sent me from Kadesh-Barnea to espy out the land - - - and now, behold, the Lord hath kept me alive, as he said, these forty and five years (1496 - 1451 B.C.), even since the Lord spake this word unto Moses, while the children of Israel wondered in the wilderness: and now, lo, I am this day fourscore and five years old (1536 - 1451 B.C.)" Joshua 14:6,7,10.

Since we know that it took Israel two years to arrive at Kadesh-Barnea (Deut. 2:14), this, in turn, leaves 38 years for them to continue to wander in the wilderness. Caleb was in his 45th year since Kadesh when he claimed his possession, therefore, leaving Joshua seven years to conquer the land. Thus, Caleb claimed his inheritance in 1451 B.C. seven years after entering Canaan. During the following year (1450 B.C.) history confirms that the Egyptians re-established their hold on Beth-shan by once again infiltrating the coastal valley of Canaan land and the valley of Megiddo, while the Israelites retained their control only in the mountainous area as the scriptures declare: "And the Lord was with Judah; and he drove out the inhabitants of the mountain; but could not drive out the inhabitants of the valley, because they had chariots of iron" Judges 1:19. "Neither did Manasseh drive out the inhabitants of Beth-Shan - - - -" Judges 1:27 (see Joshua 17:11,12).

Following our chart, Judges 2:8 states that Joshua died when he was a hundred and ten years old. What year was that? Since Joshua took seven years to conquer the land then he must have lived another 18 years since Josephus states that he was in command in Canaan land for twenty-five years: "So Joshua, when he had thus discoursed to them,

79

died, having lived a hundred and ten years (1543 - 1433 B.C. - added); forty of which he lived with Moses, in order to learn what might be for his advantage afterwards. He also became their commander after his (Moses') death for twenty-five years".[1]

Continuing with the three hundred year period, we find that there are "two adjustable periods" marked with a "rectangle". One of which falls near the beginning of the 300 years (Judges 2:10), while the other falls just before the end, preceding the united Philistine and Ammonite oppression (Judges 10:6). By correlating the intervening events with history, we believe we can come within a year of perfectly synchronizing these dates and coming to a conclusion on the length of these two "adjustable periods".

In the first "rectangle", representing an "adjustable span of time", we have 15 years, which has been suggested by Sir Charles Marston. These 15 years bring us to 1418 B.C. which is exactly 100 years since 1518 B.C., which in turn spans the very years that these young Israelites had lived, who were just under 20 years of age at the Exodus, and were not held accountable for their wilderness unbelief when entering Canaan land (Num. 32:11). The date 1418 B.C. makes that "generation" one hundred years old or younger (longevity). Very few inhabitants of Israel would have lived over that age, therefore a "new generation" had truly arisen as Judges 2:10 implies: "And also 'all that generation" (100 years) were gathered unto their fathers: and there arose 'another generation' after them, which knew not the Lord, nor yet the works which He had done for Israel".

The implication might be that a "generation" is 40 years. Their normal "promised life span" would be 70 years (Psa. 90:10 - add 30 years), while their longest possible "longevity" would be a span of about a hundred years (add 30 years). Forty years plus thirty years, and another thirty years equals 100 years. This formula still seems to hold true to this day. Thus, the "old generation" (Joshua, Caleb, and those just under twenty years of age at the Exodus - 1498 B.C.), who had seen the miracles of God, had passed away, and the "new generation" after the 100 years (1518 - 1418 B.C.) had now turned away from God and were sold into bondage to King Chushan-rishathaim of Mesopotamia (Assyria): "And the children of Israel (New generation) did evil in the sight of the Lord, and forgat the Lord their God, and served Baalim and the groves. Therefore the anger of the Lord was hot against Israel, and He sold them into the hand of Chushan-rishathaim King of Mesopotamia: and the children of Israel served Chushan-rishathaim eight years" Judges 3:7,8.

Our chronology clearly places the above eight years during the years 1418 -1410 B.C.. These eight years fall very closely within the contemporary Assyrian reign of Ashur-rim-nishushu,[2] who also reigned eight years. Perhaps, the Biblical word Chushan-rishathaim could be translat-

1. Josephus, Antiquities, V, 1:29 , **Loeb Classical Library Edition**
2. Journal of Near Eastern Studies, Vol. 8, No. 4, Oct. 1954, page 227.
   See Saggs, H.W.F., "The Greatness That Was Babylon", page 534

ed, not Chushan, but rather Shushan (Susa), which was the area of origin, some time before, of these Mesopotamian, Assyrian, Kings in 1418 B.C.. A Kassite invasion from the Zagros mountains, around Susa, had captured Babylon at one time and had greatly influenced the Mesopotamian area since 1595 B.C.. The Mesopotamian's had prefixed their kings name with their god's name (Ashur-rim, or "success god"); while the Biblical prophet would rather use the "area" (Shushan), where they had originally come from, as the prefix to the king's name. They said "Shushan" for "Chushan", or Shushan-nim-rishatha, therefore, we could compare the above name to Ashur-rim-nisheshu. Rishatha would equal Nisheshu.

1419/8-1411/10 BC 8 years(Success gods)Ashur-rim-nisheshu(contp.History)

1418 - 1410 B.C. 8 years (Judges 3:8)   Chushan-rishathaim (Bible)

The above reigns correlate with the contemporary Egyptian Kings Thutmose IV (1421 - 1413 B.C.), and Amenophis III (1413 - 1376 B.C.). The Mesopotamian King Chushan-rishathaims eight years reign overlapped both the above Assyrian and Egyptian reigns. Also, we have interesting corroborating confirmation which is recorded on the Soleb Temple columns, in Egypt, disclosing that the above King, Amenophis III, had taken captives from Mesopotamia. Contemporary history does not reveal any such attack by Egypt on Mesopotamia proper, but these Mesopotamian captives were, no doubt, taken captive in Israel by Amenophis III during the first three years of his reign, and during the last three years of Mesopotamia's rule over Israel (1418 - 1410 B.C.). Only a few years before this Amenophis II (1444 - 1421 B.C.) is recorded as attacking Shamash-Edom in his 7th year (1438 B.C) and taking 3600 Hebrews (Habiru - Apiru) captive. Again, in his 9th year (1436 B.C.), he attacked and advanced as far as the Sea of Galilee. This was only three years before Joshua's death in 1433 B.C.. Thus, these Palestinian thrusts continued during Mesopotamia's rule over Israel when Amenophis III captured either Mesopotamian soldiers or civilians, while they were subjugating the nation of Israel at that time.

There is another astonishing confirmation revealed by contemporary history which states that these foreign "Mesopotamian" invaders, who ruled Canaan Land for 8 years, were the ones who solicited help, and were rejected by Kurigalzu I. This Babylonian ruler began to rule the same year as Mesopotamia's defeat in Israel (1410 B.C.). This conclusion would give logical answer for the problem of why the inhabitants of Canaan were asking help from a country so far abroad. Naturally, Babylon would have been Mesopotamia's neighbor while she ruled eight years over Israel, and the call went out to her to help resist Israel's revolt under the leadership of Othniel.

Thus, the corroborating events in our chronological plan continue to multiply.

Now, we have God's chosen servant Othniel, the son of Kenez, Caleb's younger brother, who intervenes and brings deliverance to Israel for forty years: "And the Spirit of the Lord came upon him, and he (Othniel) judged Israel, and went out to war: and the Lord delivered Chushan-rishathaim King of Mesopotamia into his hand: and his hand prevailed against Chushan-rishathaim. And the land had rest forty years, and Othniel, the son of Kenaz, died" Judges 3:10,11.

After the death of Othniel and his forty year reign (1410 - 1370 B.C.), we have King Eglon of Moab smiting Israel, with the help of the Ammonites and the Amalekites, and possessing Jericho, the city of the palm trees (Judges 3:13, 1370 - 1352 B.C.). The children of Israel served Eglon, the King of Moab, eighteen years. These eighteen years fell during the El-Amarna period involving the Egyptian King Amenophis IV (Akhnaten: 1376 - 1359 B.C.) as John A. Wilson rightfully suggests:

"Akh-en-Aton's (Amenophis IV) preoccupation in his intellectual revolution permitted the three remaining stages of disintegration. The Hittite King Suppiluliumas (1376 - 1339 B.C. - added) moved south as a conqueror and gobbled up all of Syria. The separatism of the local princes led only to their becoming Hittite vassals. The important town Qatna was destroyed and never again was a power. Mitanni had to submit to Hittite domination (1360 B.C. - added). All of this seems to have been effected without an Egyptian sword raised in protest. With Syria lost, the disaffection rapidly spread to Phoenicia and Palestine (Moabite invasion - added). The Phoenician towns fell despite the fanatical loyalty to pharaoh of such a prince as Rib-Addi of Byblos. In Palestine, Labaya, a merchant prince who led caravans from further Asia into Egypt (they controlled the Palestine coast and the Megiddo valley - added), combined with the Habiru (Hebrews opposing Moab's rule - added) of the desert and began seizing towns for his own rule. Abddi-Khepa of Jerusalem wrote letters beseeching Pharaoh (Amenophis IV) for as few as fifty soldiers to hold the land. They were not sent. And so we see the final stage, in which the Egyptian garrisons were withdrawn from Asia and Palestine also was lost (lost control of the valleys - added). Local rebels and desert nomads (Moabites, Ammonites, and Amalekites - added) overran the vacated territory and destroyed Jericho (Eglon, the Moabite ruler reigned from there for eighteen years - added) and Tell Beit-Mirsim. The little Egyptian temple at Lachish was sacked. Most significantly, the migdol fortification at Old Gaza, which had been the center of Egyptian administration, was destroyed. Complacency, inertia, and internal distractions had lost to Egypt her vast and lucrative Asiatic Empire".[1]

Therefore, the Moabite control and the outside infringement on the peace of Israel was vanquished, and was followed by 80 years of rest for the "land": "And they slew of Moab at that time about ten thousand men (1352 B.C.), all lusty, and all men of valour, and there escaped not a man. So Moab was subdued that day under the hand of Israel. And the 'land' had rest fourscore years (80 years)" Judges 3:29,30.

The next external invasion took place 80 years later when the Midianites, with the Egyptian collaboration, conquered and swept over the land of Israel as the following quotation verifies: "And they (Midia-

1. Wilson, J.A., "The Culture of Ancient Egypt", page 230, 231; Publisher: University of Chicago Press

nites) encamped against them, and destroyed the increase of the earth, till thou come unto Gaza, and left no sustenance for Israel, neither sheep, nor ox, nor ass. - - and they entered into the land to destroy it" Judges 6:4-5.

We emphasize, it does not state that Ehud reigned 80 years (actually 20), but that the "land" had rest from the beginning of Ehud's reign for eighty years. Ussher, in his chronology, also reached this conclusion. This simple explanation solves all the seeming difficult problems of the Judges Period. Therefore, we conclude that the three intervening Israeli reigns of the Judges between the "Moabite invasion" and the "Midianite invasion" should total eighty years. In considering the intervening period in reverse, we find that Barak (Judges 5:31) judged Israel for 40 years; Jabin only harassed Israel for 20 years (Judges 4:3 - instigated by the Hititte influence as far south as the Sea of Galilee); and Ehud would then have ruled only during the "twenty years" which are left, which makes a total of 80 years. This is the simple answer of the God given Biblical problem, and we believe it is absolutely correct. Also, it is absolutely proven correct by the perfect correlation of all five events mentioned above.

Shamgar (Judges 3:31) is referred to as delivering Israel. Yet, when we consider his efforts in the light of Judges 5:6, we know that he did his exploits during the days of Jabin and Jael. Therefore, the beginning of the seven year Midianite reign, over Israel (1272 - 1265 B.C.), fell on the year following the very important date, 1273/2 B.C.. This date involved an Egyptian instigated attack by the Midianites on Palestine, with the following great Egyptian treaty with the Hittites. Thus, we conclude that the Midianite attack upon Israel was definitely instigated by Egyptian connivance. Rameses II began to reign in Egypt in 1294 B.C., and in his 21st year (1273/2 B.C.) this great treaty took place.[2] At this same time it is said that Rameses II "reduced the stronghold of rebellion in Canaan". This then refers to the Midianite invasion. That same year (1273/2 B.C.), Rameses II made a treaty with Hattusilis III (1279 - 1253 B.C.). Therefore, we conclude that the seven year Midianite rule over Israel had been instigated by Egyptian subversion and introduced by her signature to the "great peace treaty" between the Egyptian King Rameses II (1294 - 1228 B.C.) and the Hittite King Hattusilis III (1279 - 1253 B.C.) in the year 1273/2 B.C..[3] This Egyptian and Hittite treaty must have offended Assyria greatly because, 600 years later, she celebrates the anniversary of this treaty by smiting Egypt in the eighth year of Esarhaddon's reign in the year 673/2 B.C. (see page 235).

The above treaty was written on two silver tablets and a copy sent to each of the countries involved sealed under the curse of the 1000 gods of both nations. The treaty stated, "And as for these words which are written upon these silver tablets for the Land of Hatti (Hittite) and the Land of Egypt - - - whosoever does not obey them may the thousand gods of the Land of Hatti and the thousand gods of the Land of Egypt

1. Mauro, Philip, "Chronology of the Bible", page 59
2,3. Ceram, C.W., "Narrow Pass Black Mountain", page 185,187, Sidgwich & Jackson, Ltd

destroy his house, his land, and his servants".[1]

Our chronology places Jabin's incursions into Israel dating from 1332 B.C.. We also know from contemporary history that Mursillis II (Hittite: 1338 - 1310 B.C.) controlled Palestine as far south as the southern end of the Sea of Galilee during this same period. This also holds true during the second year of Seti I (1313 B.C.), at which time, Seti I attacked the Hittites.[2] Therefore, we may conclude that Jabin, the King of Hazor, was a petty ruler under the supervision of the Hittite King Mursillis II. These historic details correlate perfectly with the scripture since Jabin controlled only the area as far as the southern point of the Sea of Galilee, but he made incursions at various times further south on the main "highways" into the land of Israel. The battle (1312 B.C.) and defeat of Sisera, the General of Jabin's forces, took place near Mt. Tabor (Judges 4:6) which is, practically, in the same latitude as the southern end of the Sea of Galilee.

Further confirmation is established by an eclipse which took place in the year 1328 B.C., which inturn, is known to have taken place in the "tenth year" of Mursillis II. This heavenly phenomenon dates the first year of Mursillis II (1338 - 1310 B.C.), and clearly places him during the reign of Jabin, King of Hazor (1332 - 1312 B.C.).

During Seti I's attack against the Hittites in his second year (1313 B.C.), he left a stela at Beth-shan on which he made reference to the Hebrew (Apiru or Habiru) of a mountain district with a Semitic name.[3] This would be exactly true according to Bible chronology. Seti's attack helped weaken the Hittite forces under Jabin as Seti pushed on past Hazor to Hamath and Beirut, only one year before (1313 B.C.) Jabin's defeat by Israel in 1312 B.C. (Judges 5:31). Seti I's attack, providentially, helped to prepare the way for Barak's and Deborah's victory, the following year, over the Hittite forces under Jabin, and for their victory song recorded in Judges chapter five. The story states that Jael killed Sisera (Judges 4:22) Jabin's General.

Shamgar did his exploits (Judges 3:31) on the southern front and killed 600 Philistines during the days of Jabin's attack on Israel, and during "the days of Jael" (Judges 5:6). Shamgar's skirmishes also verifies that there were still Philistine settlements at this time, or early infiltration of the Philistine people into the land.

It is also coincidental that one of the towns which Seti I captured during his second (regnal) year was named "Bethshael" (House of Jael).[4] Could this be the home town of Jael who killed Sisera during the following year, 1312 B.C., just one year after Seti I's attack? (The Egyptian history could have been written after the two events had taken place).

Thus, we conclude, even though Israel and Egypt were enemies, they both

1. Wilson, J.A., "The Culture of Ancient Egypt", page 249, Publisher University of Chicago Press.
2. White, J.E.M., "Ancient Egypt", page 176, Allen & Unwin, England
3. Albright, W.F., "From the Stone Age to Christianity", page 277
4. Gardener, Sir Alan, "Egypt of the Pharaoh's", page 254;1961; Published by Oxford University Press

had a common enemy in the north of whom Jabin was a Petty King. Egypt was seeking to protect her trade route through the valley of Megiddo and north, while Israel was seeking to protect her land from further incursions. Providentially, Seti I's attack in 1313 B.C., inadvertently, helped Israel by weakening the Hittite hold on upper Galilee and preparing for Barak's and Deborah's success the following year (1312 B.C.). Perhaps without the Egyptian garrison at Beth-Shan, Israel may have been completely overrun by the Hittites during the days of Jabin. This is clear proof that Jabin's incursions were limited only to "border penetrations" on main roads, and that this time still remained a part of the eighty year rest for the land of Israel.

Before leaving this period we might further reiterate that the El-Amarna period fell during the period of the Moabite rule under Eglon (1370 - 1352 B.C.). Therefore, the Hebrew (Habiru) incursions during the Amarna Period were no doubt "Israel's" continual attempt to break the hold of the "Moabite rule" over the land. Egypt was naturally involved since she controlled the coastal areas of Palestine and the Megiddo valley up to this time.

The Hittite success in the north, while in the process of dominating the Mitannian Kingdom (1360 B.C.), may have ultimately helped instigate the Moabite invasion of Israel with the Ammonite, and Amelekite help a few years later.

Syria was also mentioned in the Amarna letters and they were closely associated with the place of origin of the Moabites beyond the Jordon River, and also with the Hittites from the north. For a more detailed analysis of this period, we will leave this chronology in the hands of those who have made a detailed research of the El-Amarna Letters and the contemporary period. We are assured that this chronology will meet the qualification demanded by historians, and will ultimately solve these problems.

In the next section of our chart, we have the judgeship of Gideon for a period of forty years, and followed by Abimelech, who may have been supported by the Egyptian throne and appointed as their petty king. Merneptah, King of Egypt, who reigned during Abimelech's rule (1228 - 1218 B.C.), announced that "Israel is desolate, its seed is not".[1] His reference to Israel, no doubt, refers to Abimelech's three year reign (Judges 9:22), which began in the third year of Merneptah. "It is known that during Merneptah's reign an 'Ephemeral King' by the name of 'Amenmesse' (Abimelech) is mentioned".[2] If Abimelech and Amenmesse are the same person, it may give us a wonderful clue who the Abimelech's were during the days of Abraham and Isaac. Could they have been the 12th and 13th Dynasty Kings Amenemhet I, II, and III, who also had the name (nomen) Ammenemes or Amenmesse?

1. Wilson, J.A., "The Culture of Ancient Egypt", page 255; Publisher: University of Chicago Press
2. Unger, M.F., "Archeology and the Old Testament", page 185, Zondervon

Abimelech was the son of Jerubbaal (Gideon - Judges 7:1) by his concubine that was in Shechem (Judges 8:31). This concubine is referred to as a maid servant (Judges 9:18). Since the Biblical account of this period seems to leave Shechem deleted from the list of cities conquered by Israel, we can conclude that it may have been ruled or strongly influenced at various times by the Egyptian forces that transversed the Megeddo valley and who maintained a garrison at Beth-shan. Abimelech's mother (maidservant) may have been an Egyptian, and through her family influence, Egypt may have been able to initiate Abimelech as their Petty King in Israel. Therefore, Merneptah refers to Abimelech's (Ammenemes) success and stated that "Israel is desolate and her seed is not". If Abimelech had any contact with the Egyptians personally, it would have had to have been before or during his reign, since he was killed at the end of his three year reign (Judges 9:53-54). Shechem was completely destroyed when they "beat down the city, and sowed it with salt" (Judges 9:45 - 1222 B.C.).

Following the above three years, we have Tola bringing deliverance to Israel for twenty three years, and then succeeded by Jair who judged Israel for twenty two years. This brings us to the "one year" in the "lower rectangle" which is an adjustable period. In other words the years in the "upper rectangle" could be shortened, and the number in the "lower rectangle" lengthened moving the chronology up and down to synchronize with historical events. Yet, this "one year" of sin (1177 B.C.) correlates well with the intervening historical facts, and with Judges 10:8 where it says, "that year" they (Philistines and Ammonites) vexed and oppressed the children of Israel - -". Then the scripture continues referring to the Ammonite oppression alone. "- - eighteen years oppressed they all the children of Israel that were beyond Jordan in the land of the Amorites, which is in Gilead".

Since Jephthah's 300 years ended in the 1158 B.C., therefore, the 18 years Ammonite, and the 20 years (I Sam. 7:2,14b) Philistine rule immediately, and unitedly preceded the date 1158 B.C. by 18 years beginning with the "important date 1176 B.C.", running concurrently, and pinpointing their "united" attack "that year" against Israel (1177 B.C.). Now, with the following year (1176 B.C.) of their united subjugation of Israel, clearly fixed in our mind, we can now date the first year of Rameses III's reign by confirmation screened from several historical notations.

It is known that the Philistines attacked Egypt (Rameses III) in his 9th year and 13 years later the "Philistines" were firmly settled on the coastal plains of southern Canaan.[1] If that year of firm Philistine settlement was truly the year 1176 B.C., then the test would be to simply add 13 years to the year 1176 B.C. and we then have the 9th year of Rameses III's reign (1189 B.C. - see chart). We then add nine more years and we have the first year of Rameses III (1198 B.C.). Is this date correct? For confirmation of this date, we turn to the Manual of

1. Keller, W., "The Bible as History", page 175, **Econ Verlagsgruppe Ger.**

the United Arab Republic, Antiquities Department of the Egyptian Museum, Cairo, 1963 on page 143 where we find that they date the reign of Rameses III in the same year, 1198 B.C..

It was in 1189/8 B.C. that the Philistines suffered their severe defeat at the hands of Rameses III. Thirteen years after this defeat, and retreat, the Philistines were firmly settled on the fertile brown plain between the mountains of Judah and the sea with Israel under subjection (They defeated Suppiluliuma II - Hittite - in 1190 B.C.).

The Bible lists the five cities which they possessed: Askelon, Ashdod, Ekron, Gaza, and Gath (I Sam. 6:17). Each of these cities, and the command of paid leaders, were ruled over by a "lord" who was independent and free. For all political and military purposes however the five city rulers always worked hand and hand. In contrast to the tribes of Israel the Philistines acted as a unit in all matters of importance. That was what made them so strong.[1]

While considering this very significant year (1176 B.C.), we shall not leave this subject without quoting Judges 10:7,8, and meditating on a few other important events involved around this date: "And the anger of the Lord was hot against Israel, and He sold them into the hands of the Philistines (20 years - I Sam. 7:2,14b), and into the hands of the children of Ammon (18 years). And (from) 'that year' they vexed and oppressed the children of Israel <u>eighteen years</u>, all the children of Israel that were on the other side Jordan in the land of the Amorites, which is in Gilead".

From the contents of the above verses it seems to clearly imply that the 18 years oppression had to do mainly with the children of Israel beyond the Jordan river, even though there were attacks across to the western banks (Judges 10:9). If this is so, then how long did the Philistines oppress Israel west of the Jordan river? This problem can be clearly solved. First, we should realize that there were two Philistine oppressions. One which lasted 20 years allied partially with the Ammonites (I Sam. 7:2,14b) starting with the year 1176 B.C. and lapping over into Jephthah's reign by two years. The other was the 40 year (Judges 13:1) Philistine oppression which began with Samson's 20 year judgeship and continued on into Saul's reign. Now, at the beginning of which oppression did Eli die when he heard of the victory of the Philistines and the death of his two sons (I Sam. 4:10-18)? The scriptures clearly vindicate that his death took place at the beginning of the first Philistine oppression or in the year 1176 B.C.. Detail discussion concerning this problem will be taken up later.

Second, we do not necessarily call Samuel and Eli judges. We would rather consider them both, Eli and Samuel, as spiritual prophets ministering during long periods which ran parallel to the period of the Judges in Israel. Perhaps the following chart of "The Times of Eli and Samuel" will help clarify the issue. After the death of Joshua in 1433 B.C., who no doubt acted as both judge and prophet, there seemed to be no prophet until we have the prophetess, Deborah, who ministered during

1. Keller, W., "The Bible as History", p. 175, Econ Verlagsgruppe W. Ger.

1433 B.C. - - - - Joshua dies (a Prophet and a Judge)

1312 B.C. - Barak (40 years) - Deborah the prophetess
                              :

1274 B.C. - - - - - - - : - Eli born (lives 98 years)
                              :   :

1272 B.C. - Midianites (7)- - :   :

1265 B.C. - Gideon (40)   :

1225 B.C. - Abimelech (3)   :

1222 B.C. - Tolla (23)   :

1216 B.C. - Eli's ministry begins - : - (40 years) - I Sam. 4:18

1199 B.C. - Jair (22)   :

1189 B.C. - - - Samuel born - - : - - - - - - - - - -:

1177 B.C. - Period of sin (1) defeat :   :

1176 B.C. - Ammonite(18)Philistine(20)- Eli dies and Samuel is 12 years:
                           :   I Sam. 4:15; 7:2,14b   :
       Oppression:yrs:Oppression:yrs: :

1158 B.C. - Jephthah-(6)- Ammonite: :defeat   :

1156 B.C. - Ark to Gebeah I Sam.7:1 - 17 Philistine defeat

1152 B.C. - Ibzon (7)   :

1145 B.C. - Elon (10)   :

1135 B.C. - Abdon (8)   :

1127 B.C. - Samson (20) - :--40--:

1107 B.C. - No ruler (5) : yr :   :

1102 B.C. - Saul (40) -Philistine - Samuel "old" (86 yrs) I Sam.8:1 :
                     -Oppression

1087 B.C. - - - - :-----: I Sam. 14:47,48,52   :

1084 B.C. - 18th year Saul's reign Samuel dies (104 yrs - Josephus?)   :

1065 B.C. - Scripture implies Samuel lived within 3 years of David's :*
                                 reign - Samuel 123 years

1062 B.C. - David reigns (40 years)   A.D.

* A man in South Africa has just died at the same age of 123 yrs - 1986

the days of Barak. Her 40 year ministry overlapped the birth of Eli by
two years, but Eli's ministry did not begin until he was 58 years old;
after which he laboured for 40 years (I Samuel 4:18). Samuel must have
been around 12 years old when he was brought to the temple, and when Eli
died as he heard of the death of his two sons. Therefore, Samuel was
around 12 years old in the year 1176 B.C. and he ministered to within a
few years of David's reign.

Now, let us retrace our steps and consider how the historical facts re-vealed by archaeological excavation at Jericho verifies our chronolog-ical pattern. The chief and accurate authority is J.B.E. Garstang who made excavations at Jericho early this century. Everyone of his written statements seems to fit perfectly into our chronological jig-saw puzzle. Garstang also lists various earthquakes during ancient history which affected Jericho. We have listed below our suggested slight cor-rections:

GARSTANG'S LIST[1]	OUR SUGGESTED CORRECTIONS
1. Stone Age earthquake at Jericho	1. Stone Age earthquake at Jericho
2. About 2000 B.C. destruction of Sodom and Gommorrah	2. 1904 B.C. - Destruction of Sodom and Gommorrah
3. About 1450 B.C. - Earthquake at Jericho	3. 1458 B.C. - Earthquake at Jericho (Miracle - walls fall down)
4.	4. 785 B.C. - Earthquake in the days of Uzziah
5. Earthquake about 10 B.C. im-plied by Josephus	5. Earthquake about 10 B.C. implied by Josephus

GARSTANG'S OCCUPATIONS AT JERICHO[1]	OUR SUGGESTED CORRECTIONS	
	Stone Age village	4012 - 3812 B.C.
1st City: Early Bronze Age, I 3000 - 2500 B.C.	Early Bronze Age I (Pre-Flood)	3812 - 2500 B.C.
2nd City: Early Bronze Age, II 2500 - 2000 B.C. (Noahic Flood)	Early Bronze Age II	2500 - 2356 B.C.
	Middle Bronze Age I	2356 - 1900 B.C.
150 years - History Almost Silent" -		2356 - 2200 B.C.
3rd City: Middle Bronze Age II 1900 - 1600 B.C.	(Patriarchic Age)	2003 - 1580 B.C.
	Middle Bronze Age II	1900 - 1600 B.C.
4th City, I:M. Bronze Age III 1600 - 1500 B.C.(Exodus-1498 BC)	(Early 18th Dynasty)	1580 - 1498 B.C.
	M. Bronze Age III	1600 - 1450 B.C.
4th City,II: Late Bronze Age, I 1500 - 1385 B.C.	(Exodus - 40 years)	1498 - 1458 B.C.
	Late Bronze Age I	1450 - 1385 B.C.
(No City-Jericho Israeli Storebins)		1458 - 1418 B.C.
No City: Late Bronze Age, II 1385 - 1200 B.C.	Late Bronze Age II (Mesopotamia) - - - -	1385 - 1200 B.C.
		-1418 - 1410 B.C.
No City: Early Iron Age, I - - 1200 - 900 B.C.	(destroys) (storebins)	1200 - 915 B.C.
5th City: Early Iron Age, II - - 900 - 700 B.C.	(New City of Jericho)	915 - 785 B.C.
	Uzziah's Earthquake -	785 B.C.

1. Garstang, J.B.E., "The Story of Jericho", page 128,139 , Marshall - Pickering, England

Let us first summarise the situation at Jericho. Jericho (Third City) was destroyed at the time when the 18th Egyptian Dynasty drove out the Hyksos from the land (1580 B.C.). They continued their attack well into Palestine with the Hyksos main resistance at the southern Palestinian city of Sharuhen.[1] Jericho was then rebuilt over a period of a hundred years (1580 - 1480 B.C.), during the early part of the 18th Dynasty of Egypt, which included the double walls with houses built upon them referred to in Joshua 2:15. Jericho began to prosper after it had been captured by Thutmose III in 1476 B.C. and continued as a thriving city under the supervision of Egypt for 18 years (1476 - 1458 B.C.). During these 18 years Thutmose III, as the "Hornet" (badge on his uniform), had 14 campaigns north into Palestine and Syria and finally completely subdued the territory in 1458 B.C.. That very year Joshua drove into Canaan from Mt. Nebo and Jericho fell again and was destroyed by fire (Joshua 6:24). During the next forty years, after the curse was pronounced upon it (Joshua 6:26), Jericho was only used as a place of "storebins" with only a few central places reused. No wall or foundation were attempted, and no doubt these storerooms were completely destroyed in 1418 B.C. when Mesopotamia ruled Palestine for eight years. From 1418 B.C., Jericho lay desolate in obedience to the curse, but the curse finally fell upon the builder in 915 B.C. (I Kings 16:34).

Now, let us quote verbatum from the book, "The Story of Jericho" by J.B. E. Garstang and see how wonderfully his finds correlate with our chronology: "So devastating were the effects of the destruction and burning of Jericho at the end of the Hyksos rule (1580 B.C. - added), that the new city which appeared during the next generation shrank back to the original and narrow limits upon the hill - - - The fourth city thus emerged under very difficult conditions; and it is not surprising that special attention was paid to its defences. When completed these enclosed the top of the mound with a double wall of brick, the inner member of which followed and in places actually rested on the old wall of the Second City built during the Early Bronze Age thousand years before. The outer screen wall was probably retopped at the same time, but in its present ruinous condition we can not feel certain of the period of its restoration. All this work would certainly require time, and may have been spread over a good part of the century (1580 - 1480 B.C. - added). No use seems to have been made of the old ramparts of the Hyksos period (1746 - 1580 B.C. - added), which were left in their semi-demmolished state, while the slopes which had been occupied at that time now remained uninhabited: evidently the older circuit round the brink of the mound sufficed to house the reduced population. The inception of such work, however, shows that the city was rapidly reviving; and suffering no further setbacks, it began to enjoy a measure of prosperity.

In the completed scheme the old gate tower on the eastern side became practically obliterated; indeed the line of the new wall used it as a foundation. It was replaced, however, by a new tower in the north-west corner of the enclosure, which embraced both the inner and outer walls and was 90 feet in length. Rising high above the walls at their highest

1. Wilson, John A., "The Culture of Ancient Egypt" page 165; Publisher: University of Chicago Press

point, this feature dominated not only the city but the surrounding country. Such towers, or 'migdols', where characteristic of the age, and figure on Egyptian drawings of Canaanite cities, where they look rather like a fortified 'keep' within the outer line of defense. Mural towers supported the other angles, and between these ran a **curtain of the outer wall. Within, at a distance of four or five yards, ran the main or inner wall, a massive brick structure 12 feet thick;** standing for the most part upon the disused wall of the Second City, it probably rose at least 20 feet in clear height on the outside.

This impressive appearance was, however, to some extend deceptive. Although so massive, **this new wall showed defects both of material and construction.** As usual the bricks of which it was built were merely sun-dried, not kilnbaked, and contained no binding straw. They were small and of unequal size, though an average thickness of four inches was fairly well maintained. They **are distinctly** warm in colour, hence containing possibly more earth and less clay, as compared with the solid yellow bricks of the underlying structure; and it is remarkable how much more they now show signs of wear and weakness. Moreover, the **foundations were irregular. The old wall upon which they partly rested had not been trued up continuously before the new work began, so that where there were hollows or where the two lines did not coincide, the gaps were filled (in some cases only partially) by a layer of field stones, laid upon the accumulated debris.** Such weak spots in the foundations of a wall so weighty as this would not only make it difficult for the builders to keep the courses regular, **but be likely to induce subsidence. . . A further source of weakness arose from the building of houses against or actually upon the walls. They crowded thickly against the inner face of the main wall, and on the north and west were built high above the normal town level, upon foundation provided by narrow cross walls that bonded the ramparts together, and timbers that bridged the intervening space".**

At this point it might be quite interesting to quote Joshua 2:15, "Then she (Rahab) let them down by a cord through the window: for her house was upon the town wall, and she dwelt upon the wall".

Mr. Garstang continues, "The need for building-room within the cramped limits of the new city, had evidently become acute with the increased numbers and prosperity of the community under Egyptian (Thutmose III's) protection, and in this age of apparent security overpowered military considerations; **for these dwellings would be a serious hindrance to the defenders in a siege, beside tending to put an undue stress from within upon the overloaded and untrustworthy foundation.** .... Last phase of the Fourth City (Middle Bronze Age): It was not until the 15th century B.C. had begun that a marked cultural change heralded the Late Bronze Age, which here as elsewhere in Canaan followed in the wake of the Egyptian conquests (Thutmose III) by opening the door to the expanding trade relations of the Pharaoh's empire. Hitherto, as already explained, Jericho had lain outside their imperial scheme; but about 1475

B.C. **Thutmose III** took effective control of the city. Following up the sack of Megiddo (1476 B.C. - This date is perfectly correct according to our chronology - added) by a series of irresistible campaigns he established Egypt's firm dominion throughout the country; and the association of **his predecessor Queen Hatshepsut's (Queen of the Exodus . . added) scarat with his own (in Tomb 5)** points to a date early in the series for the formal annexation of Jericho. There is no reason to suppose that the local king or people opposed this step - rather the reverse. The somewhat tardy assertion of the Pharaoh's authority at Jericho was probably due to its isolation and its relative unimportance hitherto in the imperial scheme. But Thutmose III was not only an energetic and succeessful military leader but a great organizer, and he sealed his conquests by setting up an effective system of administration: under him the Egyptians became in fact conscious of their empire, and in the new scheme the chieftains of important Canaanite cities became the Pharaoh's representatives, while Egyptian residents were established at Gaza and other chosen centres. .. To support the kings authority and no doubt (in oriental fashion) to keep an eye upon his doings, a small garrison of foreign soldiers seems now to have been established with the city (1475-1458 B.C.-added): This is to be inferred from the contents of a cremation pit, discovered on the border of the Necropolis, in which were **found a scarab of Thutmose III** and another of **later date bearing the picture of a foreign deity (Perhaps Mesopotamian since they could have used and destroyed the final storebins in 1410 B.C.) upon the back of an animal.** The indications in this case are insufficient to determine the origin of the new guards. It should be mentioned that such 'garrisons' are not to be visualized in our conception of the word; for they generally consisted of only two or three (at the most five) well armed mercenaries, usually 'sherdens'.

In the excavation of the Palace and its adjoining storerooms we noted at two different levels **the effects of earthquake and extensive burning** (His date was 1450 B.C., while our chronology corrects his slight error to the year 1458 B.C. - added) . . Here it was found that while the Palace itself had been re-established on its stout foundations, the line of the house drainage system had been completely changed and now ran over some of the Hyksos storerooms (1746 - 1580 B.C. - added) that had formerly covered the whole hill slope overlooking the spring of the east. Investigation further showed that nearly all the older storerooms had **not only suffered damage by collapsing walls and roofs, but had been largely burnt out (Joshua 6:5; 6:24)"**[1]

Let us quote Joshua 6:5,24 as clear proof from the Holy Word of God which clarifies Garstang's above conclusions: "And it shall come to pass, that when they make a long blast with the ram's horn, and when ye hear the sound of the trumpet, all the people shall shout with a great shout; **and the wall of the city shall fall down flat, and the people shall ascend up every man straight before him. .. And they shall burn the city with fire**, and all that was therein: only the silver, and the gold, and the vessels of brass and of iron, they put into the treasury of the house of the Lord" (Date of burning - 1458 B.C.).

1. Garstang, J.B.E., "The Story of Jericho", page 111-119, Marshall - Pickering, England

STORE-ROOMS AT JERICHO
(1458 - 1418 B.C.)

In obedience to the curse, Jericho was not rebuilt, yet, the Israelites did use some of the old rooms as storebins. They chose some of the central rooms on the old Tel with some reconstruction to make them usable during the days of Joshua and that generation (40 years). Mr.Garstang continues concerning his excavation in these store-rooms:

"In the reconstruction they appear to have been two-storied, but the ground, now redressed, rose above the lower storey which in some cases was more like a cellar about four or five feet high and stacked with storebins already ancient. Moreover only the central group was rebuilt, and this was done by the addition effectively of one or two new storeys upon the earlier walls now partly ruined. The visible result was a small group of ten or twelve store-rooms clustered together using some ten feet above the raised general surface of the area, just below the eastern facade of the Palace. In the restoration of these store-rooms quite a number of old vessels that had been preserved in the ruins below were now brought up to the higher level - a procedure which further complicated the archaeology of the area. For this reason amongst others we are not able to assign a precise date to this episode (1458 - 1418 B.C. - added); but it seems to have occurred toward the end of the reign of Thutmose III (1477 - 1445 - added) or just afterwards. This we infer from the contrast between the new furniture and the old".[1]

Thus, we can now clearly see that the true Biblical chronology correlates perfectly with history. The real problem lies in finding the key, which we believe that we have discovered. So now, we might all unite in correlating ancient history, the Bible, and all other events into their proper relationship according to an acceptable chronology.

THE EXODUS AND ITS EFFECT ON RAS SHAMRA

The Egyptian spoke a mixed language which consisted of words from both the Semitic and Hamitic groups. The best present day representation of this original Egyptian language is still heard among the Coptic Christians.

The new incursion into Egypt in 1713 B.C. by the family of Jacob, and co-ordinated closely with the Hyksos penetration in 1746 B.C., may have brought a new language influence into the Egyptian delta. Otherwise, by the 15th century, it also may have been the Israelites in the delta of Egypt who had only revived their own set of former religious script, which stood as the consonants of their renewed language.[2] This new Israelite religious script may not have been the result of the social influence and their association with both the Egyptians and the Hyksos people. The latter were expelled from Egypt around 1580 B.C..

This renewed religious script spread to the Sinai peninsula by Egyptian conscripted Hebrew mine workers some time between 1538 B.C. and the Exodus in 1498 B.C.. We find the main evidence of this new alphabet, and the script which was recorded around 1500 B.C., at the touquoise mines and at the temple Serabit near Mount Sinai. There, we find written evidence of the ritual customs of the Hebrews in the renewed script

1. Garstang, J.B.E., "The Story of Jericho", p..120, **Marshall Pickering**
2. Excavation at Elba dates the language now back to 2300 B.C.

93

WILDERNESS CHART
(1498 - 1458 B.C.)

	An. Hom.	B.C.	Month	Day	Text
Exodus	2514	1498/7	1	15	Ex.12:2;Num.33:3
Wilderness of Sin	2514	1498/7	2	15	Exodus 16:1
Manna & smitten Rock	2514	1498/7	2		
Ten Commandments	2514	1498/7	3	15	Exodus 19:1
Sojourn at Sinai					
Statutes & Judgments					
9½ months					Ex. 19:1;40:17
Golden Calf					
Tables broken					
Tabernacle built	2515	1497/6	1	1	Ex.40:17;Num.1:1
Numbering Israel	2515	1497/6	2	1	Numbers 1:1
Spies sent from Kadesh Barnea	2515	1497/6	2	20	Num.10:11;13:17
Miriam's death in 39th year	2553	1459/8	1		Numbers 20:1
Death of Aaron	2553	1459/8	5	1	Num.20:28;33:38
Brazen Serpent					
Defeat of Sihon & Og					Num.20:28;33:33
6 months					
Balaam & Balak					Deut. 1:3
Apostasy Baal-Peor					
Numbering New Generation	2553	1459/8	11	1	Deut.1:3;34:8
Death of Moses	2553	1459/8	12		Num.14:33;32:13
					Joshua 5:6
Entrance into Canaan	2554	1458 BC	1	14	Josh.1:1-5:10

- - - - - - - - -

(probably the religious and Biblical script of the Hebrew patriarchs). It was a linear type of writing independent both of the hieroglyphies and cuneiform (Now found at Elba, Syria, dating from 2305 B.C. to 2205 B.C. - see page 206).

Excavations at the temple library at Ras Shamra, now identified with the ancient Phoenician city of Ugarit, uncovered quantities of tablets dated back to 1400 - 1350 B.C. in eight languages and scripts (We suggest a date from 1496 to 1350 B.C.). One of these was a special "script" and "the people who wrote these inscriptions claim to be Arabs, whose forefathers came from the south of Palestine, from the district round the Dead Sea called Arabah (Joshua 3:16). And not only that, but they write of mystic rites in the same wilderness of Kadesh, where the Israelites sojourned for some time".[1]

This script used twenty-eight different characters while the original cuneiform used a larger number. After much investigation it has been suggested that this writing was alphabetical and constituted a cryptogram. To decipher, it was finally found that "Archaic Hebrew" was the original script. Not only this, but that El (Hebrew Jehovah) was their God. Also at Ras Shamra there is reference to all the Israelite sacrificial offerings as recorded in the Hebrew Old Testament. Many of the instruments of worship along with the word priest (Kohen) were also mentioned.

Deciphering this same writing at Serabit El-Khadem has proven that during, especially, the reign of Hatshepsut (which included her co-reign with Thutmose II dating - 1510 to 1498 B.C.), Palestinian slaves were conscripted to work in Sinia. These conscripts carried the knowledge of this renewed ancient Hebrew alphabet back to the scribal school at Ugarit where they incorporated it into their written records, language, and included many of the Hebrew rituals within their personal worship. Improvement and changes were gradually made down through the years and by 1000 B.C. this alphabet had finally taken the form of 22 characters. Later, it left its effect on the whole Mediterranean area by means of the Thalassocracy. The period of Phoenician dominion of the sea (824 - 786 B.C.). Perhaps the above language should be referred to as the "Isro-Syrophoenician Alphabet" in the light of its ancient Hebrew origin. It is now known to have been evident in the north Syrian city of Elba around 2250 B.C. not far from the scribal school at Ugarit.

Since we can now set the date of the Exodus at 1498 B.C., it would seem very logical, with the present information, to say that the "Isro-Syro-phoenician Script" was inherited through the Jewish nation and its tranmittal was from the areas of Elba and Haran where the Hebrew forefathers carried the script over through their faithful religious worship, and it was later nationally revived, being associated with the giving of the law at Sinai in 1498/7 B.C.. Afterwards, the Israelites wandered in the Sinai peninsula for two years (1498 - 1496 B.C.) before arriving at Kadesh. During which time, they were accompanied by released Arab Phoenician Sinai slave labours who had gathered two years of personal knowledge and experience of the miracles of the God of Israel, through their association with the Hebrew slaves, by the influence of the laws

1. Marston, Sir Charles, "The Bible Comes Alive", p.54,Eyre & Spotteswood

95

written by God supernaturally, and by the writings both on the walls of the Sinai mines and in the Sinai Serabit Temple. Therefore, we may conclude that they were also indelibly written in the minds of all captive miners who had been gathered from far and wide for forced labour by the Egyptians. The Israelites release from Egypt, no doubt, brought deliverance also to these other captive slaves in Sinai. They, freed, providentially, carried back to their own lands the Exodus news, the Godly fear of the Israelites, the stories of the miracles, the renewed script of the Hebrew worship, and the knowledge of the newly established rituals. Also, as an eternal record of these miraculous events, God, the God of Abraham, Isaac, and Jacob, spoke to Moses and said, "Write this memorial in a book" Exodus 17:14 (see Deut. 6:9; 27:2-8; Joshua 8:30-32).

Sir Charles Marston confirms the above conclusions by stating, "The archaeological evidence now suggests that the Phoenicians derived the alphabet from the Israelites. ... The obvious suggestion is that the Israelites learned to use the (renewed - added) script when in the wilderness of Sinai, and brought it with them into Palestine when they conquered the country under Joshua".[1]

Before we leave the subject of Ras Shamra, we also find that these tablets contain an account of a great battle some time in the past between Cheret, King of Zidon, and a vast host of enemies commanded by the moon god, Terach (Terah). The tribes of Zebulon and Asher are mentioned along with the Edomites (mixed multitude).

Now, under our present chronology some of these Ras Shamra tablets, may be dated as early as 1496 to 1350 B.C., and others could have been written before or after the year Joshua took Hazor at the end of the seven year offensive (1458 - 1451 B.C.). We understand by the context of Joshua chapter eleven that he smote the confederacy of kings led by Jabin king of Hazor even unto Zidon and then returned and burnt Jabin's city: "For Hazor before time was the head of all those kingdoms" Joshua 11:10. The point we want to raise here is that Joshua's attack, in the year 1451 B.C., carried him to Zidon (Joshua 11:8). Could this be the attack which the tablets are referring to involving King Cheret of Zidon? The above date of Joshua's attack correlates well with the possible date of the Ras Shamra tablets; therefore, King Cheret may have been King of Zidon in 1451 B.C.. Naturally, the tribes of Zebulon and Asher were involved in Joshua's attack upon Zidon since these tribes were located in the north. By the seventh and final year of Joshua's victories a certain remnant of Edomites, mixed multitude, and former Sinai slaves may have still remained with Joshua's forces.

As to the above statement concerning the Moon god, we now know that the city of Ur was the seat of the Moon god in the south, while the city of Nahor, in the valley near Haran, was also the northern centre of such worship where Terah finally died. Terah and his decendants, including Abraham's family and Lot, all lived in this heathenistic atmosphere at both Ur and the Haran area. Abraham was called out of this Moon god influence into faith and obedience to the true God El, the God of Abraham, Isaac, and Jacob, while their two forefathers Serug and Terah remained moon worshippers as Joshua 24:2 states:

1. Marston, Sir Charles, "The Bible Comes Alive", page 119,125, Eyre and Spotteswoode Printers Ltd

"And Joshua said unto all the people, Thus saith the Lord God of Israel, your fathers dwelt on the other side of the flood (River Euphrates) in old time, even _Terah_, the father of Abraham, and (Serug) the father of Nahor: and they (Terah and Serug) served other gods (Moon god)" Joshua 24:2.

We must not be confused by failing to realize that there are _two Nahors_. One was the son of Serug and the other the son of Terah. We must be careful to chose the right "Nahor" to solve this very simple scriptural problem.

Therefore, the god which Terah served would naturally be the Moon god. When the author (a worshipper of a polytheistic system of gods) of the Ras Shamra tablets referred to the Hebrew forces under Joshua, he spoke of them with the terms which chosely agreed with his own polytheistic method of worship. Thus, these forces were commanded by the "Moon god, Terah", or empowered by the Moon god of Terah. Then again, the phrase could mean that they were commanded by the _descendants_ of the Moon god worshipper, Terah. The heathen, even today, also can never discern between true and false worshippers; therefore, they class all religious worshippers in the same catagory.

Perhaps the above gives us added light concerning the first 18th Dynasty Egyptian King ("who knew not Joseph"), Ahmose I, whose "name" could also mean "the Moon is born".[1] Moses, a descendent of Terah, the Moon god worshipper, was born and taken into Ahmose's palace by either of his daughters Sat-Kamose or Ahotep; the latter, whom some may have confused with Hatshepsut. Perhaps, Ahmose's name indicates that the first king of the 18th Dynasty was more friendly disposed towards the Israelites than the later kings who changed their god to "Thut" and persecuted them. Although this suggestion could be contradictive, since the name "Thut" also implies Moon worship.

Both the Philistines and the Ammonites (Judges 10:8) had unitedly attacked Israel in 1177 B.C. and had won by the following year (1176 B.C.). The Ammonites ruled Israel from 1176 B.C. for eighteen years mainly beyond (East of the) Jordan until Jephthah's reign. Later, at the beginning of Jephthah's reign (1158 B.C. - 300 years after Israel's entrance into Canaan - Judges 11:26), the Ammonites beyond Jordan were defeated by him (Judges 11:32-33) and he judged Israel six years (1158 - 1152 B.C. - Judges 12:6).

At the same time of the Ammonites victory (1176 B.C.), the Philistines also were victorious "west" of the Jordan River and captured the Ark, Eli died, and Samuel was about 12 years old (Judges 10:7; I Sam. 7:2,3, 14). The Ark stayed with the Philistines only seven months (1177/6 B.C.- I Sam. 6:1) and was then returned and placed in Abinadab's house at Kirjath-jearim where it rested for twenty years (1176 - 1156 B.C.; I Sam. 7:2). Therefore, the Ark was there during the eighteen years of the Ammonite rule plus two years of Jephthah's reign or for a total of "twenty years". I Samuel chapter seven fills in the story and gives us a detailed account of the defeat of the Philistines after a "twenty year" rule over Israel (I Sam. 7:2,13). In other words the Ammonite rule of _18 years_ "beyond" Jordan ran parallel to the Philistine reign of _20 years_ "west" of the Jordan. Jephthah defeated the Ammonites beyond the

1. Gardner, A., "The Egypt of the Pharaohs" p.174,1961,Oxford Univ.Press

97

Chapter IX

LATE JUDGES PERIOD
IRON AGE
(Jephthah to Solomon's Temple)
(1158 - 1018 B.C.)

Exodus to the entrance of Canaan — — — — — 1498-1458 B.C.   40 yrs
    Deut. 2:7

Entrance into Canaan until Jephthah — — — 1458-1158 B.C. 300 yrs
    Judges 11:26

  (Jephthah defeats Amon after 18 yrs — — 1176-1158 BC Jud.10:8)
— — — — — — — — — — — — — (Philistines)— 1176-1156 BC I Sam.7:1-13)
                (20   )
Jephthah judges Israel   (years) 1158-1152 B.C.   6 yrs
    Judges 12:7   (oppression)
    (Amon defeat after 18 yrs(in Israel) - 1158 B.C.)
    (Israel.defeats Philistines after 20 yrs - 1156 B.C.)
                I Sam. 7:1-13

Ibzon judges Israel   1152-1145 B.C.   7 yrs
    Judges 12:8,9

Elon judges Israel   1145-1135 B.C.   10 yrs
    Judges 12:11

Abdon judges Israel   1135-1127 B.C.   8 yrs
    Judges 12:14

Samson judges Israel   (Philistine) 1127-1107 B.C.   20 yrs
    Jud.15:20;13:1   (40   )
                   (year   )
No King in Israel   (oppression) 1107-1102 B.C.   5 yrs
    Jud.17:6;18:1;19:1   (of   ) Jud.21:25;IK.6:1
                   (the   )
Saul King of Israel   (people) 1102-1062 B.C.   40 yrs
    I Sam.10:1;Acts 13:21 (of   )
                   (Israel)
    (Philistine oppression - (- ends -) — — — —1087 B.C.)
               I Sam.14:47-52

David King of Israel   1062-1022 B.C.   40 yrs
    II Sam.5:4;I Chron. 3:4

Solomon King of Israel until Temple 1022-1018 B.C.   4 yrs
    IK.11:42;II Chr.9:30;IK.6:1

— —
  IK.6:1 Exodus to 4th year of Solomon   — — 1498-1018 B.C. 480 yrs

Jordan at Mizpeh Gilead (1158 B.C.) and, no doubt, was involved in helping to, defeat the Philistines at Mizpeh, near Ramah, two years later (1156 B.C. - I Sam.7:11). Please notice that the Ammonites and Philistines are both mentioned in Judges 10:6-8, but in I Sam. 7:14, we find peace made with the Philistines and the Amorites. The Ammonites were defeated two years before this.

What happened to the Ark after the twenty years at Kirjath-Jearim at which time they lamented after it (I Sam. 7:2)? In the light of the King James Version quoting II Sam. 6:3, we might come to the false conclusion that Abinadab's home might have been changed to Gibeah where the Ark may have remained for another 101 years until David took it to Jerusalem. Gibeah being very close to Ramah which was the hub of Samuel's circuit among the three cities of Bethel, Gilgal, and Mizpeh (I Sam. 7:16). Mizpeh also being the place of gathering the people at the rededication, and of the defeat of the Philistines (I Sam. 7). Since the Ark was taken to battle against the Philistines 20 years before (I Sam. 4:5,6), they may have repeated this action at Mizpeh where it may have remained. But this may be an incorrect conclusion. If the Ark was taken to Mizpeh during the defeat of the Philistines, it was, no doubt, returned to Kirjath-Jearim to the area of Baale of Judah. In II Sam. 6:3 where it mentions the word "Gibeah" the Amplified translates "the hill". This translation is verified by I Chron. 13:6 where it says that David did not go to Gibeah but rather to "Baalah, that is Kirjath-Jearim, which belonged to Judah, to bring up thence the Ark of God the Lord . . ". Therefore, we conclude that Israel's lamenting after the Ark (I Sam. 7:2) did not necessarily mean a change in location but a rededication to the vows and commands (Judges 21:1-7) which the Ark represented. Either the Ark remained there continually; or, if it was taken away at times, it was returned to Kirjath-Jearim where it stayed for 101 years until David's 7th year of reign (1156 - 1055 B.C. - equals 101 years) when he brought the Ark to Jerusalem (I Chr. 13:6;II Sam. 6:3). The Amplified Version gives I Samuel 7:2 as follows which confirms our chronological measurement perfectly:

"And the Ark remained in Kirjath-Jearim a very long time (nearly 100 years, through Samuel's entire judgeship, Saul's reign, and well into David's when it was brought to Jerusalem). For it was 20 years before all the house of Israel lamented after the Lord".

The above measurement was from the time of the defeat of the Philistines when Israel lamented after the Ark until David's seventh year. If one measures from the first year of the victory of the Philistines, then we would have to add twenty more years to the measurement (101 years plus 20 years).

The Biblical text seems to imply that Samuel died about three years before the death of Saul. Samuel was around 12 years old when the Ark was first taken by the Philistines (1176 B.C.); therefore, Samuel was born in the year 1188 B.C. and died about three years before Saul died in the year 1065 B.C. (Samuel - 123 years old). If he died in the 18th year of Saul, according to Josephus, he would have been 104 years old. Samuel was "old and gray haired" (I Sam.12:1,2 - 86 years old) when he anointed Saul (see also I Sam. 28:14). Samuel's age of 123 years correlates well with Acts 13:20:

"And after that (verse 17, 'chose our fathers . . in . . Egypt') he gave unto them Judges (Exodus 18:13,25,26 - including Moses' 40 years) <u>about the space of four hundred and fifty years</u>, until Samuel the prophet (his death)".

Moses was chosen as the first spiritual judge of Israel (Exodus 18:13) in 1498 B.C. and Samuel was the last spiritual judge who died in 1065 B.C.; therefore, we have a <u>total span of 433 years</u> or "about" the <u>space of four hundred and fifty years</u> (Acts 13:20).

We continue with our chronology. Jephthah ruled six years (1158 - 1152 B.C.). Ibzon judged Israel seven years (1152 - 1145 B.C.) with Elon (1145 - 1135 B.C.) and Abdon (1135 - 1127 B.C.) following. In the year 1127 B.C., when Samson began to judge Israel, the Philistines once again became a powerful influence and oppressed Israel in Samson's day (1127 - 1087 B.C.) 40 years:

"And the children of Israel did evil again in the sight of the Lord; and he delivered them into the hand of the Philistines forty years" Judges 13:1. "And he (Samson) judged Israel in the days of the Philistines twenty years" Judges 15:20 (see Judges 16:31).

Since there is no mention as yet that they are without a ruler we conclude that Samson's twenty years coincide with the first twenty years of the forty year Philistine oppression. Following Samson's twenty years we have the period recorded for us in Judges chapters 17 - 21 which describes a time in which "there is no king in Israel" (Judges 17:6;18:1; 19:1;21:25). Some have suggested and dated this period early in the Judges period, but we disagree with this theory for these reasons:

First, Mizpeh is not mentioned as a place of seeking God until after Israel had been delivered from the first twenty year period of Philistine bondage (1176 - 1156 B.C.). Mizpeh is referred to three times during this period of <u>"no king in Israel"</u> which would synchronize this information well with the last part of the Judges Period (Judges 21:1,5,8). Also, the statement "no king in Israel" implies that it is immediately before the time that their first king, Saul, reigned. If it had been early in the Judges Period the Word of God would have rather stated, "there was no Judge in Israel". Also, it should be mentioned that Judges 21:1,5 refers to the oaths made before the Lord at Mizpeh recorded also in I Sam.7. This event recalls their lamenting after the Ark after it had been at Kirjath-Jearim for twenty years (I Sam. 7:2); their gathering at Mizpeh under the supervision of Samuel, and their rededication to the Lord:

"And Samuel said, Gather all Israel to Mizpeh, and I will pray for you unto the Lord. And they gathered together to Mizpeh, and drew water, and poured it out before the Lord (Feast of Tabernacles Ceremony), and fasted on that day, and said there, We have sinned against the Lord. And Samuel judged the children of Israel in Mizpeh" I Samuel 7:5,6.

The result of this rededication was none other than a complete deliverance from the hands of the Philistines (I Sam. 7:11). These vows are mentioned in Judges chapter 21:1,5 and may be dated 340 years after the Exodus, or in the latter part of the Judges Period.

Now, the next question is how long was Israel with out a judge or a king? First, let us reconsider what we have covered so far. We have established the year that Solomon began to build the temple (1018 B.C.). In the light of I Kings 6:1, we established the date of the Exodus 480 years earlier (1498 B.C.). Then we considered the forty years in the wilderness and the 300 years until Jephthah (Judges 11:26). We then added Jephthah's years plus the reigns of the following four Judges (Ibzon, Elon, Abdon, and Samson) which brought us to the year 1107 B.C.. Samson ruled the last twenty years of the Judges Period (1127 - 1107 B.C.), which in turn, synchronizes with the first 20 years of the final Philistine oppression of 40 years.

Now, let us return to the fourth year of Solomon (1018 B.C.) and add four years to find Solomon's first year (1022 B.C. - I Kings 11:42; II Chr. 9:30; I Kings 6:1). Before this David ruled forty years (II Sam. 5:4; I Chr. 3:4 - 1062 - 1022 B.C.), and Saul ruled for the same number of years preceding this (I Sam. 10:1; Acts 13:21 - 1102 - 1062 B.C.). This leaves exactly five years unaccounted for after the end of Samson's twenty years (1107 B.C.) and the beginning of Saul's reign (1102 B.C.). Therefore, we conclude that there was "no king in Israel" for five years (1107 - 1102 B.C.). These five years fulfill the period of "no king in Israel" and also makes up the years lacking in the 480 year span of time (I Kings 6:1) from the 4th year of Solomon (1018 B.C.) until the Exodus (1498 B.C.). See our chart on page 147.

- - - - - - -

PHASE V
BIRTH OF ABRAHAM TO THE EXODUS
2003 - 1498 B.C.
CHAPTER X
ABRAHAM'S LIFE
2003 - 1828 B.C.
(175 years)

Now that we have worked our way through the Biblical chronology to the Exodus, we have seen how marvelously all the detailed facts have synchronized perfectly. Naturally, this is what we expect if we have discovered the correct chronology which collaborates with all of Ancient History. Therefore, the ā priori argument is that all the remaining Biblical information will fit exactly into the blue-print with no genealogical gaps (except after the Noahic Flood). Besides, History itself, with all detailed information involved, will not allow long or even short gaps. In fact we shall see that history verifies the plain Biblical chronology to the letter rather than contradicting the same (see correlation of History & Bible pages 211 - 235).

The great chronologist Anstey wrote, "The chronology of the Old Testament is in strongest contrast with that of all other nations. From the creation (expulsion - added) of Adam to the death of Joseph, the chronology is defined with utmost precision, and it is only toward the end of the narrative of the Old Testament that doubts, and difficulties, and uncertainties arise(even these have since been solved - added) ... Bible chronology is an exact science. It is not built upon hypothesis and conjecture. It rests ultimately upon evidence or testimony; but it does occasionally require the use of the method of scientific historic induction".[1]

Only those who have glanced lightly at chronological study of the Bible have tended to suggest that there might be gaps in Biblical chronology. We also felt the same until we began to check the thousands of facts in a thorough investigation. The genealogical line is exact and can be used without fear of delusion from Adam's expulsion to Joseph's death and beyond. After his death you have already seen that God has provided definite measurements of time in His inspired, Holy Word which have proven dependable. Philip Mauro goes a step further. After completing his systematic study of Bible chronology he adds his confession: "The first three thousand years and more of Bible history can be reckoned with accuracy, because the scripture gives full and clear information as to the count of years ...".[2]

ABRAHAM

Just three hundred and fifty years (2355-2005 B.C.) after the "Noahic Flood" (2356 - 2355 B.C.), a period which could compare in time from the landing of the Pilgrims at Plymouth Rock (1620 A.D.) unto the year 1970 A.D., monarchal rule was finally fully re-established in the Middle East. This period, which began with world chaos, was now back to normal. The beginning of which is known by historians as a period when "History was almost silent" (Intermediate Period). During the final phase of that

1.2. Mauro, P.,"Chronology of the Bible" p.15,16,34 quote Anstey era, the 3rd Dynasty of Ur gained ascendancy for about one hundred years (2113 - 2005 B.C.). During the latter years of that dynasty, a Semitic group of peoples from Elam began to infiltrate the area around Ur. These "Elamites" were known as the "Amurru" (They are not Amorites since the Amorite brothers helped Abraham to fight the Elamites later - Gen.14: 13,24). Pressure continued through very difficult times. During the reign of Ibbi-Suen inflation had increased the cost of living to sixty times normal prices. Then suddenly under Ishbi-Erra, the last ruler of the 3rd Dynasty of Ur, the Elamites captured the city of Ur in 2005 B.C.. Noah had died the year before (2006 B.C.) and Nahor, the son of Serug, had died a few years previously (2014 B.C.), but Serug and Terah still lived beyond the Euphrates among the Moon god worshippers (Joshua 24:2). Just two years after the "Elamites" (Amurru) capture of Ur, Abraham was born there to the family of Terah in the year 2003 B.C.. Six years later (1997 B.C.) when Abraham was six years old, "Ishbi-Erra of Isin" (1997 - 1984 B.C.) regained control, expelled the Elamite garrison from Ur, rebuilt the city, and reinstalled the statue of the Moon-god Nanna, Ur's tilulary deity, which the invaders had earlier carried away to Anshan.[1] Throughout the next sixty years, during the early years of Abraham's life, peace, law, and civil order were restored in Ur, and the city was controlled from Isin by the three following rulers: Shu-ilishu (1984 - 1975 B.C.), Iddin-Dagan (1974 - 1954 B.C.), Ishme-Dagan (1953 - 1935 B.C.).

Throughout the earlier phase of these years Abraham attended school in the city of Ur. The city was located on the bank of the Euphrates River and was circumposed and bisected by a moat canal interleading to two city ports. Centrally located was the temenos (Sacred area of the Moon-god Nanna) which originally was raised above the surrounding city. The temenos was the Moon-god's terrace on which the Ziggurat stood.It is called today, "the Mound of Bitumen" which was formerly built by Ur-Nammu (2113 - 2095 B.C.), the first ruler of the 3rd Dynasty of Ur around 2100 B.C.. Ur-Nammu's name was stamped on the lower bricks of the solid mass tower, 200 feet long, 150 feet wide, and about 70 feet high (Built around 70 years after the fall of the Tower of Babel).

At school Abraham learned to write the cuneiform signs on flattened lumps of soft clay. Reading lessons were mainly hymns with lessons in writing and arithmetic. They learned the multiplication, division tables, and mastered square and cube roots.

Abraham's home, or average dwelling, measured forty by fifty-two feet, and had a side yard fifteen feet wide. The lower walls were built of burned brick, the upper of mud brick, and the whole wall was usually plastered and whitewashed. An entrance lobby led into the central court, on which all the rooms opened. On the lower floor were located the servants' rooms, the kitchen, the lavatory, the guest chamber, and also a lavatory and wash place reserved for visitors. In addition, there was a private chapel at the back of the house. Thus, all the first floor was utilized for the servants and guests. The second floor housed the family, providing five rooms for their use. The entire house of the average middle-class person had from ten to twenty rooms.[2]

1. Gardner, Sir Alan, "Egypt of the Pharaohs", page 62, 1961, publisher: Oxford University Press
2. Free, Joseph, "Archaeology and Bible History", page 49, Van Kempen Press, Wheaton, Illinois 1950

When Abraham was in his sixties, during the latter years of Ishme-Dagans reign (who set law in the land), a semi-nomadic group using donkeys for transportation (Elamites = Amurru) began to attack the countryside again near Ur and destroyed Nippur. Therefore, we conclude that around 1935 B.C., the year that Lipit-Ishtar began to rule, partially for fear of this repeated nomadic pressure, and also according to Abraham's providential call from God (Acts 7:2), Terah's family moved to Haran where it was peaceful and more secure. Lipit-Ishtar speaks of freeing citizens of certain cities.[1] This event, no doubt, refers to the "Elamite" incursion during his second year of reign (1934 B.C.).

TERAH'S MIGRATION
1935 B.C.

Why did Terah chose the area of Padan-Aram, wherein the cities of Haran and Nahor were located, as a place of migration? In the light of scriptural information surrounding the migration of Terah, we might come to certain conclusions from the following data:

1. Haran, the son Terah, died before his father in the city of Ur (Gen. 11:28).

2. Nahor, the son of Serug, died before his father, perhaps in the city of Nahor (2014 B.C.). We might conclude that the City of Nahor was named after him (Gen. 24:10). It was also located near the City of Haran. The City of Nahor was also referred to as "the abode of Bethuel" the son of Milcah and Nahor the son of Terah" (Gen. 24:10,15). Bethuel was the father of Rebekah. She was ask to be the wife of Isaac by Abraham's servant in 1863 B.C..

3. Haran was the city to which Terah migrated (Gen. 11:31), and perhaps named "Haran" because of Haran's early death.

4. Terah migrated to Haran with Abraham, Sarah, and Lot who was the son of Haran. Haran had previously died, perhaps in the City of Haran (Gen. 11:28), and Lot his son lived with his uncle, Abraham.

5. Nahor, the son of Terah, and his wife Milcah, the daughter of Haran, may not have been at Ur at the time of Terah's migration for they may have lived in the City of Nahor, since their son Bethuel lived there later (Gen. 24:10-15).

6. The City of Nahor is situated in the valley below the site of the city of Haran and is also associated with the area of Padan-Aram (Gen. 28:2,10).

7. Padan-Aram means the "Plain of Aram". Aram was the son of Shem (Gen. 10:22), therefore, Shem may have lived in that area until his death in 1853 B.C.. This could identify the location as the area of the patriarchal fathers.

"Terah", the father of Abraham, and "Serug", the father of Nahor, were the two forefathers who were heathen Moon-god worshippers beyond the Euphrates (Josh. 24:2). "Terah" lived at the southern centre of this heathen worship (Ur), while "Serug" lived at the northern centre (Nahor) near Haran. The City of Haran, where Terah settled, was probably named after Milcah's deceased father, and was also closely associated with the area of Padan-Aram. Nahor had died in the year 2014 B.C. fifty-two years

1. Saggs, H.W.F., "The Greatness That Was Babylon", page 63 , Sidgwich & Jackson Publishers

before his father Serug (1962 B.C.); therefore, they probably named the city in which some of the relatives lived, "Nahor" to honor the dead son. Nahor (not the son of Serug), the son of Terah, and Milcah his wife migrated earlier to the City of Nahor since they did not move with Terah in 1935/4 B.C. (Gen.24:10,15). Later, Isaac's servant went to the City of Nahor (Gen. 24:10) to seek the hand of Rebekah, the daughter of Bethuel, the son of Milcah and Nahor, in the year 1863 B.C., when Isaac was 40 years old (Gen. 25:20). Still later, in the year 1766 B.C., Jacob went to Padan-Aram (Haran - Gen. 28:5,10) where Laban lived. He was the son of Bethuel. Therefore, we conclude that Terah moved to Haran since it was near the city of Nahor and in the area of Padan-Aram where his forefathers and relatives lived. Also, the religious atmosphere and surroundings reminded him of Ur. So finally, around 1935/4 B.C., in the face of another nomadic incursion at Ur, and Abraham's revelation (Acts 7:2-4), the remainder of the family, Terah, Abraham, Sarah, and Lot, moved north to be in the midst of their relatives in the cities of Haran and Nahor known as the area of Padan-Aram (The plain of Aram the son of Shem - Gen.10:22).

We should remind ourselves at this point that Shem was still alive at this time and lived another eight-two years after this until 1853 B.C.. Shem may have even lived in Padan-Aram and may have had a real influence on Abraham's life while at Haran during those 7 years before the death of Abraham's father, Terah. According to Jewish tradition, some have thought that Shem may have been Melchizedek, King of Salem, priest of the most high God, who met Abraham returning from the slaughter of the kings, and blessed him (Hebrews 7:1), although this is very doubtful. Melchisedek was no doubt a theophany of God or of the Messiah. In other words he was the manifestation of the pre-incarnate form of the Messiah (the Angel of the Lord), the King of peace; without father, without mother, without descent, having neither beginning of days, nor end of life; but made like unto the Son of God; abideth a priest continually (Hebrews 7:2b,3). If not, he certainly typified the Messiah.

In 1934 B.C., perhaps only a year after Terah migrated to Haran, a coalition of nations under the "Elamite" King Chadorlaomer captured Nippur and perhaps Ur and swept down through Palestine and ruled it for 14 years (1934 - 1920 B.C.). This is recorded in Gen. 14:5. "The name Chedorlaomer is an 'Elamite name', meaning servant of the Elamite God Lagamur. Bricks in the British museum tell us that Chedorlaomer had conquered Babylon and that Eri-Aku, son of Kudur-Mabug, servant of the Elamite God Mabug, ruled Larsa. But Eri-Aku, (King) of Larsa, is Arioch of Ellasar ..."[1] (See Gen. 14:1). Our Biblical chronology pin-points Arioch's (final attack 1920 B.C.) rule during the early years of Gungunum of Larsa (See chart). Arioch was, no doubt, either a minor ruler at Ellasar, or a co-regent, or a Army General under Gungunum (1932 - 1905 B.C. - Larsa Dynasty). These events all fit together into the right period, and we are sure our chronology will correlate perfectly with a more detailed study of this historical period.

1. Anstey, Martin, "The Romance of Bible Chronology", page 131

Abraham remained there approximately seven years from 1935 B.C. until the death of his father, Terah, in the year 1928 B.C.. At which time, under the call of God (Acts 7:1-5; Gen. 12:1-4) Abraham boldly migrated into the very area of Palestine already controlled six years by his family's enemy, Chedorlaomer, the Elamite (Amurru). The final incounter with his rival came eight years later (1920 B.C.). After 14 years control over Palestine, Chedorlaomer with a coalition of nations from Elam, Ellasar, Shinar, and of the nations (Gentiles), again "smote the Rephaims in Ashteroth Karnaim, and the Zuzims in Ham (Egypt), and the Emims in Shaveh Kiriathaim, and the Horites in their mount Seir, unto Elparan, which is by the wilderness. And they returned, and came to Enmishpat, which is Kadesh, and smote all the country of the Amalekites, and also the Amorites, that dwelt in Hazezor-tamar. And there went out the king of Sodom, and the king of Gomorrah, and the king of Admah, and the king of Zeboiim, and the king of Bela (the same is Zoar); and they joined battle with them in the vale of Siddim; with Chedorlaomer the king of Elam, and with Tidal king of nations, and Amraphel king of Shinar, and Anioch king of Ellasar; four kings with five. And the vale of Siddim was full of slimepits; and the kings of Sodom and Gomorrah fled, and fell there; and they that remained fled to the mountain. And they took all the goods of Sodom and Gomorrah, and all their victuals, and went their way. And they took Lot, Abram's brother's son, who dwelt in Sodom, and his goods, and departed. And there came one that had escaped, and told Abram the Hebrew; for he dwelt in the plain of Mamre the Amorite, brother of Eshcol, and brother of Aner: and these were confederate with Abram (The Amorite brothers may have had a very large force which helped Abraham; therefore, the Amorite judgment was postponed for 400 years - Gen. 14:24; 15:13-16). And when Abram heard that his brother was taken captive, he armed his trained servants, born in his own house, three hundred and eighteen, and pursued them unto Dan (along with the Amorite forces who were confederate with Abraham - See Gen. 14:24). And he divided himself against them, he and his servants, by night, and smote them and pursued them unto Hobah, which is on the left hand of Damascus. And he brought back all the goods, and also brought again his brother Lot, and his goods, and the women also, and the people. And the king of Sodom went out to meet him after his return from the slaughter of Chedorlaomer, and of the kings that were with him, at the valley of Shaveh, which is the king's dale" Genesis 14:5-17.

Abraham broke the power of the "Elamites" in Palestine after 14 years of bondage (1934 - 1920 B.C. - Gen. 14:4,5), but their influence in the area of Ur continued until 1908 B.C., at which time Ilushuma, the Assyrian King, intervened at Ur and freed it from the "Elamites" (Amurru). This intervention was only temporary. Soon after this Sumu-abum (1892 B.C.) was made King of Babylon and he was followed by the line of very strong rulers known as the First Dynasty of Babylon (1892 - 1595 B.C.). They soon became the supreme power in the area.

The above synchronization of events agrees dramatically with suggestions presented by Dr. John Rea in his Grace Theological Seminary notes, page two, on "The Age of the Patriarchs in Palestine". Dr. Rea states: "The campaign of the four kings in Genesis 14 must have taken place after the collapse of the Ur III Dynasty, for the king of Elam took the leadership of the coalition; yet it must have occurred before Hammurabi conquered the Elamite city states in Babylonia in the early second millennium B.C.. Furthermore, Tidal King of Goiim seems to have a typically Hittite name, but is not called a Hittite. This fact can be reconciled by placing the date at a time before the invading Indo-European peoples conquered the Hattian tribes of Anatolia and assumed the name of Hittite from their defeated subjects".[1] (See confirmation on chart page 216).

Historical calculations, new archeological finds, and exploration along the shores of the Salt Sea correlate the dates of the birth of Abraham and the destruction of Sodom perfectly with our chronological system. Dr. Rea further reiterates:

"Additional proof of a date around 2000 B.C. (Our date is 2003 B.C. - added) for Abraham is to be found in the fact that early settled occupations of Transjordan continued only into the Middle Bronze I Age, but not afterwards for several hundred years. Also the site of Bab ed-Dra', on the tongue of land projecting into the eastern side of the Dead Sea, and probably to be interpreted as the high place center of worship of the five cities of the Plain, indicates by its pottery that it was not used after the twentieth century B.C. (No later than 1900 B.C.). Since Sodom and Gomorrah were destroyed during the lifetime of Abraham, the date of the abandonment of Bab ed-Dra' furnishes a positive clue for the period of Abraham".[2]

Our date of the destruction of Sodom is 1904 B.C., since Sodom was destroyed in the 99th year of Abraham, during the year of Isaac's conception, and about a year before Isaac was born (1903 B.C. - Gen. 18). Other historians have come to the same conclusion. Mr. Keller wrote, "The Vale of Siddim, including Sodom and Gomorrah, plunged one day into the abyss. The date of this event can be fairly accurately established by the geologists. ... 'probably it was about 1900 B.C. that the catastrophic destruction of Sodom and Gomorrah took place,' wrote the American scholer Jack Finegan in 1951. A careful examination of the literary, geological and archeological evidence leads to the conclusion that the corrupt 'cities of the plain' (Gen. 19:29) lay in the area which is now submerged beneath the slowly rising waters of the southern section of the Dead Sea, and that their destruction came about through a great earthquake which was probably accompanied by explosions, lightning, issue of natural gas and general conflagration".[3]

Late geological tests confirm the exact details given in the Biblical story: "The geology of the place showed a stratum of salt 150 feet thick underlying the mountains, over which much free sulphur (brimstone) was found. It is a burned-out region of oil and asphalt, where a great rupture in the strata took place long ago. Formerly, a subterranean lake of

1. Rea, Dr. John, Grace Theological Seminary notes, "The Age of the Patriarchs in Palestine", page 2
2. Rea, Dr. John, "Historical Setting of the Exodus and the Conquest", Grace Theological Seminary Notes, pages 2,3
3. Keller, Werner, "The Bible as History", page 95, Econ Verlagsgruppe, West Germany

107

oil existed beneath Sodom with a vast accumulation of gas. Apparently the gas and oil were ignited. A tremendous explosion resulted, carrying burning sulphur and asphalt into the heavens above the cities. This, mingled with the salt of the Dead Sea stratus, literally rained down upon the whole plain exactly as the Bible describes".[1]

Abraham lived 175 years until 1828 B.C.. At his death Jacob and Esau were 15 years old while Isaac, their father, was 75 years old. Eber (the word "Hebrew" comes from his name - Gen. 10:21; Luke 3:35 K.J.V.), the great, great, great, great grandfather of Abraham, outlived Abraham by four years (1824 B.C.). Being the last Patriarch of the period of longevity, his name, "Eber" (Hebrew), was chosen, thereby, designating him as the father of the physical Hebrew Race (Gen. 10:21), while Abraham was the father of their faith (Gen. 15:5,6). This was so because "Eber" lived 4 years longer than Abraham. Shem, who was born 97 years before the Noahic Flood, and conceived a son, Arphaxed, two years after the Noahic Flood (Gen. 11:10 - 2353 B.C.) in his hundredth year, lived 600 years (2453 - 1853 B.C.) and died just 29 years before the death of Eber, and 25 years before Abraham's death. During Abraham's 150th year, Abraham could have communed with Shem, his forefather of nine generations, and obtained first hand information of that which happened 90 years before the Noahic Flood. Shem, in turn, talked with Methuselah, who conversed with Adam, the friend, and the creation of God.

– – – – – – –

1. Moody Monthly, January 1967, page 59 "originally appeared in MOODY MONTHLY magazine".

Covenant's prelude (1935/4 BC) (Foretold) Acts 7:2,3 - 1st call (Ur)

Covenant promised (1928 B.C.) (Revelation) Gen. 12:1 - Abram's call
                         :   :

Covenant provision : (Prescribed) Gen. 13:14-17 :
                         :   :

Covenant pronounced (30 years) (Predicted) Gen. 15:8-19 (400 yrs) :
                         :   Probation and Purification - - (40 yrs)
Covenant promoted : (Assured) Gen. 17:4-18 :
                         :   :

Covenant postponed (<u>1898 B.C.</u>)*(Sealed-Oath)Gen. 21:27-32 - Treaty :
                         : Gen. 15:13-16 :
Covenant procured (1888 B.C.) (By faith) : Gen. 22:15-18 Isaac offered
                         :   at age of 15
Gen. 15:13 - - 400 yrs
                        :
                        :

Exodus - - - - 1498 B.C.

To reveal how dramatically the above date (1898 B.C.*) in our simple chronology is verified by Bible scholars, we turn to the twenty-first chapter of Genesis in the New Thompson's Chain-Reference Bible. There we find the date of the events in the above chapter as 1897 B.C. followed by a question mark. Now, please turn to the Scofield Reference Bible, 1945, and we find our exact date, 1898 B.C..

The above covenant, to be postponed 400 years, is predicted in the fifteenth chapter of Genesis, but the 400 year postponement was not sealed or initiated until Abraham made a future forbidden covenant (Exodus 23: 32; Deut. 7:2) with Abimelech (Amenemhet II - 1926 - 1894 B.C.), the Proto-Philistim Hamitic Egyptian King, in 1898 B.C. in his 30th year in Canaan land and recorded in the twenty-first chapter of Genesis. Thus, the later forbidden covenant, which took place in the year 1898 B.C., inaugurated the "400 year bondage" in Egypt, and ended with the Exodus in 1498 B.C..

The providential failure of Abraham in this matter, also providentially, worked for the benefit of the Gentile Amorite Race (Canaanites) by postponing their destruction 400 years, since "the iniquity of the Amorites was not yet full" Gen. 15:16. Thus, God's grace is manifested and typified through this Old Testament experience as a foregleam of the coming Dispensation of Grace for us Gentiles (Rom. 9-11; Eph. 3:1-6). But we, as Gentiles, must remind ourselves that even though destruction on the Gentile Amorites was postponed, judgment did come (Joshua 5:1; 24:8), and Israel was grafted back in again; so shall it be in these latter days at the end of the Dispensation of Grace (Rom. 11:21-27). Whether we are Jewish or Gentile the only security in these last days is the "Rock" Christ Jesus, the Jewish Messiah, who will soon return again. Are you ready? Have you faith in Him as the true Messiah, and have you person- ally repented and received Him into your heart?

By identifying Abimelech (Judges 9:22 - 1225 - 1222 B.C.) as Amenmesse (Nomen), the Egyptian Ephemeral King, and as the contemporary of Mernep- tah (1228 - 1218 B.C.), unlocks information falling in the 12th and 13th Egyptian Dynasties during the days of Abraham and Isaac (See page 204). Both Abraham and Isaac were in contact with Proto-Philistim Kings by the same name of Abimelech (Genesis chapters 21,26). Abraham's contact with a king with this name was almost 100 years earlier than Isaac's. Therefore, we conclude that there were two kings by the name of "Abimelech". By checking with our chronology, we find that there were four kings during the Egyptian 12th and 13th Dynasties who were called Amenemhet or Ammenemes (Amenmesse - See Chart) with the same name as the Ephemeral King 700 years later. The "nomen" of these four early Egyptian kings compare exactly with the same name "Amenmesse", which we believe is none other than Abimelech (1225 - 1222 B.C.) of the Judges period. In the light of the Biblical connotation, we feelthat Amenemhet (Ammenemes) I (1991 - 1961 B.C.), II (1926 - 1894 B.C.), III (1841 - 1793 B.C.), IV (1793 - 1784 B.C.) were none other than the Kings Abime- lech I, II, III, and IV. Two of whom were contacted by the two patri- archs; one by Abraham (Abimelech II - Amenemhet II), and the other by Isaac (Abimelech III - Amenemhet III). This information might possibly bring us to another conclusion. The 12th and 13th Egyptian Dynasties, during which time the above kings reigned, were none other than the Proto-Philistim group (Hamites, sons of Mizraim - Gen. 10:14) who set- tled in and ruled Egypt as far south as Gerar. This Hamite Proto-Philis- tim group were designated as Kings of the Philistines (Gen. 21:32; Gen. 26:14-16 - some may have stayed along the Palestine coast continually) since they were later driven out as rulers of Egypt by the Hyksos (1746 B.C.), and they migrated along the northern coast of Africa and to Crete where they firmly established the great Minoan Empire around 1825 to 1750 B.C.. Later, their descendents, around 1200 B.C., gained sufficient power to conquer Asia Minor and then continued their attack into Pales- tine and Egypt for revenge. Repulsed by Rameses III on the borders of Egypt (1189 B.C.), they rebound and gained pre-eminence in Israel (Wes- tern bank) in 1176 B.C. as the Philistines. There may have been earlier movements to the west, but we should remind ourselves that the main Hamite Proto-Philistim groups in Egypt did not migrate to North Africa and Crete until the dark forty year period of confusion in the middle east, caused by the Hyksos infiltration through-out the fertile crescent (1786 - 1746 B.C.), during which time, Jacob also went to Haran (1766 B.C.); sons of Heth and Ephron, the Hittites, moved to Asia Minor (1825 - 1750 B.C.); Esau to Edom (1760 B.C.); Hivites from Edom to Is- rael (1750 B.C.); and the Hyksos to Egypt (1746 B.C.). Jacob then re- turned from Haran to Palestine the final year of this period of confusion (1746 B.C. - See chart page 217). This period is known as a time of National upheaval.

Now, if our conclusion is right, and these Philistine Kings were the

Proto-Philistim rulers of Egypt, rather than just a Palestinian coastal tribe, then the status of power and influence of Abraham in the Middle East must be reconsidered in a new light. A new look must be given to Abraham's treaty in Gen. 21:27. His covenant or treaty was with the Egyptian Proto-Philistim King Abimelech II (Amenemhet II) in 1898 B.C. (30 years after Abraham's entrance into Canaan - 1928 B.C.). The treaty sealed the destiny of the Jewish people to continue in bondage to the Egyptian rulers for another 400 years (Gen. 15:13 - 1898 - 1498 B.C.). Any covenant with the heathen nations was "later" repudiated by God (Exodus 34:12,15; Deut. 7:2; Exodus 23:32). Yet, that treaty was never abrogated as we shall see.

Strangely enough Abimelech II's (Amenemhet II) treaty (1898 B.C.) with Abraham took place the "year before" Amenemhet II's son, Sesostris II, became co-regent (1897 B.C.). This implies that the Egyptian King was growing elderly and was fearful of Abraham's growing power. He wanted to protect his son who soon was to be king (four years later - 1894 B.C.) as Gen. 21:22-24 clearly indicates:

"And it came to pass at that time, that Abimelech II (Amenemhet II) and Phichol the chief captain of his host spake unto Abraham, saying, God is with thee in all that thou doest: Now therefore <u>swear</u> unto me here by God that thou wilt not deal falsely <u>with me</u> (Amenemhet II), nor <u>with my son</u> (Sesostris II), nor <u>with my son's son</u> (Sesostris III): but according to the kindness that I have done unto thee, thou shalt do unto me, and to the land wherein thou hast sojourned and <u>Abraham said, I will swear</u>" (Later a forbidden covenant).

Isaac's herdsmen's struggle (Gen. 26) with the herdsmen of Abimelech III (Amenemhet III) did not occur until some years after the death of Sesostris III fulfilling the above treaty (Gen. 21:23,24).

We should also recall that Abraham's first contact in Egypt was with an "Egyptian Pharaoh" (Sesostris I - Gen. 12:14,15). This could easily justify our view. Also, Sesostris II's co-regency with his father extended back to the time of Abraham's visit to Egypt; therefore, we agree with the following quotation, and it agrees exactly with our chronology:

"On the tomb of Sesostris II*of the 12th Dynasty, at Beni Hassen, who is thought to have been the Pharaoh at that time, there is a sculpture depicting a visit of Asiatic Semitic traders (Abraham & family) to his court, as if it were an event of some importance. This may possibly have been the very record of Abraham's visit to Egypt"[1] (The very important event was Abraham's treaty with Amenemhet II and with his co-regent son, Sesostris II, which was later forbidden, and sealed their destiny for 400 years).

If the event happened in 1898 B.C. the "year before" his co-regency (1897 B.C.) during the reign of his father, then Sesostris II would have remembered this very important event as a young prince in his father's palace while being prepared for co-regency. This important treaty protected him and his son (Sesostris III) from any false dealing by Abraham. Sesostris II could have been directly involved in the treaty since

1. Halley's Bible Hand Book,p. 98. * Some say a Noble under the king

THE MAJESTY OF EGYPT

Seated statuette of Ammenemēs III, broken from the waist downwards. Dark grey granite.
Moscow Museum

Gen. 26: Isaac was in contact with Abimelech
(III)(Amenemhet III or Ammenemes III)

it may have been the precursor of his reign; therefore, he depicted this very important event, below, of Abraham's visit to Egypt upon the wall of his tomb:

ABRAHAM'S VISIT TO EGYPT
IN 1898 B.C.

Lower: Asiatics arriving in Egypt. From a tomb painting at Beni Hassan (ca. 2000-1900 B.C.). Foreigners readily went "down into Egypt" as Abraham, Isaac and Jacob are represented as having done in Genesis. (Lepsius, **Denkmaeler**, II, 133.)

Providentially, Abraham's treaty with Abimelech II (Amenemhet II) in 1898 B.C. was no doubt the Lord's permissive will since Abraham was not allowed to possess Canaan land at that time. The sins of the Amorites were not yet full (Gen. 15:16). The postponement of obtaining the land for 400 years (1898 - 1498 B.C.) was to allow time for the sins of the Amorites to increase until they were worthy of destruction. The cup of sin had to be full (I Kings 21:26; Ezek. 23:30-35). At this place we should remind ourselves that God never makes a mistake and never judges too early or too late. He only judges nations when the cup of sin is running over. Christ Jesus, the Jewish Messiah, drank the cup of sin of substitutionary atonement once and forever for us who receive, but those who reject Christ's substitutionary atonement, whether they be of Israel, other nations, or individuals, must themselves drink personally the cup of God's "wrath" and "judgment" for sin. The Amorites proved that they were still righteous by helping Abraham recover Lot from the enemy (Gen. 14:12,13,24); therefore, God stated that He would not judge them at that time (1898 B.C.), but 400 years later (1498 B.C.):

"And He said unto Abram, Know of a surety that thy seed shall be strangers in a land that is not their's (Egypt), and shall serve them; and they shall afflict them four hundred years (1898 - 1498 B.C.)..

113

. . But in the _fourth_ generation they shall come hither again: for the iniquity of the Amorites (Amorite brothers: Mamre, Eshcol, & Aner) is not yet full" Gen. 15:13,16.

Four hundred years later, as Israel came up out of Egypt under the leadership of Moses, God ordered the Israelites to annihilate the Canaanite tribes, including the Amorites, because they were marked by God for complete destruction (See Joshua 5:1; 6:17,21; Numbers 21:2,3; I Kings 21:26; Amos 2:9,10). Their cup of sin was full. How about the nations of the world today? Nineveh repented, and obtained grace and mercy from God.

The above scripture also proves two other things: First, God still considered Egypt as the true ruling authority over Canaan. Abraham's and his son's sojourn there (215 years) was considered as a part of the Egyptian bondage, perhaps as a result of Abraham's covenant with Abimelech. The above scripture indicates that the early part of the 400 years was also known as Egyptian bondage, yet the words "fourth generation" (Gen. 15:16) limits the real bondage in Egypt proper to only half the period, or to the last 215 years. Exodus 12:40,41 adds another 30 years to the 400 year bondage clarifying that Abraham's first 30 years in Canaan, before making the covenant with Abimelech, were also included.

Second, God's inspired statement "fourth generation" proves, with out a doubt, that they were only in Egypt proper the last 215 years of a 430 year period. They were there during the time of the last four generations. There were eight generations mentioned in the Holy Script from Abraham until Gershon, the son of Moses, during whose life the Exodus took place. The eight generations were Abraham, Isaac, Jacob, Levi, Kohath, Amram, Moses, and Gershon. Therefore, if they were brought up after the "fourth generation" it would naturally be after the last four. Thus, Jacob went to Egypt proper in 1713 B.C., after the birth of Levi in 1756 B.C. according to the scripture, and was there 215 years, or for "four generations". They came out under the guiding hand of Moses in the year 1498 B.C..

The average time between each Biblical generation during these 430 years was 61 years. The scriptures prove that Abraham had his first child in the Biblical line when he was 100 years old; Isaac, when he was 60 years old; and Jacob, when he was 88 years of age. This long span of time just happened to be peculiar within the Biblical line through the providential working of God, while other Biblical lines or heathen lines may have had triple the number of generations during the same period. To us this is no problem especially when we know that Moses had to have been 41 or 42 years old when Gershon was born. So we add up the years of the four known generations and we have 290 years accounted for out of the 430 years. This leaves exactly 140 years for the other four unknown generations which would be exactly 35 years between the remaining four generations. This does not seem an extreme amount of time since it was not always the first born that was chosen by God. Certainly, it leaves no room for long gaps in the Biblical Chronology.

Therefore, we conclude that it was exactly 430 years between the time of the call of Abraham (1928 B.C.) and the Exodus (1498 B.C.) as Exodus 12: 40,41 confirms:

"Now the sojourning of the children of Israel, which they sojourned in Egypt, was four hundred and thirty years. And it came to pass at the end of four hundred and thirty years, even the self same day it came to pass, that all the hosts of the Lord went out from the land of Egypt".

The Septuagint's translation of Exodus 12:40 also confirms our view:

"And the sojourning of the children of Israel while they sojourned in the land of Egypt and in the land of Chanaan was four hundred and thirty years".

Josephus also adds his testimony to the list and verifies our decision:

"They left Egypt in the month Xanthious, on the fifteenth day of the lunar month: four hundred and thirty years after our forefather Abraham came into Canaan, but two hundred and fifteen years only after Jacob removed into Egypt".[1]

Last, but not least, the New Testament ratifies our conclusion in Galatians 3:17:

"And this I say, that the covenant, that was confirmed before of God in Christ (Gen. 12:1,2,3,7 - 1928 B.C.); the law, which was four hundred and thirty years after (1498 B.C.), can not disannul, that it should make the promise of none effect".

At this point we should remind ourselves that according to I Kings 6:1 it was 480 years between the fourth year of Solomon (Temple begun - 1018 B.C.) and the Exodus (1498 B.C.). Also, it has been established above that it was 430 years from the Call of Abraham (1928 B.C.) unto the Exodus (1498 B.C.). Now, add the two together and we have a span of 910 years. This period's length is verified by the most astounding confirmation of all found in the following quotation:

"Students of Hebrew Chronology may be interested in this new date for the foundation of the First Dynasty of Babylon. The shortest of the Hebrew chronologies dates the migration of Abraham as occurring 910 years before year four of Solomon".[2]

CANAAN LAND IN ABRAHAM'S DAY

To picture the conditions in Canaan land during the early years of Abraham's life, let us turn to the delightful, and often repeated Egyptian story of Sinuhe, a nobleman in the Egyptian court.

Contemporary Babylonian history of the early days of Sinuhe reveals that Ur of Chaldees had been recaptured from the Elamites by Ishbi-Erra of Isin (1997 - 1984 B.C.) and a period of peace followed. During the reign of the following ruler, Shu-ilishu (1984 - 1975 B.C.), when Abraham was

1. Josephus, "Antiquities of the Jews", Bk 2, Ch. 15,2,p.83,Loeb Classic
2. Glanville, S.R.K., "The Legacy of Egypt", page 13, Edited, Oxford, University Press, London   115

only 32 years old and still in Ur (1971 B.C.), the events concerning Sinuhe took place in the courts of Egypt. Approximately thirty-six years later (1935/4 B.C.), perhaps caused by the Elamite (Amurru) invasion of Palestine, and Abram's initial call (Acts 7:1-5), Abram moved to Haran and Sinuhe returned to Egypt for his last days.

Amenemhet I (Abimelech I - 1991 - 1962 B.C.), the first ruler of the Twelveth Egyptian Dynasty and the grandfather of Amenemhet II (Abimelech II - 1927 - 1894 B.C.), whom Abraham later contacted, was growing old. Therefore, he was considering Sesostris I, his son, as his co-regent in 1971 B.C., only nine years before his death in 1962 B.C.. During this very occasion Sinuhe, a nobleman in attendance in the Egyptian Court, became involved in a political intrigue. The whole affair was revealed to the king, therefore, Sinuhe feared for his life and fled north towards the "Princes' Wall" and Retenu (The Sandrambler's country of Palestine). The story continues:

"As I headed north I came to the Princes' Wall, which was built to keep out the Bedouins and crush the Sandramblers (Wilderness Wanderers). I hid in a thicket in case the guard on the wall, who was on patrol at the time, would see me. I did not move out of it until the evening. When daylight came . . . and I had reached the Bitter Lake I collapsed. I was parched with thirst, my throat was red hot. I said to myself: This is the taste of death! But as I made another effort and pulled myself on to my feet, I heard the bleating of sheep and some Bedouins came in sight. Their leader, who had been in Egypt, recognized me. He gave me some water and boiled some milk, and I went with him to his tribe. They were very kind to me".

'Sinuhe's escape had been successful. He had been able to slip unseen past the great barrier wall on the frontier of the kingdom of the Pharaohs which ran exactly along the line which is followed by the Suez Canal today. This "Princes' Wall" was even then several hundred years old. A Priest mentions it as far back as 2650 B.C. (During the fourth Egyptian Dynasty - Wall was destroyed later during the Noahic Flood - 2356/5 B.C. - added): The Princes' Walls are being built to prevent the Asiatics forcing their way into Egypt. They want water . . . to give to their cattle. Later on the children of Israel were to pass this wall many times: there was no other way into Egypt. Abraham must have been the first of them to see it when he immigrated to the land of the Nile during a famine (Gen. 12:10)'. (This famine would have taken place right after Abraham entered Canaan Land in the year 1928/7 B.C., or shortly after, making it fall during the last years of Sesostris I reign - added).

Sinuhe continues: "Each territory passed me on to the next. I went to Byblos, and farther on reached Kedme (Syria east of Damascus) where I spent eighteen months. Ammi-Enschi, the chief of Upper Retenu (Palestine) made me welcome. (Ammi-Enschi - a western Semitic name, an Amorite a descendant of Ham - Gen. 10:16) He said to me: 'you will be well treated and you can speak your own language here'. He said this of course because he knew who I was. Egyptians who lived there had told him about me".

"Ammi-Enschi said to me: Certainly, Egypt is a fine country, but you ought to stay here with me and what I shall do for you will be fine too. He gave me precedence over all his own family and gave me his eldest daughter in marriage. He let me select from among his choicest estates and I selected one which lay along the border of a neighbouring territory. It was a fine place with the name of Jaa. There were figs and vines and more wine than water. There was plenty of honey and oil; every kind of fruit hung on its trees. It had corn and barley and all kinds of sheep and cattle. My popularity with the ruler was extremely profitable. He made me a chief of his tribe in the choicest part of his domains. I had bread and wine as my daily fare, boiled meat and roast goose. There were also desert animals which they caught in traps and brought to me, apart from what my hunting dogs collected . . . There was milk in every shape and form. Thus, many years went by, my children grew into strong men, each of them able to dominate his tribe. Any courier coming from Egypt or heading south to the royal court lived with me. I gave hospitality to everyone. I gave water to the thirsty, put the wanderer on the right way, and protected the bereaved.

When the Bedouins sallied forth to attack neighbouring chiefs I drew up the plan of campaign. For the prince of Retenu for many years put me in command of his warriors and which-ever country I marched into I made . . and . . . of its pastures and its wells. I plundered its sheep and cattle, led its people captive and took over their stores. I killed its people with my bow thanks to my leadship and my clever plans.

At length in his old age he began to yearn for his homeland. A letter from his Pharaoh, Sesostris I, summoned him to return (Sesostris I became co-regent during the ceremony when Sinuhe fled) . . . 'make ready to return to Egypt, that you may see once more the court where you grew up, and kiss the ground at the two great gates . . . Remember the day when you will have to be buried and men will do you honour. You will be anointed with oil before daybreak and wrapped in linen blessed by the goddess Tait (Embalming). You will be given an escort on the day of the funeral. The coffin will be of gold adorned with lapis-lazuli, and you will be placed upon a bier. Oxen will pull it and a choir will precede you. They will dance the Dance of the Dwarfs at the mouth of the tomb. The sacrificial prayers will be recited for you and animals will be offered on your alter. The pillars of your tomb will be built of limestone among those of the royal family. You must not lie in a foreign land, with Asiatics to bury you, and wrap you in sheepskin'.

Sinuhe's heart leapt for joy, he decided to return at once, made over his property to his children and installed his eldest son as 'Chief of his tribe'. This was customary with these Semitic nomads, as it was with Abraham and his progeny. It was the tribal law of the patriarchs, which later became the law of Israel. 'My tribe and all my goods belonged to him only, my people and all my flocks, my fruit and all my sweet trees. Then I headed for the south.

I found his Majesty on the great throne in the Hall of Silver and Gold. The king's family were brought in. His Majesty said to the Queen: "See, here is Sinuhe, who returns as an Asiatic and has become a Bedouin". She gave a loud shriek and all the royal children screamed in chorus. They

117

said to his Majesty: "Surely this is not really he, my Lord King". His Majesty replied: "It is really he".

I was taken to a princely mansion, in which there were wonderful things and also a bathroom . . . there were things from the royal treasure house, clothes of royal linen, myrrh and finest oil; favourite servants of the king were in every room, and every cook did his duty. The years that were past slipped from my body. I was shaved and my hair was combed. I shed my load of foreign soil and the coarse clothing of the Sand-ramblers. I was swathed in fine linen and anointed with the finest oil the country could provide. I slept once more in a bed. Thus, I lived, honoured by the king, until the time came for me to depart this life".[1]

Thus, when Abraham came into Canaan land and experienced the famine (1927 B.C.) during the last three years of Sesostris I's reign (1928 - 1926 B.C.), he "went down into Egypt to sojourn there; for the famine was grievous in the land" (Gen. 12:10). At that same time Sinuhe was already spending his last years in ease in His Majesty's court. The Biblical text continues the story of Abraham's experience with Pharaoh Sesostris I, who desired Sarah at the age of 65 (1928/7 B.C. - this had to be Sesostris I - Gen. 12 - since "later" Abraham's wife, Sarah, at 89 years of age - died at 127 years - was also desired by the following King Amenemhet II - Abimelech II, recorded in Gen. 20, 1904 B.C. at Gerar, the border city between Egypt and Israel):

"And it came to pass, that when Abraham was come into Egypt, the Egyptians beheld the woman (Sarah - age 65 years) that she was very fair. The princes also of Pharaoh saw her, and commended her before Pharaoh (Sesostris I - he died two or three years later in 1926 B.C.): and the woman was taken into Pharaoh's house. And he entreated Abraham well for her sake: and he had sheep, and oxen, and he asses, and menservants, and maidservants, and she asses, and camels. And the Lord plagued Pharaoh and his house with great plagues because of Sarah Abraham's wife.

And Pharaoh called Abraham, and said, 'What is this that thou hast done unto me? Why didst thou not tell me that she was thy wife? Why saidst thou, She is my sister? So I might have taken her to me to wife; now therefore behold thy wife, take her, and go thy way'.

And Pharaoh (Sesostris I) commanded his men concerning him; and they sent him away, and his wife, and all that he had. And Abraham went up out of Egypt, he and his wife, and all that he had, and Lot with him, into the south. And Abraham was very rich in cattle, in silver, and in gold. And he went on his journeys from the south even to Bethel, unto the place where his tent had been at the beginning, between Bethel and Hai; unto the place of the altar, which he had made there at the first: and there Abraham called on the name of the Lord" Gen. 12:14 - 13:4 (Sesostris I may have fallen ill as a result of the plague, or shortly after, since he died in the third year after Abraham entered Canaan, and Amenemhet II - Abimelech II reigned soon after in 1926 B.C., whom Abraham and Sarah later - 1904 B.C. - contacted in Genesis chapter twenty).

1. Keller, Werner, "The Bible as History", page 75-78, Econ Verlagsgrupp West Germany

CHRONOLOGY OF THE PERIOD

We now have established the date of the Exodus as 1498 B.C.. Therefore, the measurement of 430 years beyond the above year, according to Exodus 12:40,41; Galatians 3:17, extends our chronology back to the call of the patriarch Abraham out of Haran (1928 B.C.). Abraham's first call came in Ur and his second call came when he was 75 years old after the death of his father Terah at the age of 205 years in Haran (Acts 7:4; Gen. 11:32; 12:4). Thus, Abraham was 75 years old in the year 1928 B.C., therefore, he was born during the 130th year of Terah's life.

Isaac, Abraham's son, was born when Abraham was 100 years old (Gen. 17:1,17,23). This would make Isaac born in the year 1903 B.C.. Isaac, in turn, was blessed with a son, Jacob, when he was 60 years old (Gen. 25:26), which brings us to the year 1843 B.C. (Jacob's birth). Then Jacob went to Egypt when he was 130 years old in the year 1713 B.C. (Gen. 47:9).

Therefore, with the above information, we can compute the year Jacob went to Haran. Joseph was thirty-nine years old the year Jacob was 130 years old (Gen. 45:3,4,6) when he entered Egypt (1713 B.C.). By adding the 39 years of Joseph's life to the year 1713 B.C. when he came to Egypt, we find the year of Joseph's birth (1752 B.C. - Jacob was 91 years old). We also know that Jacob had served Laban exactly 14 years when Joseph was born (Gen. 30:25). Therefore, we add another 14 years to 1752 B.C. which brings us back to 1766 B.C., the year Jacob came to Haran at the age of 77 years. After Jacob worked 14 years and Joseph was born, he ask Laban to allow him to return to Israel (1752 B.C. - Gen. 30:25). But after a discussion, Jacob decided to serve Laban another six years for his flocks (until 1746 B.C. - Gen. 31:38-41), which made Jacob 97 years old completing the total of twenty years in Haran (1766 - 1746 B.C. - Gen. 31:41).

Since Joseph was born in 1752 B.C. (Jacob's 14th year in Haran), we know that it was in the year 1735 B.C., at the age of 17, that Joseph was sold into Egypt (Gen. 37:2). We also know that Joseph was thirty years of age when he was released from two years imprisonment in Egypt (1724 - 1722 B.C. - Gen. 41:1) at the beginning of the seven years of plenty (1722 - 1715 B.C.):

"And Joseph was thirty years old when he stood before Pharaoh King of Egypt (Salitos - Rameses II - Exodus 1:11). And Joseph went out . . . . throughout all the land of Egypt. And in the seven plenteous years (1722 - 1715 B.C.) the earth brought forth by handfuls" Gen. 41:46-47.

Thus, we know that Joseph was seventeen when he was sold into Egypt (1735 B.C.), twenty-eight when he went to prison for two years (1724 B.C.), thirty when he was released at the beginning of the seven years plenty (1722 B.C. - Gen. 41:27,54), thirty-seven years old when the years of plenty ended (1715 B.C.), and in the second year of the famine (1713 B.C.), nine years after being released from prison, Joseph was thirty-nine years of age when revealed to his brethren (1713 B.C. - Gen. 45:3,4,6). Jacob came down into Egypt that same year (Gen. 45:9-11) at the age of 130 years (Gen. 47:9). Jacob died seventeen years later in the year 1696 B.C. (Gen. 47:28).

Joseph died after 110 years in the year 1642 B.C. (Gen. 50:26). Joseph's death was exactly 2370 years after the expulsion of Adam from Eden, if we allow one year for Noah's Flood (4012 B.C. until 1642 B.C. equals 2370 years).

Thus, Jacob went to Haran in 1766 B.C. at the age of 77 and returned to Canaan twenty years later in the year 1746 B.C. at the ripe old age of 97 years. It is also known historically that 1746 B.C. was the year that the Hyksos forces invaded Egypt. Also, this may have been the good reason for Jacob deciding to return to Israel, since the Hyksos had infiltrated the Middle East for all those years while Jacob was in Haran.

Now, that we have established both the dates of Joseph's death (1642 B.C.) and the Exodus (1498 B.C.), it is comparatively simple to find the number of years that transpired between Joseph's death and Moses' birth. At the Exodus (1498 B.C.) Moses was exactly 80 years old. This would make his birth date in 1578 B.C.. We subtract Moses' birth date, 1578 B.C., from the date of the death of Joseph, 1642 B.C., and we have exactly 64 years, on which all Bible chronologists agree. Please see chart on page 121 for a clear picture of our chronology during the years of the patriarchs.

– – – – – – –

EIGHT GENERATIONS

ABRAHAM'S 2003 BC	ISAAC'S 1903 BC	JACOB'S 1843 BC	LEVI'S 1956 BC	KOHATH'S 1906? BC	ARMRAM'S 1841? BC	MOSES' 1578 BC	GERSHON'S 1530? BC

Proph.
Gen.15:13   4 Generations   4 Generations
Proph.
Seal'd
Gen.21:27

1898 BC   ___   400 years   ___   :   ___   "Out Egypt after 4 generations"   ___
                Gen.15:13   Gen. 15:16
                                                             Ex. 6:16-20
Abram's 75th   Jacob   Num. 26:57-59   Exodus
years   :   into   & the
1928 BC :   Egypt   Law

BC : ___ 215 years ___ ___ -1713 BC   According 215 years LXX, ___ ___ -1498 BC
                                           and the Samaritan version
1991:   1786 BC

Pro-Philistim 12,13th Dyn: Jacob-Haran
                           Heth-Hittites   ___ Period of Confusion 40 yrs. ___   BC   BC
2003 Abram 100th Isaac   Esau-Edom   1580 BC   1316   1199   1105
      _____   Phili-Hittites-Crete   ___ -:Seth worship:
      1903 BC M.E.invas.-Hyksos-Egypt   ___ 18th Dyn ___ :   19th & 20th Dyn
                           :1746 BC Hyksos-Seth Worship :- Egyptian 15th to 17th Dyn
Isaac's 60th yr Jacob   :- 64 years -:
_____
1843 BC   Joseph   Moses
                        1713 BC   dies   born
Jacob's 91st year Joseph revealed   1642 BC   1578 BC
_____ 1752 BC
1898 BC
                        39th year ___ -- 110th year ___ -- 80 yrs --
30 yrs
Joseph's 1st yr

in Canaan: Salitos(Rameses II)   ___ ___ ___ ___ ___ ___ 1294 BC Rameses II
                  2nd Hyksos King   :adds to bldgs of Joseph

                  ___ ___ 430 year ___ ___   & Salit 480(Rameses III)- 1018 BC
Abram's Covenant Ex.12:40,41;Gal.3:17   ___ 480 yrs ___   I Kings 6:1
w/Abimelech, Egyptian
Philistim King Amenemhet II   B.C.:

1898 BC ___ ___ 400 years of Gen. 15:13 & Acts 7:6 ___ ___ -1498   Celebrate|Celebrate|Celebrate|Celebr.
                                                                         2nl Seti 5th|Joseph 5t|19th yr
                                                                                   1313 BC 1213 BC 1113 BC
                  1713 B.C. ___ ___ 400 year Stella at Tanus - 400th yr - 400th yr 500th yr 600th yr
1928 B.C. ___ ___ ___ Jewish traditional Short Chronology 910 years from Abram's call to 4th of Solomon—1018BC

"And they placed over them (Israel) princes of tribute to number, to inform, to answer the king according to their burden bearing, and rebuilt the collapsed store-cities of Pithom and Raamses for Pharaoh (18th Dyn)" Exodus 1:11.

Historically, we know that the Hyksos invaded Egypt in 1746 B.C. and demolished most of their cities while conquering the land. After 16 years of Army rule under Rameses I, they finally re-established monarchal rule by putting Salitos (Rameses II) on the throne in 1730 B.C.. He inturn began to rebuild the cities of Egypt. Now the question concerning Exodus 1:11 is this. Does the scripture refer to the Hyksos rebuilding (1727/6 B.C.), or the rebuilding during the 18th Dynasty (1568 - 1498 B.C.)?

We can now come to certain conclusions within our chronological framework. In our chart "Middle East Kingdoms Correlated", we notice that in 1736 B.C. Dinah was defiled by the Hivite which caused Jacob to move to Mamre (Arbah or Hebron) to live. Rachel dies during that year on the way near Bethlehem (Gen. 34,35). Thus, we conclude that Jacob began to build "Hebron" at Arbah during the following year 1735/4 B.C. (Numbers 13:22; Gen. 35:27).

In 1730 B.C. Salitos or Sethos (Rameses II) was enthroned in Egypt, and Joseph, at the age of 22 years, was there at that time. If we say that Jacob had the city of Hebron well built by his third year after arrival at Arbah (Hebron, 1735 - 1733 B.C. inclusive), then according to Numbers 13:22, Pithom, or Zoan, was built seven years later around 1727/6 B.C., which was also the third year of Salitos (Rameses II). Dr. John Rea suggests,"Since the Nineteenth Dynasty rulers (1316 to 1199 B.C. - added) were in all probability "descended" from the Hyksos (1746 - 1580 B.C. or Dynasties 15 through 17 - added), the name "Rameses" may have been used in Hyksos times".[1] This, we believe is the exact truth and the correct answer. The same god (Seti or Seth) was worshipped during both periods as the Stella of Tanus confirms, and celebrated at three different times later and dating back to the former period.

Isaac died at Hebron in the year 1723 B.C. (Gen. 35:27-29), and the following year Joseph was released from prison in Egypt in his 30th year (1722 B.C.). That year Joseph was authorized by Pharaoh Salitos (Rameses II) to also make store-cities out of Pithom and Raamses. This would have been only four or five years after Salitos began to rebuild these cities. In fact, at the time of Joseph's release, they may have been still involved in the re-construction. Joseph, during the years of plenty (1722 - 1715 B.C.), probably completed the building work of the cities by finishing large granaries to hold the stored grain kept for the coming famine years (1715 - 1708 B.C.).

Therefore, Joseph, as vizier and second to Pharaoh, was involved completing their construction at this time. The family of Jacob only arriv-

1. Rea, Dr. John, "The Historical Setting of the Exodus and the Conquest", page 6, Grace Theological Seminary notes

ed two years later in 1713 B.C. during the second year of the famine (Gen. 45:6). During that year, the pinnacle of the power of Rameses II, the Stella of Tanus (Raamses, Avaris, or Zoan) was unveiled and the god Seth (famine god) was exalted.

Now, when considering the interpretation of Exodus 1:11, we must also survey the context of the chapter and realize that the information covers a period of 133 years from the going down of Jacob into Egypt (1713 B.C.) until the beginning of the 18th Dynasty (1580 B.C.). Since the real persecution probably did not radically start until the second king of the 18th Dynasty (Amenophis I - 1557 - 1536 B.C.) during whose reign Moses escaped to Midian (1538 B.C.), we should add at least another 30 years to the above span of time. Only the 18th Dynasty Kings "knew not Joseph" (Exodus 1:8), since Joseph died in 1642 B.C. under the Hyksos Dynasties.

Now, our question is this. When were Pithom and Raamses actually store-cities? Were they being built as store-cities by the pharaoh's of the 18th Dynasty just before the Exodus, or were they store-cities years before during Joseph's days and now being rebuilt by the 18th Dynasty and simply being "designated" as such? The latter is no doubt true. Joseph helped build (Numbers 13:22; Gen. 41:48,49) and made Pithom and Raamses into store-cities during the seven years of plenty (1722 - 1715 B.C.) under the second Hyksos King Sethos (Suta and sometimes called "Salitos". He is probably the first Rameses II of the Tanite family - Gen. 47:11). Salitos, as the first Rameses II, put his name on the buildings during this time. Later, a descendent, Rameses II (1294 - 1228 B.C.), simply added buildings under the same name and received the glory for them all. The real problem is to decide which buildings were built by whom?

Josephus refers to the Hyksos and claims that he quotes directly from Menetho in the following statement. The Hyksos "burnt down our cities, and demolished the temples of the gods . . . At length they made one of themselves King, whose name was Salatis . . . he found in the Saite Nomos (Sethroite) a city very proper for the purpose, and which lay upon the Bubastic channel, but with regard to certain theologic notion was called Avaris, this he rebuilt".[1] Later, he also refers to Joseph as speaking to the Egyptian King.

Therefore, we conclude by the above statement and by chronological computation that Joseph was second to the second Hyksos King Salitos (Sethos) who rebuilt cities and established store-cities under the name of Rameses II and stored food in preparation for the seven years of famine foretold by Joseph (Gen. 41:48,49 - 1715 - 1708 B.C.).

Now, under the 18th Dynasty, approximately 150 years later, these same cities had been once again burnt during the expulsion of the Hyksos (around 1580 B.C.). Kamose (Ahmose I) said, "I razed their (Hyksos) towns and burned their places, they being made into red ruins forever on account of the damage which they did within this Egypt".[2]

1. Josephus against Apion, Book I, 14, **The Loeb Classic Library Edition**
2. Gardiner, Sir Alan, "Egypt of the Pharaoh's, page 167, 1961, publisher: Oxford University Press

Thus, under the three Pharaoh's of the Israelite bondage, Amenophis I (1557 - 1536 B.C.), Thutmose I (1536 - 1510 B.C.), and Thutmose II (1510 - 1498 B.C.), these store-cities were being rebuilt again. The Hebrew word "store" or "treasure" (misknot) could have a connotation referring to a collapsed, lanquored, or weakened condition. Therefore, in the light of the time involved between the two periods of reconstruction, and the disaster which the Hyksos brought upon the cities' in 1580 B.C., we would translate Exodus 1:11 as follows: ". . and rebuilt the collapsed store-cities of Pithom and Raamses for Pharaoh".

THE PROBLEM OF DINAH'S BIRTH

The problem concerning Dinah's birth is to have her born late enough in the chronology for Leah to have all her other children first (Gen.30: 21). Second, we must have Dinah born early enough for her to be old enough to be desired as a wife for the Hivite before Joseph was seventeen years old. We also know that Joseph was born the 14th year of Jacob's sojourn in Haran (Gen. 30:25; 31:38-41).

To accomplish the above within the framework of the known chronological facts, we must suggest that Rachel was taken as a wife for Jacob at the same time as Leah. This solves the problem. Jacob worked seven years before cohabiting with Leah, while he worked seven years after he cohabited with both Leah and Rachel. This answers the problem, and is also indicated in Gen. 29:30:

"And he (Jacob) went in also (at the same time as Leah) unto Rachel . . and served with him (Laban) seven other years" (See Gen.29:27,28).

There are several proofs which correlate with the above interpretation:

1. Jacob went in unto both Leah and Rachel at the same period since Leah's first born son (Reuben) was born as a result that God saw that Leah was hated and Rachel loved (but barren), therefore, God opened Leah's womb to conceive Reuben (Gen. 29:31,32).

2. Second, we know that Joseph was born during the 14th year at Haran (Gen. 30:25,26; 31:38-41). We also know that Rachel's non-conception and her maid's, Bilhah's, conception took place during the second seven years before Joseph's birth (including Asher), since the births are listed in order.

3. Third, all of Leah's children up to Issachar were all conceived by the 14th year in Haran. Issachar and Joseph were both conceived within one year of each other during the period of 1752/1 B.C., since Rachel bought Reuben's mandrake (Reuben was seven years old) during the wheat harvest (June) and conceived (Joseph in Jacob's 14th year - 1752 B.C.), while Leah, also, was allowed to lay with Jacob but conceived only by her maid, Zilpah, the following year 1751 B.C. - Gen. 30:14-18). Joseph's and Issachar's births were the result of the selling of the mandrake.

In 1750 B.C. Leah conceived and brought forth Zebulun (Gen. 30:20), and Dinah was born of Leah the following year (1749 B.C. - Gen. 30:21). Therefore, Dinah was 13 years old in 1736 B.C. when the Hivite cohabited with her. That year Jacob went to Hebron with his family. On the way Benjamin was born (1736 B.C.) of Rachel at Bethlehem and Rachel dies (Gen. 34,35). The following year in 1935 B.C., when Joseph was 17 years old, Joseph was sent to Dothan where he was sold into Egyptian bondage by his brethren (Gen. 37:2).

Please turn to our chart "Middle East Kingdoms Correlated" for a complete chronology of the birth of Jacob's family (p. 211).

PROBLEMATIC NUMBER ENTERING EGYPT

"All the souls of his sons & his daughters were 33"	"And these she bare unto Jacob even 16"	"And all souls were 14"	"And all the souls were 7"
Gen. 46:46	Gen. 46:18	Gen. 46:22	Gen. 46:25
Leah's progeny number 33	Zilpah progeny number 16	Rachel's progeny number 14	Bilhah's progeny number 7

Leah's progeny	Zilpah progeny	Rachel's progeny	Bilhah's progeny
1. Reuben 1	34. Gad 7	50. Joseph 9	64. Dan 11
2. Hanoch 1	35. Ziphion 28	51. Manasseh 42	65. Hushim 54
3. Hezron 2	36. Haggi 29	52. Ephraim 43	66. Naphtali 12
4. Phallu 3	37. Shuni 30	53. Benjamin 10	67. Jahzeel 55
5. Carmi 4	38. Ezbon 31	54. Belah 44	68. Guni 56
6. Simeon 2	39. Eri 32	55. Becher 45	69. Jezer 57
7. Jemuel 5	40. Arodi 33	56. Ashbel 46	70. Shillem 58
8. Jamin 6	41. Areli 34	57. Gera 47	
9. Chad 7	42. Asher 8	58. Naaman 48	
10. Jachin 8	43. Jimnah 35	59. Ehi 49	
11. Zohar 9	44. Ishuah 36	60. Rosh 50	
Uncounted-Shaul	45. Isui 37	61. Muppim 51	
Canaanite woman	46. Beriah 38	62. Huppim 52	
12. Levi 3	47. 1.Heber 39	63. Ard 53	
13. Gershon 10	48. 2Malchiel 40		
14. Kohath 11	49. Serah 41 (Sister)		
15. Merari 12			

16. Judah 4
17.1.Er(Died)13
18.2.Onan " 14
19. Shelah 15
20. Phares 16
21.1.Hezron 17
22.2.Hamul 18
23. Zarah 19
24. Issachar 5
25. Tola 20
26. Phuvah 21
27. Job 22
28. Shimron 23
29. Zebulun 6
30. Sered 24
31. Elon 25
32. Jahleel 26
33. Dinah 27(Sister)

1. Shaul, the son of Simeon, born of the Canaanitish woman is not counted (Servants symbolic of the mixed multitude).

2. "All the souls that came with Jacob to Egypt, which came out of his loins, besides Jacob's sons wives, all the souls were threescore and six" Gen. 26:26.

These 66 souls refer to the living men souls who arrived in Egypt. Therefore, we subtract the names of Er and Onan, who died, and the names of the two daughters, Dinah and Serah, from the above seventy.

3. "And the sons of Joseph, which were born him in Egypt, were two souls, all the souls of the house of Jacob, which came into Egypt, were threescore and ten" Gen. 46:27. These

seventy are listed and numbered above: Leah had 33 children (Shaul not counted), Zilpah had 16, Rachel had 14, and Bilhah had seven children which makes a total of seventy. Since Serah (daughter) is numbered among Zilpah's children making a total of 16, therefore, Dinah must be numbered among Leah's children totaling 33 children. This leaves Shaul born of the Canaanitish woman out.

4. "Then sent Joseph, and called his father Jacob to him and all his kindred (wives), threescore and fifteen souls" Acts 7:14. Now if we add Jacob's name and the names of his four wives, Leah, Zilpah, Bilhah, and Rachel to the above seventy, we have seventy-five. The three dead (Er, Onan, and Rachel) are counted, yet Shaul, the son of the Canaanitish woman, is still not listed

The number of the descendants of Jacob who migrated to Egypt given to us in Gen. 46:8-27 lists the 66 sons and counts only two daughters mentioned in verses fifteen and seventeen (Serah and Dinah their sisters). Therefore, we conclude that the total number of females have not been tabulated. Only the men are numbered (Num. 1:2-4). Yet, verses six and seven says that Jacob brought all his seed with him; "His sons, and his sons' sons with him (They are listed above - 12 sons, 52 grandsons, 4 great grandsons, and two granddaughters), his daughters, and his son's daughters (They are not listed except for Dinah and Serah), and all his seed (men and maid servants included in his household - Shaul and his family) brought he with him into Egypt".

For an example, Abraham had a large household after only 8 years in Canaan land (1928 - 1920 B.C.) when it is said that he "armed his train-ed servants, born in his house three hundred and eighteen . . ." Gen. 14:14. If Abraham had a household of men numbering 318, Jacob and his twelve sons could have had a much larger household after 46 years? Therefore, Jacob's family group, including household servants, could have added to a formidable number.

If Jacob had 64 sons, grandsons, and great grandsons, then it seems log-ical to state that he may have had at least 64 daughters, granddaught-ers, and great granddaughters. In 46 year (see charts) at least two thirds of them could have been married. This means that at least 42 daughters and granddaughters could have been married into some other family. Say, in 32 years, these couples could have had four children each, which is a conservative estimate. Each of the 84 families may have had a man and wife for servants, and they, in turn, had child-ren. In computing these figures, we could have a total of 1000 people or more, including the mixed multitude, who went down into Egypt with Jacob in 1713 B.C.. Multiply this number eleven times in 215 years and you have 2,048,000 people. There were outcasts in Egypt who also gather-ed with Israel in the land of Goshen verified by history. This would certainly be a sufficient number to provide 600,000 fighting men accord-ing to Exodus 12:37,38. The number of a thousand or more would have given suitable reason for Pharaoh Salitos or Sethos (Rameses II - the 2nd Hyksos King) to have said to Jacob and his sons, "the land of Goshen let them dwell" Gen. 47:6. The scripture continues and states:

"And Joseph placed his father and his brethren, and gave them a possess-ion in the land of Egypt, in the best of the land, in the land of Rameses (II, 1730 - 1711 B.C.), as Pharaoh had commanded" Gen. 47:11.

We are convinced that there were two Rameses Kings during the Hyksos period. One known as the first King (Army General - Rameses I), and a second Hyksos King named Salitos (Sethos) who no doubt, was called Rameses II, and Jacob and his descendants settled in the <u>land of Rameses</u>". If they were placed in the <u>land of Rameses</u>, then it seems logical to think that there was a "Rameses" King that owned and ruled the land at that time. Joseph, under his supervision, prepared the recently built (some were rebuilt) cities into store cities during the period of plenty (1722 - 1715 B.C.) in which grain was stored in behalf of the coming famine (1715 - 1708 B.C.). Jacob came to Egypt during the second year of the famine (1713 B.C. - Gen. 45:6).

Some 416 years later their descendants, under the reign of another Rameses II (1294 - 1228 B.C.), added buildings to the Egyptian Empire under the same name. He became some what of a plagiarist by taking the honour for much of his architectural ability originating from the works of a earlier Rameses II, who really deserved credit for building many of them during the days of Joseph, as a King with the same name, Rameses II.

At the end of the 13th Egyptian Dynasty, that Dynasty collapsed in 1781 B.C., while a woman, Sebknefrure, was ruling for a short time (1784 - 1781 B.C.). She introduced a period of forty years (1786 - 1746 B.C.) which is known as the "Period of Confusion". During that period, in the year 1766 B.C., Jacob went to Haran, while the "Hyksos" were making strong penetrations into the Middle East. They were from a Indo-European Kassite origin from the Zagros mountains, and were known as the "Asiatics" (They may have subdued the Proto-Hittites of Asia Minor around 1775 B.C. before infiltrating Palestine). During this confused period, not only did the Egyptian Hamitic Proto-Philistim move to North Africa and to Crete, but Jacob went to Haran; Esau went to Edom; the seed of Ephron, the Hittite, went to Asia Minor (the beginning of the late Hittite Empire), while the Hivites became prominent in Palestine from Edom. At the end of which (1746 B.C.), Jacob (Israel) and his 12 sons returned to Israel.

Therefore, in 1746 B.C., the very year that Jacob returned from Haran, the Hyksos left the Fertile Crescent and invaded Egypt land. Since the 12th and 13th Egyptian Dynasties, residing in Egypt before the Hyksos invasion, were Proto-Philistim (Hamite) group, many of them migrated from Egypt, before or during 1746 B.C., along the North African coast and on to the Island of Crete (This was no doubt the beginning of the late Great Minoan Kingdom which lasted to around 1190 B.C.). There descendants of these same Proto-Philistim people of Crete, five centuries later, rebounded into Israel as the "Philistines" after their victorious invasion through the Hittite country (Asia Minor) while heading for revenge on Egypt, their former home land, in the year 1190/89 B.C.. As they continued through Israel, and attacked Rameses III (Egypt) in his eighth year of reign (1190 B.C.), they were defeated and returned settling, and later ruling over western Israel in 1176 B.C.. (Probably, this attack against Rameses III was attempted revenge against the "Rameside Family" for their "Hyksos" invasion 556 years before, whose first King (Army General) also went under the same name "Rameses").

After their defeat, the Philistines (Former Proto-Philistim Hamitic peoples - Gen. 10:13,14) rebounded from the shores of Egypt into Israel where they gradually gained control of western Israel in the 22nd year of Rameses III (1176 B.C.) as the Philistines. According to Gen. 10:14; Amos 9:7; Jer. 2:23, the Island of Crete (Caphtor - Kaptara) is from "whence came forth the Philistines" and they are synonymous with the Cherethites according to Zeph. 2:5; Ezek. 25:16.

After the Hyksos invasion of Egypt in 1746 B.C., there followed a period of 16 years (1746 - 1730 B.C.) in Egypt of consolidation and rebuilding under their first enthroned King (Army General - Rameses I), and then under their second King, Salitos or Sethos (Rameses II), who began to rule in 1730 B.C., the Hyksos Kingdom was well established in Egypt. These were the Hyksos Kings and approximate dates of their reign in Egypt land:

HYKSOS KINGS IN EGYPT FOR 166 YEARS
(1746 - 1580 B.C.)

Hyksos invasion of Egypt - - - - - - - - - - - - - - - - 1746 B.C.

Hyksos consolidation (Military Ruler Rameses I)   1746 - 1730 B.C.(16)

   (Joseph to Egypt under Rameses I   1735 B.C.)

Salitos (Rameses II - Joseph second in power)   1730 - 1711 B.C.(19)

   (Jacob to Egypt - Joseph revealed   1713 B.C.)

Apepi I   1711 - 1705 B.C.(6)

Apephis   (Ruled over more than Egypt)   1705 - 1644 B.C.(61)

   (Khayan co-regency perhaps 10 years from - - - 1654 B.C.-Palestine)?

Khayan (Khian - alone)   1644 - 1614 B.C.(30)

   (Joseph dies - 1642 B.C.; Khayan takes Thebes 1640 B.C.)

Aa-seh-re   1614 - 1611 B.C.(3)

Apepi II   1611 - 1580 B.C.(31)

The Hyksos capital was a fortress city on the shore of the Egyptian Delta called Avaris, a massive-walled military encampment garrisoned by heavily armed troops from which they plundered upper Egypt. Since they invaded Egypt in 1746 B.C., they brought devastation and destruction, but afterwards settled down, and assumed Egyptian names and adopted Egyptian customs.

Avaris, their capital, no doubt rebuilt by them, was on the extreme east of the Delta, close to the borders of Asia. According to Breasted, it was so that they might rule both Egypt and their Asiatic domminions. Gardiner suggests that there was a local Palestinian chief during this time with the same name as the fourth Hyksos King (Khayan or Khian). Perhaps he was the Palestinian Chief during his ten year co-regency?

Their first military ruler (Rameses I) gradually established peace out of chaos during the first sixteen years, and then Salitos (Rameses II) was finally enthroned as Pharaoh about 1730 B.C.. He ruled 19 years. During the time that the Hyksos forces were consolidating their rule under Rameses I, Joseph was sold into Egypt in the year 1735 B.C. at the age of 17 years (Gen. 37:2). Joseph was 30 years of age as he stood before Pharaoh, Rameses II, at the beginning of the seven good years (1722 B.C. - Gen. 41:46). Since he had been in prison two full years (Gen. 41:1), Joseph was 28 years old when he was imprisoned in the year 1724 B.C. in the sixth year of Salitos (Rameses II).

Since we are interested in the year 1722 B.C. as the first year of the seven years of plenty, we might include the following quotation:

"The signs of the zodiac were certainly in use among the Egyptians 1722 years before Christ. One of the learned men . . . found upon a mummy-case in the British Museum a delineation of the signs of the zodiac, and the position of the planets; the date to which they pointed was the autumnal equinox of the year 1722 B.C.. Professor Mitchell was chosen, to ascertain the exact position of the heavenly bodies belonging to our solar system on the equinox of that year. This was done, and a diagram furnished by parties ignorant of his object, which showed that on the 7th of October, 1722 B.C., the moon and planets occupied the exact point in the heavens marked upon the coffin in the British Museum".[2]

Who ever this was that was in this mummy-case, he died the very year Joseph was released from prison and should illustrate the exact burial case they used in the days of Joseph. Gen. 40:20 states that the chief butler died two years before this date (1724 B.C.) during the ceremony commemorating the pharaoh's birthday.

During the Hyksos period, we have several important correspondences with the Genesis story:

"The capital was near the north-eastern frontier; there were special circumstances to account for the hearty welcome accorded to the patri-archs; Yaqob scarabs have been found among Egyptian relics of the Hyksos era; the embalming process compares with Joseph's Genesis story; and horses were first introduced by them.

There are other details which are also common to this period. As vice-roy, Joseph rides in Pharaoh's 'second chariot' (Gen. 41:43). That im-plies the time of the Hyksos. These 'rulers of foreign lands' (Asiatics) were the first to bring the swift war-chariot to Egypt. We know too that the Hyksos rulers were the first to use a ceremonial chariot on public occasions in Egypt. Before their day this had not been the practice on the Nile. The ceremonial chariot harnessed to thoroughbred horses was in those days the Rolls Royce of the governors. The first chariot belonged

1. Breasted, "History of Egypt", page 218
2. Donnelly, Ignatius, "Atlantis: The Antediluvian World",p. 187 quoted Goodrich, "The Sea & Her Famous Sailers", London, 1859

to the ruler, the 'second chariot' was occupied by his chief minister".[1]
Therefore, Joseph rode in that chariot (Gen. 41:42,43).

In the tomb of Baba at El-Kab, was an inscription of the time of
Se-Kenen-Ra-Taa III, a vassal king of Upper Egypt under the early Hyksos
rulers. There we find a close identification between the story of Baba
and the Biblical story of Joseph:

"I (Baba) collected corn, as a friend of the harvest god I was watchful
at the time of sowing. And when a famine arose, lasting many years,
I distributed corn to the city each year of famine".[2]

Baba lived sometime during the reigns of the first three Hyksos rulers
and may have been one of the regional authorities chosen by Joseph for
collecting grain.

Another wonderful correlating factor is that history verifies our
theory that Rameses II (Salitos) built the cities of Raamses and Pithom.
The city of Raamses was none other than Tanis (Zoan) and is known as Tel
Rotab.

Avaris, the capital known as Pithom or Succoth was also built by Rameses
II. The Greeks knew it as Ero or Heraopolis. The Tel is known by the
Arabic name, Tel-el Maskhulah. There seems to be some disagreement over
the identification of which one of the above cities was actually known
as the city of Raamses. The fact is there is abundant evidence that
Rameses II either built or rebuilt all of these cities and that
he was none other than Salitos (Rameses II) of the Hyksos period. Each
of the cities could have been referred to as the city of Raamses since
he was the builder or rebuilder, yet when we consider Exodus 1:11 and
Numbers 33:3-6, we realize that only one city must have been identified
by that name.

The site of Pithom was identified in 1883 by Prof. E. Naville as Tel-el-
Maskhulah. the "Mound of the Statue" cut out of red granite in the form
of a chair where three figures are sitting, known as Rameses II between
the solar gods Ra and Tum (these could have been the gods they worshipped before 1713 B.C. when they turned to the worship of the Seth god -
added). Prof. Naville found a number of inscriptions which show
not only that the site represents an ancient city whose religious name
was Pi Tum ("the abode of Tum"), while its civil name was Thuka (Succoth), but also that the founder of the city was Rameses II.[3] He continues, "The founder of the city, the king who gave to Pithom the extent
and the importance we recognize, is certainly Rameses II. I did
not find anything more ancient than his momuments. . . . It is he who
built the enclosure and the storehouses. He (Rameses II - Salitos or
Sethos - added) is the only king whosename appears on the naos and on
the monuments of Ismailiah".[4] Prof. Kyle was able to confirm the Hebrew
slavery in Egypt with his following words: "Every point in the story of

1. Keller, Werner, "The Bible as History", p.103, Econ Verlagsgruppe,Ger.
2. Griffith, J.S., "The Exodus in the Light of Archaeology", p. 38-43
3. Ibid , p. 43-45 , Robert Scott, Roxburghe House, London
4. Ibid

the insurrection is written upon the ruins at Pithom. The place was called Pithom (Exodus 1:11), the bricks were laid of mortar (Exodus 1: 14), contrary to the usual Egyptian method of brick-work; the bricks in the lower courses were fitted with good clean straw, those of the middle courses were made with stubble mixed with weeds and all pulled up by the roots (Exodus 5:10-12), while the bricks of the upper courses were made of Nile mud".[1]

The latter quotation by Prof. Kyle naturally refers to the 18th Dynasty when these store cities were reconstructed during Israel's slave bondage (1548 - 1498 B.C.).

Seth worship was first introduced during the ancient Egyptian Dynasty II during King Perabsen's (Sakhemit) reign around 2913 B.C.(Seth, the second in the genealogical line from Adam, died in 2970 B.C.). Seth worship was revived by Salitos or Sethos (Rameses II) 1200 years later during Joseph's period of glory (a descendant of Seth) and during the "dedication" of the city of Tanus (Zoan - foundation laid around 1727/6 B.C.) in the year 1713 B.C.. Also Joseph was revealed that year, and during which time King Salitos or Sethos (Rameses II) gained full possession of all the cattle; of all the lands; and the servitude of all the people (Gen. 47:13-26). That year (1713 B.C.) was the height of his personal glory, and of the Kings dedicated glory of the city of Tanus by the unvailing of the Stella of Tanus.

Subsequently, the same Ramesside Royal Kingly line (Sethroite-Saite Nomos) revived "Seth worship" once again during the XIX Dynasty in the second year of Seti I (1313 B.C.).[2] This celebration of the "Repetition of the birth of Seth" (Rebirth of the "worship" of Seth - added) fell on the 400th anniversary fulfilling the 400 year Stella at Tanus. Just one hundred years later the 500th anniversary was celebrated during the fifth year of Seti II (1213 B.C.), when the story which compares to Joseph's was retold.[3] The final recorded anniversary (the 600th) was held in the 19th year of Rameses XI or in the year 1113 B.C.(the 600th year of the "Repetition of the birth - worship - of Seth"; the 600th year since Joseph was revealed in Egypt; the 600th year since the dedication of the city of Tanis - Zoan - in the year 1713 B.C.; and the 600th year anniversary of Rameses II's highest pinnacle of power).

The following quotation confirms our theory quite clearly: "The Ramesside royal family was certainly in possession of the best local sources, and at the time of the celebration of the fourth centenary of the town of Tanis, they must have accepted the date 1700 B.C. as 'approximately' that of the foundation of the town and the introduction of the worship of Seth; that is at the time of the Hyksos invasion or shortly before (from Sethe, AZ, 65, 85 f. - the so-called Stella of year 400 from Tanis).[4] Notice that the quotation states that the Royal Ramesside family possessed the best local source of information concerning the celebration 400 years before this time. In other words Seti I was a descendant of King Rameses II (Sethos or Salitos) during Joseph's day, and Seti I, in his second year (1313 B.C.), celebrated the 400th

1. Ibid   (er: Oxford University Press
2. Gardiner, Sir Alan, "Egypt of the Pharaoh's",p. 249,304,1961,Publish-
3. Halley's Bible Hand Book, page 104, 1948
4. Kees, Hermann, "Ancient Egypt", page 198,Publisher:Faber & Faber Ltd

anniversary of the "Repetition of the rebirth of the worship of Seth".

Seth was the "son" of Adam in whose days men began to seek God (Gen. 4:26); while the pantheistic worship of Egypt designated 'Seth' (son of Adam) as the 'sun god' (son god)[1] or 'Typhon' god who had power of 'drought', or of the 'pitiless sun'. Since seven years of "drought" was involved in Joseph's (son of Seth) days, and he brought great deliverance and sufficient supplies during the 7 years of famine, it seems logical, from a physical standpoint, that they would have renewed "Seth" worship, the god of drought, and celebrated and commemorated it with the unvailing of the Stella at Tanis in the year 1713 B.C..

Finally, may we ask why a story like unto Joseph's was repeated during the 500th anniversary of the city of Tanus (Zoan)? The implications are clear and plain, and the details only once again confirm our simple Bible chronology which has been hidden all these years in the rocks of archeaology and in the "Rock" of the Holy Word of God, the Lord Jesus Christ.

- - - - - - -

1. Gardiner, Sir Alan, "Egypt of the Pharaoh's", page 9, 1961, Publisher Oxford University Press

PHRASE VI
POST-FLOOD PERIOD
(2356 - 1928 B.C.)
428 YEARS
CHAPTER XIII
CHRONOLOGICAL DATA

This portion of our chronology is based on the simple genealogy found mainly in Gen. 11:12-32; 12:1-4. There is only one problem in this portion of the genealogy found in the following quoted scripture, which compares exactly with the problem found in Gen. 5:32. Yet, both difficulties can be solved with the help of other correlated scriptures. In Gen. 11:26, we have the following quotation: "And Terah lived seventy years, and begat Abram, Nahor and Haran".

The problem above is this? In what year of Terah's life was Abram born? This question and its answer is very important to further our chronological computation. Were they all born in his seventieth year? No. The simple answer to the above problem is that the first of these three sons was born in his seventieth year, and the other two were born in different years after that year. When we compare this problem with Gen. 5:32, we prove that the three sons were not the same age (Gen. 9:24;10:21), and the three brothers could be listed in reverse order when correlated with the above verses. Thus, we conclude the possibility of the same in Gen. 11:26. Besides, we have clear substantiation that Abram was not born in Terah's seventieth year, but much later, when we consider several other scriptural verses. Abram was no doubt the youngest, but mentioned first. Let us first consider Gen. 11:32; 12:1,4 since the verses follow in order:

"And the days of Terah were two hundred and five years: and Terah died in Haran. Now the Lord had said unto Abram, Get thee out of thy country, and from thy kindred, and from thy father's house, unto a land that I will shew thee:. . . So Abram departed, as the Lord had spoken unto him; and Lot went with him, and Abram was 'seventy and five years old' when he departed out of Haran".

The implication of the above scriptures is that Abram left immediately after the death of Terah at the age of 75 years. This is clearly confirmed in Acts 7:4 as quoted below:

"Then came he (Abram) out of the land of the Chaldaeans, and dwelt in Charran (Haran); and from thence, 'when his father was dead', he removed him into this land, wherein ye now dwell".

Thus, our conclusion is this. If Terah was "two hundred and five years old" when he died, and Abraham was "seventy-five years of age at his death", then Abraham was born in Terah's 130 year. This solves our problem and completes our genealogy, thereby totaling 428 years from the first year of the Flood to the call of Abram.

Some may question the elderly age of Terah at the birth of Abram since he was 130 years old. But we must remember the longevity of the patriarchal lives following the Noahic Flood. The first three in the genealogical line following Noah after the Flood lived over 400 years, while four of the next five lived well over two hundred years of age. Terah, himself, lived another 75 years after Abraham was born. We should also recall that each one in the genealogical line also "begat sons and daughters" Gen. 11:11,13,15,17,19,21,23,25.

The changed climate and the tilt of the earth at the time of the Flood had a gradual shortening effect on man's life span until we get to the time of Joseph's death when it regulated itself when death came around 100 years of age. This process "may" be soon reversed in its order as we come to the end of this age (6000 years) near the end of this century. The next thousand years may encompass a great many changes described in Isa. 65:17,18,20,25 as follows:

"For, behold, I create new heavens and a new earth: and the former shall not be remembered, nor come to mind. But be ye glad and rejoice for ever in that which I create: for, behold, I create Jerusalem a rejoicing, and her people a joy. . . . There shall be no more thence an infant of days, nor an old man that hath not filled his days: for the 'child' shall die an 'hundred years old' (if a child is as a hundred, then a man will live for 900 years); but the sinner being an hundred years old shall be accursed. . . . The wolf and the lamb shall feed together, and the lion shall eat straw like the bullock: and dust shall be the serpent's meat. They shall not hurt nor destroy in all my holy mountain (kingdom), saith the Lord".

Since Abraham's call took place in the year 1928 B.C. (already confirmed), we add 428 years to that date and we have 2356 B.C. for the first year of Noah's Flood, and 2355 B.C. as the date when the Noahic Flood

. . . .

Noahic Flood lasted	1 year (Compare Gen. 7:11;8:13,14)
Arphaxad born	2 years after the Flood (Gen. 11:10)
Arphaxad was	35 yrs old when Salah born (Gen.11:12,13)
Salah was	30 yrs old when Eber born (Gen.11:14,15)
Eber was	34 yrs old when Peleg born (Gen.11:16,17)
Peleg was	30 yrs old when Reu born (Gen.11:18,19)
Reu was	32 yrs old when Serug born (Gen.11:20,21)
Serug was	30 yrs old when Nahor born (Gen.11:22,23)
Nahor was	29 yrs old when Terah born (Gen.11:24,25)
Terah was (not 1st born)	130 yrs old when Abram born (Gen.11:32;12:1-4; Acts 7:4)
Abraham was	75 yrs old at his call from Haran (Gen.12:1-4)

Total 428 years - 1st year Flood to call of Abram

. . . .

ended. Arphaxad was born two years after the Flood (Gen. 11:10 - 2353 B.C.) in the year that Shem was one hundred years old. Thus, Shem was born in the year 2453 B.C., and was 97 years old when the Flood began. Since Noah was 600 years old when the Flood began (Gen. 7:11), and 601 years old when the Flood ended (Gen. 8:13), we can solve the problem of the age of Noah when Shem was born. Add 600 years to the first

134

year of the Flood, and we have 2956 B.C. for the birth of Noah.

In the light of Gen. 5:32, we might conclude that all three sons were born during his 500th year. In actuality only the eldest was born then, since we know, according to Gen. 10:21, that Japheth was the eldest, and Ham was the youngest (Gen. 9:24). So, when was Shem born? First, the eldest was born in Noah's 500th year, so we subtract 500 years from the known year of Noah's birth (2956 B.C.) and we have 2456 B.C. (Japheth's birth). Thus, Shem was born in Noah's 503rd year (2453 B.C.), since Shem was 100 years old when Arphaxad was born in 2353 B.C., which was two years after the Noahic Flood "ended" (Literally 3 years after the "beginning", and two years after the "end" of the Flood).

Some have questioned whether the 600th year means 599th year or the 600th. We admit there is this possibility found in Gen. 7:6: "Noah was a son of 600 years (600th), and the Flood of waters was on the earth". According to the rendering (600th) in the Hebrew, it could logically be interpreted either way. Yet, we believe that the Hebrew rendering of Gen. 7:11 settles the matter and puts the event in the year six hundred (the 600th year had been completed). We must also remember the Flood ended in Noah's 601st year (Gen. 8:13). In Ivan Panin's book "God Counts", page 22, he says, "This peculiar expression, 'In year six hundred years', whatever else it may mean, does not mean the six hundredth year (or 599th year). It is meant, in fact, to guard against this very error". Besides, we have seen how this one extra year makes the whole chronology fall into a beautiful mathematical pattern.

Therefore, we conclude that Japheth was born as the eldest son in Noah's 500th year (2456 B.C.), and Shem, the second born, was begotten in the 503rd year of Noah (2453 B.C.), and Ham, the youngest, was born a few years later. If we allow three years for weaning a child, we would come to 2450 B.C. for the birth of Ham, or any number of years later. Ham's birth is the only uncertain date.

Most chronologists make the mistake of having Shem born in Noah's 502nd year by "not" allowing one year for the Noahic Flood. We must realize that Arphaxad was born in 2353 B.C. two years after the end of the Flood (2355 B.C.) rather than two years after the beginning of the Flood (2356 B.C.). This one extra year, added to our computation, makes Joseph's death fall in the 2370th year after Adam's expulsion rather than the 2369th.

FLOOD AFTER-MATH

The great catastrophe of the World Wide Flood had just begun to subside. As Noah, Shem, Ham, ánd Jepheth, with their wives (perhaps their servants), opened the door of the Ark, the storm clouds were melting away on the horizon. The sky above had a deep blue which they had not known before. This was strange? In the Old Kingdom, from which they had suddenly emerged, there had always been a mist going up from the earth which watered the plants in the former world wide tropical climate (Gen. 2:6). But this was a new world with a difference. It was spring time (Gen. 8:14), and an appropriate time for the earth to begin its

135

renewal. Those terrible, foreboding clouds which had brought this first torrent of rain on the earth (Gen. 2:5) were receding over the horizon. Would those dark ominous masses return and repeat this horrible disaster?

Noah and his sons hurriedly built an alter of thanksgiving to their one true Triune God (Gen. 8:20), who had not only warned them of the disaster, but had protected and delivered them over into a new sphere of life, and seemingly, into a new world. Suddenly, Noah, and his family saw a new, startling display, a spectacle in the heavens against the backdrop of clouds, which was fantastically beautiful. A rainbow! What did it mean? The first to be seen on earth. As they sang their praises to God, while the incense of their sacrifice went up to God, the rainbow became the capstone of glory announcing a new beginning.

Suddenly, from the heavens above Mount Ararat, whereon they stood, the voice of God spake unto Noah, and to his sons saying:

"And I, behold I establish my covenant with you, and with your seed after you, and with every living creature that is with you, of the fowl, of the cattle, and every beast of the earth with you; from all that go out of the Ark, to every beast of the earth. And I will establish my covenant with you; neither shall all flesh be cut off any more by the waters of a flood; neither shall there any more be a flood to destroy the earth . . . I do set my bow in the cloud, and it shall be for a token of a covenant between me and the earth" Gen. 9:8-11,13.

Little did Noah and his sons realize, as they looked out over this beautiful spring scene from their mountain pinnacle, that, even at that very moment, the northern and southern hemispheres, which had been tropical up to this time, were going through their death throes. The earth had tilted by the Hand of the Almighty 23½ degrees causing sudden frigid temperatures. Even as the waters receded, the howling arctic winds newly created, were trapping mountains of water, and turning them into an arctic waste. Tropical animals, and mastodons were being frozen instantly where they stood upright with the tropical growth still in their mouths. Little did they realize, as they stood enjoying the spring weather of Mount Ararat, that seven or eight months later the first ice and snow would form on the very slopes on which they stood for the first time, and seal the Ark (built for 120 years - Gen. 6:3 - 2476 - 2356 B.C.), beneath a glacier of ice; as a deep freeze to preserve the Ark as a last day testimony to our last generation of this dying world 4345 years later in 1990 A.D.

"According to an Associated press despatch from Istanbul, Turkey, a team of research scientists of the Scientific Exploration and Archaeological Research Foundation (SEARCH) found some pieces of wood which they believe to be about 4,000 years old. They were found on top of a 17,000 foot mountain in Eastern Turkey in the vicinity of Mount Ararat on July 31st and August 2nd, 1969".[1]

The following year a report came from the scientists in Spain, "Scientists from Spain have determined the wood to be at least 5,000 years old and containing residues of pitch".[2]

1. "Midnight Cry", page 229, 1969. 2. "Christian Beacon", Mar. 5, 1970

According to our chronological calculations the Noahic Flood took place in the year 2356 B.C., and if we allow 120 years for Noah to build the Ark, and an additional 50 years or more for trees to grow to a size to be usable in some parts of the Ark, then we would arrive at a date for the age of the wood in the Noahic Ark at 4,500 years, or older.

"The dendrochronological laboratory at the University of Arizona recently discovered a stand of still older trees in the white Mountains of California, a group of bristlecone pines. Their discoverer says: 'Only recently we have learned that certain stunted pines of arid highlands, not the mammoth trees of rainy forests, may now be called the oldest living things on earth. Microscopic study of growth rings reveals that a bristlecone pine tree found last summer at nearly 10,000 feet began growing more than 4,600 years ago and thus surpasses the oldest known sequola by many centuries . . . Many of the neighbors are nearly as old; we have now dated 17 pines 4,000 years old or more' . . ."[1]

Noah turned and looked at the place of sacrifice made from timbers taken from the bow of the Ark not knowing that an expedition 4,272 years later would still be able to describe it in such picturesque way as follows:

"The expedition found, on the peak of the mountain, above the ship burned remains of the timbers which were missing out of one side of the ship. It seems, that these timbers had been hauled to the top of the peak and used to build a tiny one room shrine, inside of which was a rough stone hearth like unto alters (Gen. 8:20) which the Hebrews used for sacrifices, and it had either caught fire from the altar or had been struck by lightning, as the timber were considerably burned and charred over and the roof was completely burned off".

The report continues in its description of the Ark from an earlier view from a plane:

"I looked and was amazed. A submarine? (Partially submerged in a mountain lake) No, for it had stubby masts, but the top was rounded over with only a flat catwalk of about five feet wide down the length of it. What a strange craft, built as though the designer had expected the waves to roll over the top most of the time, and had engineered it to wallow in the water like a log, with the stubby masts carrying enough sail to keep it facing the waves. We flew down as close as safety permitted and took several circles round it. We were surprised when we got close as safety permitted and took several close looks at the immense size of the thing, for it was as long as a city block and would compare very favorably in size with the modern battleship of today. It was grounded on the shore of the lake with one-fourth under water. It had been partly dismantled on one side near the front, and on the other side there was a great doorway nearly twenty feet square, but with the other door gone. This seemed out of proportion as even today ships seldom have doors even half that size".

Later, the expedition by foot reported: "The Ark was found to contain hundreds of small rooms, and some rooms were very large, with high ceilings. The unusually large rooms had fences of great timbers across them, some of which were two feet thick, as though designed to hold beasts

1. Whitcomb, Dr. John, "The Genesis Flood", page 393, The Presbyterian & Reformed Pubishing Company   137

much larger than elephants. Other rooms also were lined with tiers of cages somewhat like one sees today at a poultry show, only instead of chicken wire, they had rows of tiny iron bars along the front. Everything was heavily painted with a waxlike paint resembling shellac, and the workmanship of the craft showed all the signs of high type of civilization.

The wood used throughout was oleander - the 'gopher' of Genesis 6:14 - which belongs to the cypress family and never rots, which of course, coupled with the fact of its being frozen most of the time, accounted for its perfect preservation".[1]

Naval architects who have worked out the displacement of the Ark from the specifications given in the Bible state that it was between 30,000 and 40,000 tons - as large as most of our great liners of today.

Another expedition of the 9th August, 1883 stated also: "Effecting an entrance into the structure, which was painted brown, they found that the admirality requirements for the conveyance of horses had been carried out, and the interior had been divided into partitions fifteen feet high. Into three of these only could they get, the others being full of ice, and how far the Ark extended into the glacier they could not tell'.' . . . "Even though Bible references say that eight souls went into the Ark it has been suggested by some that the servants were never counted: only those of the family that were in the blood line were counted, and the number named. There were always many servants of these families who shared the destiny of their masters (Gen. 7:1) where it states, 'And the Lord said unto Noah, come thou and all thy house into the Ark'. This would account for a great deal ethnologically after the Flood, and it is a fact that ethnologists have failed to take into account in their efforts to trace the origin of nations and peoples".[2]

POST-DILUVIAN KINGDOM

Just before the Noahic Flood, "Lugalzaggisi (2381 - 2356 B.C.) obtained supremacy of the Sumerian Kingdom by defeating Urukagina (2388 - 2381 B.C.), and attempted to gain direct control of the whole of Babylonia by subjugating Kish, once again the predominant city in north Babylonia".[3] During this period Noah and his family were completing the finishing touches to the "Ark", after 120 years of labour, as a last testimony to that generation of coming judgment. Then suddenly it came. In the 25th year of Lugalzaggesi (2381 - 2356 B.C.), in the 969th year of Methuselah (3325 - 2356 B.C.), and "in the 600th year of Noah's life (2956 - 2356 B.C.), in the second month, the seventeenth day of the month (2356 BC), the same day were all the foundations of the great deep broken up, and the windows of heaven were opened. And the rain was upon the earth forty days and forty nights. In the selfsame day entered Noah, and Shem, and Ham, and Japheth, the sons of Noah, and Noah's wife, and the three wives of his sons with them, into the Ark" Gen. 7:11-13.

1. Christian Digest, 1960's, page 10, quote Evangelical Christian
2. Ibid
3. Saggs, H.W.F., "The Greatness that was Babylon", page 48, Sidgwick & Jackson Publishers

14   Head of Sargon of Agade
Perhaps Ham or son "Cush" father of Nimrod

Contemporary history relates the destruction of the Old Kingdom, the Sumerian civilization and their King, Lugalzaggisi with this simple statement, "Uruk (Erech or Warka) was smitten with weapons (God's weapons - 'God's four sore judgments' - Ezek. 14:19-21 - added), its kingship was carried to Agade (Accad or Akkad)".[1] The Cambridge Ancient History page 144, I Part 2 states this concerning the end of Lugalzaggisi's reign, "With utmost distinctness it ends an age".

Thus, the Sumerian civilization was smitten by the Noahic World Wide Flood and it kingship was carried to Accad by the "Hamitic"(Sargon) branch of the Noahic family.

There is a real possibility, which is astonishing, that "Sargon", the first ruler of Akkad (Accad - Agade) after the Flood, may have been none other than Noah's son, "Ham", who was in the Noahic Ark. This is clearly suggested by the following contemporary historic statement:

"Sargon the mighty king, the king of Akkad (Agade or Accad), am I . . . My humble mother conceived me; she bore me in secret, placed me in an Ark of bulrushes, made fast my door with pitch and gave me to the river (Flood) which lifted me up and carried me to Akki the irrigator . . .".[2]

Thus, this traditional, historical statement concerning the first great tribal King after the Flood, Sargon (Sharrun-Kin of Agade), states that he survived the great river or Flood in a vessel caulked with pitch. Could King Sargon have been "Ham", the son of Noah; or a son of Ham (Cush), born during the Noahic Flood, or possibly conceived in the Ark and born shortly afterwards? There is real confirmation that "Sargon" and "Ham" may be the same person, and that their grand-sons may be the self-same person, Nimrod or Naram-Sin. The following quotations, one from the Bible, and the other from history clearly verifies this:

"The sons of Noah, Shem, Ham, and Japheth. . . And the son of Ham, Cush . . . and Cush begat Nimrod (Grand-son of Ham)" Gen. 10:1,6,8.

"Literary tradition has attached itself strongly, not only to Sargon, but also to his third successor and grand-son, Naram-Sin".[3] (Nimrod)

```
             Noah                                        -
              :                                          :
             Ham    - (Ham and Sharrum- Kin)   -    Sargon
              :       (both survived in  a )             :
              :       (pitched vessel     )              :
             Cush   -        (sons)            -         ?
              :                                          :
(G. Kingdom) Nimrod-    (Both Grandsons)       - Naram-Sin (G. Kingdom)
```

Now, we have found that both Ham and Sargon had experiences in a caulked, pitched, vessel; they were contemporary; they both had grandsons with great kingdoms; and finally both kingdoms were involved with the same city state which included Accad (Agade), Babel (Babylon), Erech

1. Ibid, page 49
2. Finegan, J., "Light from the Ancient Past", p.38,Princeton Univ.Press
3. Saggs, H.W.F., "The Greatness that was Babylon", page 49,55, Sidgwick & Jackson Publishers

(Uruk or Warka), and Calneh (Nimrud - Gen. 10:8-10). Naram-Sin, Sargon's grandson, who ruled from Agade (Accad), "mentioned that precious wares flowed into Agade from all quarters, whilst city adminis-trators sent in their monthly and yearly tribute. It was probably Naram-Sin's imperialism which, in Sumerian eyes as well as in fact, led to the downfall of his dynasty. Literary tradition is ambivalent toward Naram-Sin: he is treated not only as a great hero but also as an ill-fated ruler".[1]

It was after the 25 year reign of Sharkalisharri, son of Naram-Sin, and after the final three years which included the reigns of the following last four rulers, that the kingdom, which included Babylon, was destroy-ed by the Gutians, therefore dispersing the inhabitants of the Great Fertile Crescent. The tower of Babel was also destroyed at that same time, which was inturn, the providential intervention of God. The last four rulers of the 1st Dynasty of Agade ruled for only a total of three years. Their unstable reigns illustrates the chaos involved before Babylon fell in 2169 B.C.. The 1st Dynasty of Agade had lasted 146 years (2315 - 2169 B.C.), and was destroyed 186 years after the Noahic Flood. The correlating proof that Babylon was destroyed at this very time is set forth later by studying the meaning of each of the post-diluvian names of the descendants of Noah, and correlating dates of their births with the above historical chronology.

Since Sargon, the first ruler of the 1st Dynasty of Agade succeeding the Flood, first dwelt at the city of Kish and then some moved to Accad, we therefore conclude that Noah and his family probably settled around Kish in the land of Shinar (Gen. 11:2).

As we all know, the city of Kish was a well known name among various city-states vying for power before the Flood in the Sumerian civilizat-ion, and, according to the following historical data, we know that Kish was one of the first tribal villages to be rebuilt after the Flood. The date of the restoration of Kish correlates exactly with our Bible chron-ology following our Flood date and falls rightly during Sargon's reign.

". . . Sumer in the third millennium B.C. (3000 -2356 B.C. - added) con-sisted of a number of city-states vying for supremacy over the land as a whole. One of the most important of these was Kish, which according to Sumerian legendary lore, had received the 'kingship' from heaven im-mediately after the 'Flood'".[2] (Kish was located between Babel and Erech).

Now, how do we establish the date for Nimrod's Kingdom as portrayed in the following scripture?

"And the sons of Ham (Hamitic peoples); Cush (Negro races), and Mizraim (the Hebrew name for Egypt), and Phut (Libyia), and Canaan (Palestine) . . . and Cush begat Nimrod (Naram-Sin): he began to be a mighty one in the earth (around 140 years after the Flood - added) . . . and the be-ginning of his kingdom was Babel (Babylon), and Erech (Uruk - Warka), and Accad (Akkad or Agade), Calneh (Nimrud) in the land of Shinar" (Gen. 10:6,8,10).

1. Saggs, H.W.F., "The Greatness that was Babylon", p.51,Sedgwick Jackson
2. Kramer, S.N., "History begins at Sumer", page 67, Jarrold & son, Ltd

By means of the above scripture, we know that <u>Nimrod</u> was the one who organized "Babel" and other cities into a Kingdom of city-states. Also, Naram-Sin's contemporary reign also marked the re-emergence of the system of centralization of government.[1] He was known as the "King of the Four Regions". Therefore, we conclude that they and their kingdoms were one.

Josephus gives a means of dating Nimrod's Kingdom by stating that it began 1903 years before the attack on Babylon by Alexander the Great in 331 B.C., which takes us back to 2234 B.C.. Therefore, we suggest the following chronology for the 1st Dynasty at Agade, which correlates with our Bible chronology perfectly and is based on Josephus' correct date.

1ST DYNASTY OF AGADE

B.C.

2356 - Ham as "Sargon" preserved by the Ark (See picture - page 139)

2355 - Noahic Flood ends

2350 - Noah and family settles around Kish

2320 - Sargon leaves Kish and goes to the area of Accad (Agade)

2315 - Sargon's (Sharrun-Kin) tribal rule at Agade (57 years)

2258 - Rimish's tribal rule at Agade (9 years)

2249 - Manishtushu's tribal rule at Agade (15 years)

<u>2234*-</u> <u>Nimrod (Naram-Sin)</u> (Erech,<u>Calneh(Nimrud)</u>,Akkad,Babel - Gen.10:10)

2197 - Sharkalisharri city rule (Son of Naram-Sin - 25 years)

2172 - Iqiqi, Nanum, Imi, and Eluln (total of 3 years)

2169 - Fall of Babel - Gutian invasion from Lagash(confusion of tongues)

Since we know the date that Nimrod organized the Kingdom of Babel into a confederation of city-states (2234 B.C.*), we add the years reign of each ruler before that date, and we fix Sargon's tribal reign in the year 2315 B.C., which is exactly "forty years" after the Flood. We designate his reign as "tribal" since the Bible confirms that there were no organized city kingdoms before Nimrod's reign, and also because of the limited population at that time. During the first 40 years, they probably lived in tents around Kish, then during Sargon's 57 years of reign, he began to organize military out posts, villages, and towns supervised from his headquarters at Agade (Accad), while later, Nimrod, his grandson, founded the kingdom of city-states of Erech, Babel, <u>Calneh</u> <u>(Nimrud)</u>, and Accad 121 years after the Flood. Nimrod and the following rulers built the tower of Babel and temples to Enlil, but the Gutian invasion destroyed them all in 2169 B.C., 186 years after the Flood.

"The later civilized people of Babylonia remembered the period of Gutian domination with abhorrence, as a time of barbarism when the gods were not respected and temples (including Babel - added) were plundered and neither women nor children spared".[2]

1. Saggs, H.W.F., "The Greatness that was Babylon", page 55
2. Ibid, page 53,54 , **Sidgwick & Jackson publishers**

These attacks by the Gutians scattered the tribal groups, and the former ruling Hamitic race were forced southward into Palestine, Egypt, Arabia, and Africa. Thus, during the next 154 years (2169 - 2015 B.C.), or during the latter years of Peleg's life (2254 - 2015 B.C.), "was the earth divided" (Gen. 10:25). This verse is explained in Gen. 10:32b where it states, ". . and by these were the nations divided in the earth after the Flood". The earth was re-inhabited into their tribal areas during Peleg's days. This was the means that God providentially used to re-establish what the Noahic Flood had destroyed as a result of their sinfulness. This "National Migration Period" falls during the time of the "Sumarian Legendary period" known as the "Heroic Age" (or during Peleg's days - 2254 - 2015 B.C.).

Thus, the Hamitic branch of Noah's family under Nimrod's confederacy controlled the Fertile Cresent until 2169 B.C. when the Gutians (perhaps from Lagash) conquered and destroyed Babylon. This resulted in the dispersal of the races.

Nimrod was the son of Cush (Black person), and Cush was the son of Ham. This proves that Nimrod was only two generations from the time of the Noahic Flood, and it could also imply that Sargon was either of Noah's family, or of the first generation after the Flood. If Nimrod was the grandson of Ham, and if Nimrod can be identified as Naram-Sin, who was, in turn, the grandson of Sargon, then Sargon could be none other than "Ham" the son of Noah.

Ham's son's name was "Cush", and the connotation of the word itself connects it with the "negro" or "black" races. After living in Israel for almost seven years we know that an Israeli, in his Hebrew language, call the "black" or "negro" races "Cushim" (plural of "Cush"). Nimrod was a son of "Cush", therefore he must have been some what colored in complexion as was his father Sargon (see page 139).

The name "Nimrod" means "Rebel". He rebelled against the nomadic way of life only 122 years after the Flood, and organized the first post-Flood Kingdom. It is also known that one of his son's names was "Tammuz" (Dumuzi, Daonos - implies revival of the evil worship - fertility cult - originating during the days of Jered in the Old Kingdom), therefore, they were turning away from God so soon after such terrible destruction.

During the Old Kingdom, the descendants of Seth began to inter-marry with the sons of Cain, who had God's "mark" upon them. Thus, as a result of this inter-marriage during the days of "Jered" (means to go down - spiritual degradation), Noah had mixed genes within his loins. Therefore, after the Noahic Flood, we have the "three racial groups" coming forth from the loins of Noah by which he replenished the earth. They were the "Semitic" (Jewish race), "Caucasian" (White race), and the "Hamitic" (Black race) racial groups.

Let us recall and emphasize that the Old Kingdom was completely wiped out, except for Noah's families and perhaps servants, by the Noahic disaster. The Sumerian civilization was destroyed by the Flood and not by an invasion. The Sumerian, Akkadian, and Egyptian languages were carried through to the Semitic period by Noah's family, by literature, personal knowledge, and perhaps, by servants. Most historians place the destruction of the Sumerian civilization between the dates of 2375 - 2350 B.C.. Simple Bible chronology properly pin-points the exact year of the division of the two civilizations (Sumerian and Semitic) by the Flood in 2356/5 B.C.. Historians also state that the Inter-mediate Kingdom began immediately after this exact time.

The first seven great Egyptian Dynasties were all rulers over thriving civilizations before the Flood, and the seventh, whose first ruler was a woman, fell into complete chaos, ending with seventy rulers in seventy days (Manetho) as the Noahic Flood destroyed them. This should be a good reminder today, that when women come into pre-eminency, it inevitably leads to destruction.

It is quite coincidental that the Westminister Historical Atlas Map, 1945, dates the disappearance of the Ante-Diluvian Chinese Dynasties, and their last ruler (Laou) in the year 2357 B.C.. History states that during "Laou's days there was a world wide Flood". The Holy Bible agrees:

"'In 2357 B.C. his son Yau (Laou) ascended the throne, and it is from his reign that the regular historical records begin. A great Flood, which occurred in his reign, has been considered synchronous and identical with the Noahic Deluge, and to Yau (Laou) is attributed the merit of having successfully battled against the waters'.

There can be no question that the Chinese themselves, in their early legends, connected their origin with a people who were destroyed by water in a tremendous convulsion of the earth. Associated with this event was a divine personage called Nin-va (Noah?)".

We also know that Chinese history reappears some 200 years later (2150 B.C.). M. Anstey, in his reference to Du Halde's writings on China, states:

"There is, therefore, nothing in the high antiquity of China to conflict with the conclusion arrived at by Du Halde, whose admirable work on China stands unrivalled for the copiousness and correctness of information it contains, that 'two-hundred years after the Deluge the sons (descendents) of Noah arrived in North-West China'" (2155 B.C. - 14 years after Babel fell).

This clear confirmation reiterates our premise that the World Wide Noahic Flood took place at this exact time (2356/5 B.C.). History provides no other gap in chronology for such a universal catastrophe.

In turn, the Egyptian Dynasties did not reappear until about 2225 B.C.. This leaves a gap in chronology in Egyptian history of 130 years. May we again ask why? Let us be honest and sincere in our approach. Our very destiny, and the destiny of thousands of others, whom we may influence are hanging in the balance. Let us recall that the Holy "Words of the Lord are pure Words: as silver tried in a furnace of earth, purified seven times" Psa. 12:6. Seven is the number of God, or of perfection. To doubt, or question the veracity of the Holy Word of God is to "wrest the scriptures unto their own destruction" II Peter 3:16.

1. Donnelly, Ignatus, "Atlantis: The Antediluvian World", page 208
2. Anstey, M., "The Romance of Bible Chronology, Vol. I, page 103

Again, we know that monarchal rule was instituted in the Old Kingdom approximately 3200 B.C. in Egypt, Babylon, and Assyria, yet, this high standard of civilization suddenly disappeared in 2356/5 B.C., when their rich cultural heritage relapsed into tribal nomadic life (See chart on page 209). The Assyrian historians list "sixteen tribal rulers" after 2350 B.C., during the Akkadian period, identified as "tent dwellers", or "tribal nomads". It was only in the days of Apiasal, the seventeenth ruler (approximately 2100 B.C.), that they left their nomadic life and became city dwellers again. Strong monarchal rule, as a whole, in the Middle East was not firmly re-instituted until around 2050 B.C., or shortly before. Why was this true? Only the Noahic Flood answers the question.

While attending the University of Cape Town, South Africa, we had the privilege to hear a notable French lecturer and archaeologist comment on his finds during his exploration of the Sahara desert in North Africa. He stated that around 2000 B.C., or earlier (at the time of the Flood), a great change began to take place in the Sahara region. The previous period included evidence of civilization, heavy growth of vegetatian, farming, tools, chariot wheels, and settlements were evident in many places in the Sahara area. After this period of civilization, the sea seemed to have covered it all with clear evidence of fossils, of plants, sea shells, and other evidence of oceanic life. Thus, we know that the whole region had been inundated by waters. A.M. Behwinkel declares that there are the same evidences in other regions all over the world when he states:

"Fossils of plants and man-made implements found in the Sahara show that this great African desert was at one time covered with luxuriant vegetatian and was inhabited by man. Similar remains have been found in the Gobi desert of China and in the great desert areas of northwestern India".[1] This is another definite proof that there was a world Flood.

After the Flood disaster, there began a slow deterioration in the Sahara region, and it gradually grew more arid from that time until approximately 1000 B.C. (the days of Solomon), when the Sahara desert arrived at its present day desolate state. Additional confirmation is found in this following quotation:

"It is evident that the people who lived in the lush Sahara region in those early times left no written records of great cities with temples, palaces, fortifications, and feats of engineering as in ancient Babylon or Egypt. They did have a degree of civilization, and a wonderfully fruitful area in which to live. Evidence suggests that they inhabited this area after the second (and short) inundation of the earth. But they were driven from the area (or destroyed by the Flood - added) by a great and mysterious catastrophe - the drying up of that vast region of Northern Africa. But while they lived there they enjoyed the knowledge of pottery making, metal working, the use of the wheel for transportation, domestication of crops and animals, and a form of religion which is thought to border on magic. There were similarities in cultural knowledge between the Saharan and the early Sumerian civilizations (Pre-Flood - added). That they took with them certain knowledge when they

1. Behwinkel, A.M. "The Flood", 1951,p.4 (See Wilson, John A., "The Culture of Ancient Egypt",p.20 - Note from S.A. Huzzayin,"The Place of Egypt in Prehistory",Vol.XLIII,1941) Concordia Publishing House. Used by permission

migrated westward and southward into the fertile Sahara region is log-ical (We say they were destroyed by the World Wide Flood).

Why did this area dry up? It is evident that the Sahara has been a true desert for only about 3000 years, and that <u>between the time of the di-vision and scattering of the human race</u> and the time of Solomon a <u>progressive drying process took place</u> which caused these people to migrate southward into Central and Southern Africa".[1]

Let us now consider this same period in the light of the known biblical information and genealogy (See charts). Arphaxad was born two years after the Flood (Gen. 11:10), and Peleg his great-grandson, was born in 2254 B.C.. Gen. 10:25 states the following: "And unto Eber ("Hebrew" - born 2288 B.C.) were born two sons: the name of one was Peleg; for in his days was the earth divided; and his brother's name was Joktan".

Peleg lived 239 years (2254 - 2015 B.C.). In his days the earth was divided (re-divided), or re-inhabited by different tribal groups. Peleg's life began 100 years after the World Wide Noahic Flood, which gave the people some time to begin to fulfill God's command: "And God blessed Noah and his sons, and said unto them, be fruitful, and multiply, and replenish the earth" Gen. 9:1.

The earth had to be replenished with mankind and re-divided amongst tribal groups according to God's foreordained will (Gen. 1:28). How did God accomplish this in the days of Peleg? Peleg was born in the 4th year of Rimush's tribal rule, or in the days of Ebrum of the City of Elba, in Syria. During the reign of their last ruler of the 1st Dynasty of Agade, Elulu, the Gutians invaded and scattered the people. They destroyed Nimrod's idea of a city state (Nimrod destroyed Elba in 2205 B.C.), and the tower built unto heavens where they worshipped the Moon god (towers to the Moon have been rebuilt in these last days). This scattering divided the tribal groups in different areas. This soon con-fused their languages (2169 B.C.), which was God's providential will to prevent them introducing then, the present scientific age of a United One World Order which will soon bring world destruction.

Another problem is the seeming discrepancy in God's statement, "And the whole earth was of one language, and of one speech" Gen. 11:1. It is a historical known fact that several pre-Flood languages were carried over after the Noahic Flood. The only possible answer is that even though two or three languages were truly carried over from before the Flood by Noah, by his family, by literature, and perhaps, by his servants, they unitedly knew one language (Akkadian). They were so closely knit in group unity, until the Gutian invasion 186 years after the Flood, so that they all were able to communicate together in one language. Never-theless, two or three other languages were kept alive during that per-iod, since the life span was still very long.

The situation could be compared with Israel's today. The Israelites today know scores of languages mastered during their world wide dispersal, but they converse together in one language (Hebrew). But if they were again scattered, many would soon revert back to their mother tongue of their youth. In a few generations their personal understanding of Hebrew lost, and their languages would be confused once again.

1. "The Midnight Cry", Feb. 1967,p.30, London, England

Thus, the Gutian invasion, the destruction of Babel, and the scattering of the people under these cruel rulers drove them to every part of the earth during the remainder of Peleg's life time (2169 - 2015 B.C.), thereby reviving these various original languages and starting other dialects. This historical explanation does not deny the possibility of a supernatural intervention by God in this matter.

What other Biblical confirmation do we have which prove that our chronological calculations are right? In a portion of the following chart, we have the meaning of all the names of the Godly line following Peleg. Notice how these names express the political situation which was prevalent during that period. This span of time agrees historically with the "National Migration Period", and the "Sumarian Legendary Period", known as the "Heroic Age".

Historians confirm that this period is known as a period of migration as verified in this following statement:

"In the entire fertile crescent nomadic pressure seems to have reached its height in the period between 2200 and 2000 B.C., when it also penetrated into Egypt. From about 2000 B.C. on, the nomads seem to have become more interested in settling down than in making fresh raids, so sedentary occupation began to expand, and with it the arts of Civilization".[1]

Now, we lay the Biblical chronology, including their dates and meaning of names, over the chronology of the "1st Dynasty of Agade" as follows:

CHRONOLOGY OF 1ST DYNASTY OF AGADE AND THE BIBLE
(The Biblical chronology is underlined)

B.C.

2356 – Ham (Sargon) preserved by the Ark as Flood begins

2355 – Noahic Flood ends

2350 – Noah and family settles around Kish?

2320 – Sargon (Ham) leaves Kish & goes to Accad (Agade)?

2315 – Sargon's (Sharrun-Kin) tribal rule at Agade (57 years)

Lk.3:35 K.J.
"Heber" – – 2288 – Eber born means "Hebrew"died 1824 BC 4 yrs after Abram

| 2258 – Rimish's tribal rule at Agade (9 years)

(Son)| – 2254 – Peleg born means "divide"(his days earth re-divided)

m 2249 – Manishtushu's tribal rule at Agade (15 years)

i *2234 – Nimrod (Naram-Sin)(Erech,Calneh,Akkad,Babel-G.10:10)

g 2224 – Reu born means "friendship"(peaceful co-existence)

r 2197 – Sharkalisharri city rule(25 yrs – son of Naram-Sin)

a 2192 – Serug born means "firmness" (Gutians infiltrate)

t 2172 – Iqiqi, Nanum, Imi, and Eluln (total of 3 years)

e 2162 – Nahor born means "slayer"(Gutian attack 2169 BC-Babel)

| 2133 – Terah born means "wandering"(people scattered-migrate)

:– –2015 – Peleg dies – "In his days the earth was divided".

1. Albright, W.F.,"From the Stone age to Christianity",p.163,164, The Johns Hopkins University Press, Baltimore/London, 1957

Thus, the Sumarian Heroic Age" (2250 - 2000 B.C.) coincides with the period of "National Migration" (Peleg's days). It was the beginning of an adolescent and barbaric cultural stage and the exploits of the pre-Flood Sumerian characters were told and retold. The 350 year span of time between the end of this period and the Flood would compare to our present time back to the days of the pilgrim fathers in America.

During the latter part of this same period, the 3rd Dynasty of Ur arose, which introduced the first code of law under Ur-Nammu (2113 - 2095 B.C.). Later, the death of Noah took place in 2006 B.C. (350 years after the beginning of the Flood at the age of 950 years - Gen. 9:28,29), and three years later (2003 B.C.), Abraham was born as the "Father of a Multitude", or the "Father of our faith", who, in turn, introduced the "Patriachal Age".

As a last proof and testimony of the sudden disappearance of the Sumarian peoples and the catastrophic eclipse of civilization in the year 2356/5 B.C., these following **"twenty quotations"** are given. These quotations should impress upon the general reader that there is clear evidence of a lapse in history which began in 2356/5 B.C., which was caused by the **"World Wide Noahic Flood"**. Therefore, as a result, there was approximately 50 to 75 year interregnum with a complete lapse of civilization, and clear evidence of world wide nomadic life following (tent dwelling period): then a gradual recovery to full monarchal rule by 2100 B.C. (see Assyrian chronology - page 209):

TWENTY QUOTATIONS

"When the darkness of the **First Intermediate Age** descended upon Egypt in the 22nd century B.C. **high culture suffered temporary eclipse,** old traditions were neglected, and the ancient customs fell into desuetude. As often again in the recorded history of man, it seemed to thinkers that all worth-while possessions of humanity had been lost in the **general catastrophe of civilization . . ."[1]** (Noahic Flood).

"Amon-Re came to the fore during the **First Intermediate Period** (2300 - 2065 B.C.) the **era of confusion** between the fall of the **Old Kingdom** and the establishment of the **Middle Kingdom"**.[2]

"Akkad was the name given to northern Babylonia from the city of Agade which Sargon brought into great prominence as the capital of a **new Semitic empire** dominating the Mesopotamian world from about 2360 - 2180 B.C.".[3]

"Around 2350 B.C. there comes to power in Mesopotamia the first Semitic dynasty which is known to us from direct sources: the Dynasty of Accad (Akkad or Agade), founded by Sargon the Great".[4]

"But after the 'sixth' (Egyptian) Dynasty had well begun, a definite **decline and retrogression had set in.** We find ourselves then **groping in a dark age** wherein were no arts and no written history".[5]

1. Albright, W.F., "From the Stone Age to Christianity" page 182 **(Ibid)**
2. White, J.E.M., "Ancient Egypt", page 26, **Allen & Unwin, England**
3. Unger, M.F., "Archeology and the Old Testament", page 88 , **Zondervon**
4. Moscati, S., "The Semites in Ancient History", p.49, **Univ. Wales Press**
5. Rimmer, H., "Dead Men Tell Tales", page 45 , **Hart-Kolportasie-bekery**

"Though Manetho includes Dynasties I-XI in his section for the **Old Kingdom**, the monuments prove that the **Old Kingdom came to an end with the VI Dynasty** (Egyptian)".[1]

"The Delta and Middle Egypt were plunged into **fearful chaos.** Their nobles were dispossessed, **a reign of terror commenced,** palaces and temples were destroyed. For nearly 300 years, from the **closing years of Pepi II** (Dynasty VI) to the foundation of the Middle Kingdom (2300 - 2065 B.C.), **anarchy ruled in Egypt**"[2] (Pre-Noahic Flood Chaos & afterwards).

"When did the **bell toll** for the Sumerians? . . . For at the zenith of their civilization, **about 2350 B.C.,** the Sumerian city-states were succeeded by **nomadic tribes of a Semitic people** who had settled in the region of Akkad. ."[3] (The bell toll came with a World Wide Flood).

"The evidence is impressive that **a sudden flood** overwhelmed the human race and that all mankind on earth today has descended from the one surviving family.

Anthropologists have learned from ancient religions literature as well as from primitive tribes that some **form of a flood story** occurs all around the world. The oldest known uninspired account, as the Mesopotamian, written on clay tablets, at least, as early **as 2000 B.C.** . "[4].

"**Sargon** the mighty king, the king of Akkad, am I . . . My humble mother conceived me; she bore me in secret, placed me in an **Ark of bulrushes, made fast my door with** pitch and gave me to the river which did not over whelm me . . .".[5]

"Whatever his origin may have been, Sargon became King about 2277 B.C. and ruled Agade (Accad), a city which he built and where he founded a dynasty . . . ".[6]

"We have seen how between 2800 B.C. and 2270 B.C. a succession of powerful kings - the first pyramid builders - ruled a prosperous Egypt, and that **after 200 years interregnum,** another dynasty rose and reunited the kingdom, reigning over it throughout the **Middle Kingdom** (2100 - 1900 B.C.)".[7]

"About the **22nd century B.C.** or even a **little earlier, we find a break** in the continuity of occupation; fewer and fewer towns were inhabited and such centers as Ai were destroyed and abandoned".[8]

"If Sargon's advent to power is not, so far as we know, the result of a Semitic invasion from without, can we say that it is at least a **victory**

1. Budge, Sir E.A.W., "Egypt", page 80, J.M. Dent & sons Ltd,London 1914
2. White, J.E.M., "Ancient Egypt", page 152, Allen & Unqin, England
3. Lissner, Ivan, "The Living Past", page 35. A.P. Putnam sons(Jan. 1967.
4. Coder, Dr. S.M., "Archaeology confirms the O.T.", Moody Monthly p. 5
5. Finegan, J., "Light from the Ancient Past", page 38,Princeton Univ.P.
6. Swain, J.W. "The Ancient World", page 70
7. Cottrell, L., "The Anvil of Civilization", page 104, Trinity Press L.
8. Albright, W.F., "From the Stone Age to Christianity", page 163, The Johns Hopkins University Press, Baltimore/London, 1957

of the Semitic over the Sumerian element in the population? It is widely accepted as such a victory; but this view likewise does not stand the test of careful scrutiny; for it is a remarkable fact that neither the Semitic conqueror Sargon, nor his Sumerian adversary Lugalzaggisi, represents himself as the leader of a people; both their titles are those of kings of cities, and then in more generic terms 'of the land'. Moreover, the Sumerian Lugalzaggisi does not hesitate to put up a Semitic inscription in the temple of his god, while within the royal family, that of Kish, Sumerian and Semitic names succeed one another, and so become devoid of any ethnical significance. Neither side in its invocations of the gods makes any distinction between Sumerian gods and Semitic ones, and finally, there is not, in the extensive literature of later periods, the slightest trace of any national animosity"[1] (Only the Noahic Flood can solve these problems).

"From the 6th to the 10th (Egyptian) Dynasty, history is almost silent"[2] (The 6th & 7th were the last pre-Flood dynasties).

"And when the 6th (Egyptian) Dynasty ended in 2270 B.C. it was followed by an age of confusion. Historians call this the period of the 7th to the 10th Dynasties, the First Intermediate Period".[3]

"Then, about 2350 B.C. the Semites come to power with the advent of Sargon the Great, who unites Mesopotamia under his rule and presses into Syria (Elba - added) and Asia Minor".[4]

"Sargon had been a high official under king of Kish, and about 2340 B.C. he founded a city of his own, Akkad".[5]

"The Sumerians and all other in Babylonia were subdued by the Semitic Sargon I in the Old Accadian period (c. 2360 - 2180 B.C.)"[6] (The World Wide Noahic Flood brought this about).

"It was not until about 2300 B.C. that the centralized power of the Old Kingdom faded and gave place to feudal anarchy".[7]

". . . Sumer, in the third millennium B.C., consisted of a number of city-states vying for supremacy over the land as a whole. One of the most important of these was Kish, which according to Sumerian legendary lore, had received the "Kingship" from immediately after the FLOOD"[8] (Historic confirmation of the Noahic Flood).

THE RUINS OF TROY AND THE FLOOD PERIOD

"During its 7th and final phase, in 2500 B.C. or thereabouts, Troy II suffered destruction by a fire of such great intensity that many of its stones were calcined by the heat. Evidence abounds that the inhabitants fled in such haste that they were able to make little or no attempt to rescue their household goods and other belongings.

1. Moscati,S., "The Semites in Ancient History"p.50, Univ. Wales Press
2. Murray, M.A., "Splendour that was Egypt", p.20,Sedgwick & Jackson Ltd
3. Cottrell, L., "The Lost Pharaohs", pages 20,21,Pan Books Ltd., London
4. Moscati, S., "The Face of the Ancient Orient"p.59,Vallentine Mitchell
5. Frankfort, H., "The Birth of Civilization in Near East"p.72(Bibliogr)
6. Harper Bible Dictionary, page 710, Harper & Row Publishers Inc., N.Y.
7. The Egyptian State Tourist Administration, pages 22,23
8. Kramer, S.N., "History begins at Sumer", page 67, Jarrold & sons, Ltd

Troy III (4): The occupation which followed the fire is marked by an unusual abundance of deer bones, the results, it has been suggested, of a sudden influx of[1] game, or of the devising of improved methods of hunting, or both" (A result of the Flood).

INDIA'S CIVILIZATION & NOAH'S FLOOD

"On the Indus River before 2500 B.C., a people of mixed origin and diverse ethnic composition combined to produce a bronze age civilization with a well developed art and architecture probably matched by an equally advanced religion, . . . who was in touch but differed markly from the Sumerians, **vanished completely in some unknown catastrophe"**[2] (Noahic Flood).

To deny the plain implications of the above **"Twenty Quotations"**, and to doubt that Ancient History and the Bible correlates is nothing short of intellectual folly. We rest our case on the authority of the Eternal Word of God, on the facts of history, and on the testimony of these twenty eminent scholars.

The Eternal Word of God continues to cry out; the stones of the centuries cry out; the voices of men through various means cry out; and finally, the Spirit of the Lord doth cry out as a last day testimony at the close of this age. Are you prepared to meet your **Maker?**

- - - - - - -

1. Cleator, P.E., "The Past in Pieces" p.97, Geo. Allen & Uniwin Ltd
2. Noss, John B., "Man's Religions", pages 123,124 , Macmillan Co.,1966
   New York

Ante-Diluvian world, God, creation, and the beginning of all things! How? When? Listen, we have just gleaned from very ancient manuscripts a new look at creation:

In the beginning was God (Elohim); in the beginning was His Word (God's Son); and in the beginning was His Spirit (God's Holy Spirit). God the Father (Godhead bodily) planned all things (Creator - Gen. 1:1 - Operation Center); God the Son (the living Word - Finger, Hand, Arm, & Bosom of God) directed by His Right Hand all things (Isa. 41:20; Eph. 3:9 - Administration Center); God the Spirit accomplished all things by "moving" with "creative" power (Gen. 1:2 - Manifestation Center).

In the beginning heaven was as the earth, so was earth as heaven. A day was as a thousand years in God's time, and a thousand years was as a day.

God was, God is, and God shall be infinite, and time, as we know it, was not, for time was as infinite as God. In the infinite time, our infinite God created all things perfect. "For Jehovah created the heavens and the earth and put everything in place, and He made the world to be lived in, not to be an empty chaos or a waste" (Isa. 45:18). He made the Cherubim with four wings, and the Seraphim with six wings. He made the angels also. The Cherubim were chosen as the anointed one that covereth and protected the very throne of God in His heavenly sanctuary (Ezekiel 28:18) which was in the mountain (Kingdom) of God (Ezekiel 28:16). The earthly Paradise was then known as the Garden of Eden of God (Ezekiel 28:13 - heaven on earth) where the Cherubim walked, and where God walked in the cool of the day (Gen. 3:8). There, God fellowshipped with His created beings, Adam and Eve, who were made after His "likeness" and His "image" (Gen. 1:26 - a triune person).

Then suddenly there was chaos in the midst of His creation, and darkness was upon the face of the waters of the deep (Gen. 1:2,3). Lucifer, the highest exalted Cherub that covereth (Ezekiel 28:14; Heb. 9:5), and one third of the angels of heaven, revolted against God in heaven (Rev. 12:3,4 - "Behold a great red dragon - Lucifer - and his tail drew the third part of the stars - angels - in heaven"), but God cast him and his angels out of heaven (Rev. 12:10b). "And Jesus said unto them, I beheld (personally) Satan as lightning fall from heaven" Luke 10:18. Since then, Lucifer is the prince of the power of the "air" (second heaven - Eph. 2:2), and his fallen angels represent the principalities, powers, rulers of darkness of this world, and "spiritual wickedness in high places (air)" Eph. 6:12.

"How art thou fallen from heaven, O Lucifer (day star, or angel who became the angel of darkness) son of the morning! How art thou cut down to the ground, which didst weaken the nations? For thou hast said in thine heart, I will ascend into heaven, I will exalt my throne above the stars (angels) of God: I will sit also upon the mount (kingdom) of the

congregation, in the sides of the north (heaven): I will ascend above
the heights of the clouds; I will be like the most High (Isa. 14:12-14).
"Yet thou shalt be brought down to hell to the sides of the pit" (Isa.
14:15; Rev. 19:20).

"Therefore rejoice, ye heavens, and all ye that dwell in them (heaven
cleansed of Lucifer and his angels). Woe to the inhabiters of the
earth and of the sea. For the devil (Lucifer) is come down unto
you, having 'Great Wrath' because he knoweth that he hath but a
short time" (Rev. 12:12 - Lucifer has a short time to gather his
sons and final followers on earth during the last half of the Tribulation Period).

"Art thou not He that hath . . wounded the dragon (Lucifer)?" (Isa.
41:9). "And He (Christ) said unto them, I beheld Satan (Lucifer) as
lightning fall from heaven" Luke 10:18. "He that committeth sin is of
the devil (Lucifer); for the devil sinneth from the beginning. For this
purpose the Son of God was manifested, that He might destroy the works
of the devil (Lucifer), . . In this the children of God (through Christ)
are manifest, and the children of the devil (Lucifer) . . " I John 3:8,
10a. "The thief (Lucifer) cometh not, but for to steal, and to kill, and
to destroy: I (Christ Jesus) am come that they might have life, and they
might have it more abundantly" John 10:10.

Thus, one of God's created beings brought darkness, destruction, and
havoc upon God's first creation. Yet, in the process of time God will
cleanse the world and heaven of the Father of lies, and God will put
Lucifer, his angelic fallen creatures, and all followers in the
pit (Rev. 20:1,2; Isa. 27:1). In the mean time God first wants to
test the descendants of Adam with an opposition force to see, by free
will choice, whether they are sons of God (I John 3:10), or sons of
their father the devil (I John 8:44). Whom will ye serve? In whom are
you pinning your faith?

God reassures us that He has providential control of all things and
His plan and blue print are working out perfectly according to His foreordained will from the foundation of the earth. For He states, " . . I
am the Lord, there is none else. I form the Light (sent Christ the Light
into the world born of a virgin), and create darkness (God's created being, Lucifer, who fell by free choice and disobedience which brought
darkness), I make peace (sent Christ the prince of peace), and create
evil (created the being which brought evil, who was a free angelic being
with a free choice of will, as man has today): I the Lord do all these
things (Isa. 45:6b,7).

Therefore God's Spirit moved upon the face of the waters (of chaos
to re-create). And God said, Let there be Light (Christ manifested
as the Angel of the Lord) and there was Light (Christ in the midst of
His new creation). And God saw the Light (Christ), that it was good
(Christ said unto him . . "there is none **good** but one, **that is God**");
and God divided the Light (Christ - sons of Light) from the darkness
(Lucifer - sons of darkness). And God called the Light (Christ) **Day,** and
the darkness he called **Night** (Satan). And the evening and the morning
was the first day (of re-creation) Gen. 1:3-5. "But ye brethren, are not

in **darkness** (not of Lucifer) that the day should overtake you as a thief. Ye are all the children of **Light** (Christ) and children of the **Day**: we are not of the **night nor of darkness**" II Thess. 5:4,5.

Thus, in the beginning God created the heavens and earth,[1] and in the beginning was the Word, the Word was with God, and the Word was God; and the Word was made flesh (Christ) and dwelt among us;[2] rejoicing in the habitable part of the earth; and His delight was with the sons of men,[3] walking in the garden in the cool of the day.[4] (Man) beheld His Glory,[5] the Glory as the only begotten of the Father, full of grace and truth,[5] and the Word was with God, and the Word was God. The same was in the beginning with God.[6] Then I (Christ) was by Him (God), as one brought up with Him; and I was daily His delight, rejoicing always before Him.[7]

Before the mountains were settled, before the hills, was I (Christ-Wisdom-I Cor. 1:24) brought forth, while as yet He had not made the earth, nor the fields, nor the highest part of the dust of the world. When He prepared the heaven, **I (Christ) was there**; when He (God) set a compass upon the face of the depth, when He established the clouds above; when He strengthened the fountains of the deep; when He gave to the sea its decree, that the waters should not pass his commandment, when He appointed the foundation of the earth.[8]

Thus, in the beginning God created the heaven and the earth,[9] and through faith, we understand that the worlds were framed by the Word (Christ) of God, since all things were made by Him; and without Him was not anything made that was made;[10] so that things which are seen were not made of things which do appear.[11]

And the Spirit of God moved upon the face of the waters, and God said, Let there be Light.[12] That was the true Light, which lighteth every man that cometh into the world. He (Christ Jesus) was in the world, and the world was made by Him, and the world knew Him not. He came unto His own (Jewish people), and His own (people) received Him not. But as many as receive Him (whether Jewish or Gentile), to them "gave He power" to become the sons of God, even to them that believe on His Name.[13] (He was the one) by whom also He made the worlds; - being the brightness of His Glory, and the express image of His person, and upholding all things by the **Word (Christ Jesus) of His power.**[14]

Therefore, Lord, thou hast been our dwelling place in all generations. Before the mountains were brought forth, or ever thou hadst formed the earth and the world, even from everlasting to everlasting, thou art God.[15] But thou, Bethlehem Ephratah, though thou be little among the thousands (of villages) of Judah, yet out of thee (Bethlehem - meaning house of bread) shall He come forth unto me that **is to be Ruler in Israel**; WHOSE GOINGS FORTH HAVE BEEN FROM OF OLD, FROM EVERLASTING **(Eternal One)**.[16]

1. Gen. 1:1   5. John 1:14   9. Gen. 1:1   13. John 1:9-12
2. Jn. 1:1,14   6. John 1:1,2   10. John 1:3   14. Heb. 1:2,3
3. Prov. 8:31   7. Prov. 8:30   11. Heb. 11:3   15. Psa. 90:1,2
4. Gen. 3:8   8. Prov. 8:25-29 12. Gen. 1:2,3   16. Micah 5:2

And (the) God (Head boldly) said, Let US make (create) man in OUR image, and also after OUR likeness [1] (this first man was a failure, so 4000 years later God, Himself, was manifest as man in His own image); thus, man had a **BODY**, after the image of the **Father,** which made man **world-conscious**; man had a **SOUL**, as the **Son of God** became a living soul, which made man **self-conscious**; and man was endued with a **SPIRIT**, as the God Head moved by the Spirit, which made man **God-conscious**. Thus, the **Triune God Head** (Elohim) completed forming the **triune man in His own image** within the sixth day of recreation, and that point of time in the sixth day is unknown. Thus, man rejoiced in the presence of God with full knowledge of all things. Man walked in perfect fellowship and communion with God continually in this delightful Paradise until man sinned against God and was sent forth from the Edenic Garden. The man, Adam, and his mate, began to die by gradually aging from that day. That point in the sixth day of recreation in which man departed from God's presence and from Paradise was the year 4012 B.C., therefore, the Jubilee year fell on that date every 50 years after that time until zero B.C. (see Appendix B, page 188 on the subject, "The Sabbatic and Jubilee Years").

THE ADAMIC CURSE

Before his fall and departure from God, Adam found himself occupying the most beautiful, central, and productive part of the Edenic Universe. It was located in the midst of the flourishing Great Fertile Crescent of the Middle East near the delta where four great rivers meet (Gen. 2:10-14). Before their fall, God's new man arrived on the scene during the Palaeolithic Era (Old Stone Age), but he had little need of implementing the meaning of the term, since he lived in the midst of God's perfect Paradise with a landscape of tropical fauna. This, Paradise, with its lavish productivity, resulted in a food-gathering way of life with no immediate need of implements (Gen. 2:15).

Historians state that during this era "the whole area from the Atlantic coast of North Africa and across the Near East to Iran enjoyed a far more abundant rainfall (mist - Gen. 2:6). Thus, at that time, the whole region was a kind of vast park, (a Garden of Eden - added), with plant life sufficient to support a wide range of plentiful supply of animals; among this extensive fauna was a rare food-gathering biped - the creature we know as Palaeolithic Man".[2]

This fabulous, original, perfect Paradise was known as "heaven on earth", since man and his Creator were in such close communion and fellowship, so that they walked and talked together in the cool of the evening (Gen. 3:8).

Suddenly, as in the former age, this era made a quick sprint towards catastrophe, because of the disobedience of man towards his Maker (Gen. 3). As a result, man was immediately driven forth from the presence of God and from Paradise in the year 4012 B.C. (Gen. 3:23,24). Man began to age from that date. The Palaeolithic beauty began to deteriorate and gradually fade away, and so it happened with man.

1. Gen. 1:26
2. Saggs, H.W.F., "The Greatness that was Babylon", page 4 , **Sedgwick & Jackson Publishers**

"And the Lord God said unto the serpent (Lucifer), because thou hast done this, thou art cursed above all cattle, and above every beast of the field; upon thy belly shalt thou go, and dust shalt thou eat all the days of thy life (the serpent symbolizes the evil of Lucifer): and I will put enmity between thee and the woman, and between thy seed (Lucifer's followers) and her seed (seed of the Messiah); it (He - Messiah) shall bruise thy head, and thou (Lucifer) shalt bruise His heel"(Messiah's).

Unto the woman he said,"I will greatly multiply thy sorrow and thy conception; in sorrow thou shalt bring forth children; and thy desire shall be to thy husband, and he shall rule over thee".

And unto Adam he said,"Because thou hast hearkened unto the voice of thy wife, and hast eaten of the tree of which I commanded thee, saying, Thou shalt not eat of it: cursed is the ground for thy sake; in sorrow shalt thou eat of it all the days of thy life; thorns also and thistles shall it bring forth to thee; and thou shalt eat the herb of the field; in the sweat of thy face shalt thou eat bread, till thou return unto the ground; for out of it wast thou taken: for dust thou art, and unto dust shalt thou return" Gen. 3:14-19.

The Adamic curse had fallen upon man (Rom. 3:9), as God had forewarned, and upon all creation (Rom. 8:22). Both man and all creation began their travail and their journey towards death and desolation.

For a while Adam and Eve, his wife, and their descendants (Gen. 5:4)[1], managed quite well in the food-gathering way of life, but each year crops grew worse. Strange non-food bearing plants began to entangle themselves with the others. They were literally forced to labour to survive in a continuous warfare with these obstinate plants. The scientific application of the word "Palaeolithic" had begun. Pieces of stones and sticks worked well at first to kill these few foreign invaders, but as the perfect era faded with time, the battle became more intense after only a few decades. They were forced to improve their implements to ease the burden of labour. Flint stones were proven to chip easily, thereby providing sharpened objects so they would more readily destroy these "plants out of place" (weeds).

Viewing this early attempt to make implements, archeaologists, almost six thousand years later, called that period the Neolithic Era (the New Stone Age). Subsequent to this time some of Adam's relatives (he begat sons and daughters - Gen. 5:4)[2] began exploration hikes each week in various directions. One week they extended their journey. On their return, they reported that they had travelled far up the Euphrates river and testified to have seen a beautiful mountain toward the south west, which they named Sirion (Senir). Centuries later it was called Mt. Hermon. At its height the tropical mist had seemed to form a white cap on its top, and they were determined to investigate. This first expedition led to a settlement near Jericho where they found that they could enjoy the clear, clean, spring waters which flowed from the base of the snow capped mountain. This settlement was known by later archeaologists as the Jermo culture.

1,2. Josephus, Bk. I, Ch. 2, par. 3, note, "The number of Adam's children were 33 sons and 23 daughters"

The population in the Fertile Crescent continued to increase, expand, and migrate. One group reported that lightning from God had started a grass fire and that they had managed to keep the flame going continually by adding brush and wood. The black bitumen from the sea south of Jarmo was found to make good fuel, but was difficult to transport or to handle. Perhaps it could be used for something better. It was soon discovered that melted animal fat provided the best fuel and could be contained in a cupped rock. It also gave a delightful glow at night and provided light. Some time later one tribal family received a report from the descendants of Cain that they had discovered a strange green rock formation, and after putting the substance in the fire it melted and cooled into something very hard and pliable. They suggested that tools might be made from the material to help fight the invasion of weeds. As to the date it was rumoured that Adam, their great, great grandfather, was about 235 years old. The archeaologists, later, called their discovery the Chalcolithic culture (when copper and flint was used together - 3812 B.C.).

This was about the time that Enos was born, Seth's son, in the year 3777 B.C.. It was during Enos' days that many began to call on God's name once again (Gen. 4:26). Prayer was certainly needed at that time. Although the discovery of brass and copper had easied the burden of labour, it was rumored that weapons had been made with the metal by which family feuds were being waged amongst the Proto-Iranian culture (descendants of Cain) further to the east. With such an advantage, would they try to return to the Euphrates valley and cause havoc?

In the meantime, another family group migrated further to the west. They had discovered a great blue sea, and had then followed its coast south until they also discovered a large, beautiful river which flowed in the opposite direction than the Euphrates or Jordan rivers. This was strange. What did it mean? They named it the Nile. The family group that settled there was later designated as the Badarian peoples (culture).

Perhaps it should be recalled that Adam's first son, Cain, did not turn out so good. He became jealous over Abel, his brother, because of difference in views concerning atoning sacrifices which their father, Adam, had taught them to bring before God according to God's instructions. Cain began to deviate from his father's words to bring a live animal to be sacrificed for "the atonement for their sins". Cain just did not seem to have the spiritual discernment to understand the significance of a "substitutionary atonement" for sin, which in turn, provided salvation and deliverance from the Adamic curse which came by sin. Adam, his father, had explained that the life of an animal had to be given as a substitute for the life of man in payment for the debt of sin. This implied nothing less than that God demanded both physical and spiritual death in payment for sin. Therefore, the lamb was slain as a substitute sacrifice for sin in behalf of the person who offered it as atonement (atonement = at-one-ment with God). Yet, Cain argued and insisted that it was terrible to kill animals and brought only fruit of the field instead. Naturally, God only accepted Abel's sacrifice according to the God given instructions which Adam had passed on to his children.[1] This

1. This lamb sacrifice was symbolic of atonement pointing to the coming Messiah who atoned, with one sacrifice, once, for the world's sin, if believed and received

angered Cain. In spite of the clear explanation, with his dread and dislike of blood sacrifice, he became so jealous and angry with his brother, Abel, so that he sinned even worse. He killed and shed his own brother's blood, rather than sacrifice an animal, which is not a redeemable creature. A real picture of man's fallen nature and rebellion from the beginning.

God had not talked much with man since the Edenic relationship, but He readily intervened at that time and spoke directly to Cain. As a result, Cain had a strange mark put upon him (Gen. 4:15), and he and his family were forced to "separate" from the Semitic group. They moved east of Eden to the land of Nod and built the city of Enoch. Some parts of the family later migrated towards India and other eastern areas during those early days, ultimately creating another racial group known as the Proto-Iranian culture. This racial group was "marked for separation" by the eternal Hand of God. Breaking this God given command of separation of tribal groups would only be fully reaped in these present latter days.

But now, according to the early Semitic fears, these Proto-Iranian racial groups began to rebound and re-enter the Fertile Crescent from the East. They brought with them the tomtom music and the organ and the harp in the person of Jubal (Gen. 4:21); the knowledge of metallurgy (industry) and weapons of warfare in the person of Tubal-cain, the artificer in brass and iron (Gen. 4:22); the beauty queen of the daughters of men in the person of Naamah (beautiful - Gen. 4:22; 6:2) for "they were fair" to look upon; the fugitive and wandering spirit of Cain (Gen. 4:12-20); the Cain spirit of nakedness that was evident in Ham (Gen. 9:22); the spirit of polygamy of Lamech (Gen. 4:19); the spirit of bloodshed and debauchery also in the person of Lamech (Gen. 4:23), and the degrading life of the city (Gen. 4:17).

During the days of Jered (sixth generation from Adam) the descendants of Cain began to return to the Middle East bringing their worldly corruption and mark with them forcing the Semitic race to inter-mingle either by arms or treaty. They soon inter-married with the Semitic peoples. Spiritual decay soon set in which led directly to the Noahic destruction (see II Peter 2).

Josephus (1950 years ago) gives us a good spiritual picture of the corrupt, and evil characters of the descendants of Cain:

"However, he (Cain) did not accept of his punishment in order to amendment, but to increase his wickedness; for he only aimed to procure every thing that was for his own 'bodily pleasure', though it obliged him to be injurious to his neighbours. He augmented his household substance with much wealth, by 'rapine and violence'; he excited his acquaintance to procure 'pleasures and spoils by robbery', and became a great leader of men into 'wicked courses'. He also introduced a change in that way of simplicity wherein men lived before; and was the author of measures and weights. And whereas they lived innocently and generously while they knew nothing of such arts, he changed the world into cunning craftiness. He first of all set boundaries about lands: he built a city, and fortified it with walls, and he compelled his family to come together in it; and called that city Enoch, after the name of his eldest son Enoch. . . . Nay, even while Adam was alive (until 3082 B.C.), it came to pass that the posterity of Cain became 'exceeding wicked', every one successively dying, one after another, more wicked than the former. They were 'intolerable' in war, and 'vehement in robberies'; and if any one were slow to 'murder people', yet was he bold in his profligate behaviour, in acting unjustly, and doing injuries for gain".[1] ("As it was in the days of Noah so shall it be in the days of the coming of the Son of Man - Jesus Christ" Matt. 24:37).

Therefore, this inter-marriage, in turn, created a third racial group during the days of Jered (3552 - 2590 B.C.) known as the "Sumerian Culture". This suggestion is well confirmed since the theory has been propagated by later historians as the following quotation verifies: "Note that Sumerian achievements were actually the product of at least three ethnic groups . . . the Proto-Iranian, the Semitic, and the Sumerian".[2] This is why the pre-Sumerian period is referred to by some as the Irano-Semitic civilization (4012 - 3552 B.C.). Thus, the Sumerian civilization, the result of inter-marriage, lasted for only a period of around 1200 years (3556 - 2356 B.C.) resulting in the Noahic disaster.

Jered and his sixth generation soon deteriorated spiritual in that they began to inter-mingle with these corrupt races through inter-marriage. Thus, God raised up a witness against them in the seventh generation in the person of Enoch (seven is God's perfect number). Enoch lived 65 years, perhaps in unbelief (3390 - 3325 B.C.), during the early years of Jered; then Enoch had a tremendous life-changing revelation from God during the birth of his son, Methuselah. This revelation revealed that on the day of the death of his new born son, Methuselah, a "World Wide Flood" of destruction would come on the earth. Therefore, in the light of this heavenly vision, Enoch walked with God in the midst of evil chaos for 300 years, ". . and he (Enoch) was not because God took him (Rapture)" Gen. 5:24. Futhermore, Enoch's message, appling both to that generation, and also to our last-day generation, is found in Jude 14-16 as follows:

"Behold the Lord cometh with ten thousands of His saints, to execute judgment upon all, and to convince all that are ungodly among them of all their ungodly deeds which they have ungodly committed, and of all their hard speeches which ungodly sinners have spoken against Him. These are murmurers, complainers, walking after their own lusts; and their mouth speaketh great swelling words, having men's persons in admiration because of advantage".

TRADITIONAL CREATION STORIES

As we analyze the Sumerian concept of the historical events of creation, there is no room for an evolutionary historical process. Everything happened through the providential working of the Omnipotent One, predetermined by His divine decree. They had a created world view with the world hemisphere coming in view fully created with no slow evolutionary process of environary interaction involved. Primitive monotheism, which agrees with Genesis, can be gleaned from earliest data. The degen-

1. Josephus, "Antiquities of the Jews", Bk. I, Ch. II, par. 2, Loeb Libr.
2. Kramer, S.N., "History begins at Sumer", page 281, Jarrold & sons Ltd
   Norwich, Britian

eration of this early monotheistic view followed resulting in forty names of other deities by the 29th and 28th centuries at Ur in southern Babylon, while 700 gods are listed at Shuruppak (Tell Turah) by 2600 B.C. in central Babylonia.[1]

The Sumerian's view concerning their flourishing civilization, which included all branches of their economy, was that it always had been more or less as it was from the beginning. Later, this concept degenerated into a pantheon of gods as mentioned above. The various branches of the economic structure had existed since God had planned and decreed them. This strong theological conviction amongst the early Sumarians was no doubt traditionally transferred to them from Adam (Semitic line) to the Proto-Irano line through Cain's family, whom God separated from the Semitic group and caused them to move east, "And **Cain went out from the presence of the Lord,** and dwell in the land of Nod, on the east of Eden (Iran)" Gen. 4:16.

These Proto-Irano-Semitic groups (Cainite) were the earliest known inhabitance during the Obeid period and their remains are evident everywhere immediately above virgin soil. Later, and in the early stages of the Uruk period, these two groups began to inter-mingle in the days of Jered (3552 - 2590 B.C. - added).[2]

This very inter-mingling, according to the book of Enoch, of the "Proto-Irano-Cainite" culture with the separated "Semites" resulted in the creation of another cultural group, the "Sumerians". Thus, by the beginning of the monarchial period (3275 B.C.), the Sumerians were growing in numbers, and by the late Uruk period they were a group to be reckoned with in lower Mesopotamia. The area of future Babylonia was where the largest amalgamation of racial names were found.

"It is in the later stage of the Uruk period that we find the introduction of the cylinder seal as well as the first inscribed tablets. And since, according to present indications, the language represented on these tablets, in spite of the largely pictographic character of the signs, seems to be Sumerian".[3]

Inscriptions are predominately in Sumerian language with a few votive tablets in Semitic. These votive "Semitic tablets" should remind us that Adam lived 930 years and died in 3082 B.C. during this period. His presence would certainly provide first hand information concerning the one true God, the same God as the God of Abraham, Isaac, and Jacob.

Sanchoniathon was an early Phoenician historian, and according to Philo Byblius, was a native of the city of Berytus. He was a contemporary of Semiramis, the Queen of Assyria, who reigned a few years with Adad-Nirari III (855 - 827 B.C.). In her days King Menuas of Urartu (Armenia) was making a flattering detente with the Queen by excavating a canal and naming it "Shamiram" after her name.[4]

According to our chronology this would make Sanchoniathon, also the contemporary of the Phoenician King Pygmation (871 - 840 B.C.); of Joash

1. Albright, W.F.,"From the Stone Age to Christianity", page 170,190
2. The Book of Enoch, page 34 (**Above Johns Hopkins Univ. Press 1957**
3. Kramer, S.N., "History begins at Sumer"p.272, **Jarrold & sons, Ltd.**
4. Ibid, page 277

(Judah - 880 - 841 B.C.); of Jehoahaz (Israel - 859 - 842 B.C.); of Sheshank III (Egypt - 872 - 833 B.C.); and living during the time of the founding of Carthage (864 B.C.). He gave an interesting resume of creation.

Sanchoniathon Phoenician history may be regarded as one of the most authentic memorials of the events which took place before the Flood, to be met with in heathen literature. It begins with a legendary cosmogony. It relates how the first two mortals were begotten by the wind (Spirit). His wife's name was Baau (Darkness, perhaps because of the temptation - added). It refers to the Fall, the production of fire, the invention of huts, clothing, the origin of the arts of agriculture, hunting, fishing, navigation, and the beginnings of human civilization. He gives an interesting account of the descendants of the 'line of Cain'. His history of the descendants of the 'line of Seth' reads like a heathen version of the account in Genesis. The whole system is an unintelligible jargon as though sorted from the mythologies of Egypt and Greece. Fragments of Sanchoniathon's writings have been preserved to us in the writings of Eusebius.[1]

Berosus, a historian of Belus of Babylon, who obtained his information from the temple Belus, wrote during the period 331 - 268 B.C. concerning history from the time of Creation unto his time. Fragments of his writings have been passed to us by later historians such as Apollodorus (144 B.C.), Polyhistor (88 B.C.), Abydenus (60 B.C.), Josephus (37 - 103 A.D.), Africanus (220 A.D.), and by Eusebius (265 - 340 A.D.).

His story of Creation confirms that there were ten generations before the Flood, and ten generations following agreeing with the Massoretic text. In the Babylon creation story Alorus corresponds to Adam while Xisuthrus, the tenth, corresponds to Noah. In Xisuthrus reign the great Deluge destroys the world while he and his family escapes in a vessel which he built. The inclusion of animals, birds, the abating waters, the sending forth of the birds, the location Armenia, and the dispersal to Shinar all agree with the Biblical text.

In Shinar, Nimrod, aspiring to world rule, builds the Tower of Babel, but the builders are dispersed and the Tower destroyed.

Some historians may have inadvertently confused quotations on "restoration" after the Flood with God's Adamic recreation. These mistakes must be carefully culled out. For example it would be natural to miss-apply the following quotation to God's Adamic recreation rather than applying it to the period after the Noahic Flood:

"Among a number of allusions to creation, one, dated c. 2350 B.C., describes the act of the god Atum who brought forth gods on a primeval hill above the waters of chaos".[2]

The primeval hill could easily be Mt. Ararat, and the waters of chaos could well be the world wide Noahic Flood. The above date would correlate correctly with our Biblical Flood date of 2356 B.C.. It is also interesting to note that the god Atum (only as Ra), in the

1. Anstey, M., "Romance of Bible Chronology", page 17, Vol. I
2. New Bible Dictionary, page 273, InterVarsity Press, 38 De Montfort St Leicester LEI 7GP

above quotation, is not listed among the ten pre-Diluvian Egyptian gods, but is listed among the representative gods during the Post-Diluvian and Patriarchal period.

Besides this, there are unanimity among various ancient legends, collected from many nations, that there were ten Ante-Diluvian patriarchs, and the tenth patriarch was involved in a World Wide Flood.

The Sibylline books speak of ten ages which elapsed between creation and the Flood; the Hindus enumerate ten fathers; the Chinese speak of ten emperors; while the Biblical, the Egyptian, the Babylonian, and the Assyrian historians actually list the names of the Ante-Diluvian patriarchs:

B.C.	Characters in the Bible	Rulers of Assyria	Rulers of Babylonia	God's of Egyptian	
4012-3082	Adam(Father)	Alulim	Aloros	Ptah	(Father)
3882-2970	Seth(Son)	Alalmar	Aloparus	Ra(Re)	(Sun-Son)
3777-2870	Enos(Worship)	Emenluanna	Almelon	Su(Shu)	(Spirit)
3687-2777	Cainan	Kickunna	Ammenon	Seb(Geb)	
3617-2722	Mahalalad	Enmengalanna	Amegalarus	Hosiri	
3552-2590	Jered	Dumuzi(Tammuz)	Daonos	Set(Seth)	
3390-3025	Enoch	Sibzianna	Eudoraohus	Horus	
3325-2356	Methuselah	Emenduranna	Amenpsinos	Tuth	
3138-2361	Lamech	Uburratum	Otiartes	Ma	
2956-2006	Noah	Zinsuddu	Xisuthros	Hor	

They are also unanimous in confirming that there were ten Post-Diluvian patriarchs from Noah to Abraham listed as follows: Shem, Arphaxad, Salah, Eber, Peleg, Reu, Serug, Nahor, Terah, and Abraham. The tenth person of both lists were great figures in the history of the Bible and of Israel. Noah, the tenth of the Ante-Diluvian world, initiated the means of escaping world judgment in the physical sphere (Flood), while Abraham, the tenth of the Post-Diluvian world, initiated the means of "faith" (Father of faith), whereby all nations may have access to a way of escape from judgment, while Christ Jesus under the New Covenant has initiated the atonement by His death as the Lamb of God chosen from the foundation of the world in the spiritual sphere (Gospel).

As the cry of warning went out to that last generation just before the World Wide Noahic Flood through Enoch and Noah, so the Mid-Night Cry is going out today. Are you ready? Has your sins been atoned for once and forever by Jesus Christ the Jewish Messiah, Saviour and Redeemer (Emmanuel - God with us)?

- - - - - - -

CHAPTER XV

ANTE-DILUVIAN HARMONY

We have already discussed the question that arises from Gen. 5:32. What year was Shem born in Noah's life? We found that Noah was 500 years old when Japheth (eldest - Gen. 10:21) was born, while Shem was born in Noah's 503rd year rather than 502nd year, since we allow one year for the Flood, and two years after the **end** of the Flood (not the beginning) for the birth of Arphaxad in the 100th year of Shem (Gen. 11:10,11).

As we consider Gen. 6:1-7, the next question is, who are the "sons of God"? Do they represent the Semites (sons of Seth - the Godly) who, later, inter-married with Cain's descendants (the Proto-Irano race), or shall we accept the so called "Book of Enoch's" interpretation? The Book of Enoch suggests that these "sons of God" were the fallen angels, the children of the heaven, who co-habited with the "daughters of men".

The co-habitation of angels (spirits) with humans is not such a problem (known in the occult demon world), but the reference of the reproduction of physical seed is difficult to accept. Luke 20:34-36 implies that angels neither marry or are given in marriage, therefore, reproduction of physical seed is impossible. Although there may be a great difference between good angels and fallen angels. Whether a different application could be applied to the one-third of the angelic forces (Rev. 12:4,9), who now have been cast out of heaven into the air (Eph. 2:2;6:12), and will be cast down to the earth along with their leader, Lucifer, according to the scripture is the question (Rev. 12:10,13)? Such a theory is very doubtful. In the spirit world of the "demon possessed" there seems to be some indication of "spirit co-habitation". (It was reported in 1983 that spirit beings co-habited with native women along the Amazon River and caused pregnancy. This has not yet been confirmed). Yet, no reproduction of physical seed is presently known. Verse six in Jude could imply such an angelic fall: ". . but (they) left their own habitation". Although it is not necessary to think that when they left their habitation that they came to earth in physical forms. Actually, they left their spiritual position as good angels and became evil; therefore, God cast them out of heaven into the air (Eph. 2:2), but they are still spirit beings.

After reading the book of Enoch, we also reject the total inspiration of the book and would clearly place it with the Apocrypha. Therefore, we doubt the interpretation found in that book, unless verified in these last days of evil.

Personally, we would rather interpret the verse as the inter-marriage between the line of Cain (Proto-Iranian) and the daughters of the Messianic Godly line (Seth). Cain and his descendants were cursed (Gen. 4:11) with the promise of being a fugitives and vagabonds (Gen. 4:14), and a "mark" was put upon him (Gen. 4:15). If the Godly line inter-married with the line of Cain which was during the days of Jered (meaning "to go down" - spiritual corruption) then automatically this curse and the effects of this mark would obviously be transferred into the Godly line and passed on from the days of Jered through the loins of Noah.

Perhaps this is why such a variety of nations, colors, varied religious groups, and even giants (Philistines) sprang from the loins of Noah, whose descendants have in turn multiplied and inhabited the whole earth since the Noahic Flood (racial curses are only taken away in Christ Jesus).

The book of Enoch also implies that the wicked descendants of this inter-marriage (Sumarians) were seeking longevity, but Enoch makes a pronouncement against them as follows: "And no request that they make of thee shall be granted unto their fathers on their behalf; for they hope to live an eternal life, and that each one of them will live five hundred years" (Longevity was cancelled with the Noahic Flood - added).[1]

When we compare the above statement with Gen. 6:1-7, we would conclude that the wicked desired to live another 500 years, but God intervened and gradually limited their demands to just one hundred and twenty years (Gen. 6:3) by the days of Joseph. Since the Flood took place in 2356/5 B.C., the above scripture would date the revelation to Noah in the year 2476 B.C.. The first revelation of the coming judgment came through a personal revelation to Enoch in the Godly line. This happened 969 years before the Flood at the time of the birth of his son Methuselah (3325 B.C.). The name "Methuselah" means "judgment at his death". With Enoch's personal revelation of coming judgment, when his son died, caused him to walk with God three-hundred years before being taken (raptured).

According to an Egyptian tradition they also had a revelation of the coming judgment three-hundred years before the Flood, or in 2656 B.C.. Thus, we have three known recorded revelations of the coming Flood through the Bible and extant history (There may be others - Therefore, the Flood had to be subsequent to the above dates).

B.C.

Enoch's revelation   - 3325   (969 years before the Noahic Flood)

Egyptian revelation   - 2656   (300 years before the Noahic Flood)

Noahic revelation   - 2476   (120 years before the Noahic Flood)

Therefore, we conclude that the full revelation of the coming judgment of the Noahic Flood was between 3325 B.C. to 2476 B.C.. The last 120 years (2476 - 2356 B.C.) Noah warned the whole world by personally building the Ark before their eyes as the last witness. We, today, are also forewarned that "as it was in the days of Noah so will it be in the days of the coming of the Son of man". The total span of time of God's Flood warning included the time from the First Egyptian Dynasty through the Seventh Dynasty (the whole Pre-Flood monarchal period). The Seventh Egyptian Dynasty, according to Manetho, had 70 rulers in as many days, which pictured the chaos in Egypt just before the Noahic Flood. It was during the Third Egyptian Dynasty, about 2815 B.C. (Carbon 14 date + - 2770 B.C.), under the leadership of their King Zoser (Djoser - in his days - Day Star - the Sothic Cycle Calendar was begun by Imhotep between 2785 - 2773 B.C.) when the first Pyramid was built.

1. "The Book of Enoch", page 37

It is coincidential and astonishing that there are historical records which inform us that these Egyptian Kings were building these mighty pinnacles (Pyramids) to protect their dead bodies from the coming judgment (Noahic Flood) and world destruction. There is every indication that they succeeded through the providential protection by means of tombs and Pyramids, thus preservering their bodies through the coming Noahic Flood as a witness from antiquity. Even though the Flood brought catatrophic changes in some parts of the world, God could have providentially protected Egypt and many other places in the Fertile Crescent as a witness of pre-Flood history. We should recall that the dove brought back an olive leaf from a tree located some where in the Fertile Crescent which indicates that the tree survived the Noahic Flood intact (Gen. 8:11).

There may have been a general knowledge of the coming Flood either by message or personal revelation amongst all nations. Certainly, news of Noah's witness as a ship builder in the midst of a wilderness would have spread far and wide during 120 years of waiting. Mankind was certainly without excuse. The following Arab accounts were taken from the appendix to Vyse and Perrings great work, "The Pyramids of Gizah" and were quoted by L. Cottrell in his book "The Mountains of Pharaoh". They, in turn, were found in the Bodleian Library. The Arab account of Abou Ma'sher Ja'fer Ben Mohamed Balkhi who wrote in 894 A.D., stated the following concerning the Pyramid builders of Egypt:

"The wise men, previous to the Flood, foreseeing an impending judgment from heaven either by submersion or by fire, which would destroy every created thing, built upon the tops of the mountains and in Upper Egypt many Pyramids of stone, in order to have some refuge against the approaching calamity. Two of these buildings exceeded the height, being four hundred cubits high, and as many broad and as many long. The length and breadth of each stone was from eight to ten cubits square; and they were as well put together that the joints were scarcely perceptible . . .".[1]

The Arab account of Masoudi, 967 A.D., states according to Coptic tradition that the revelation came "three hundred years" before the Noahic Flood in the year 2656 (our date), and that he was of the sixth generation (Jered was also of the sixth generation - 3552 - 2590 B.C.; therefore, the 300th year - 2656 B.C. - would have fallen during the fourth Egyptian Dynasty shortly after Mycerimus, pictured on page 173, a Pyramid builder):

"That Surid, Ben Shaluk, Ben Sermuni, Ben Termidum, Ben Tedresan, Ben Sal, one of the kings of Egypt **before the Flood**, built two great Pyramids; and, notwithstanding they were subsequently named after a person called Sheddad Ben Ad (our chronology lists either King Dadef-Ra or Shepses-Kaf - page 203), that they were not built by the Adites, who could not conquer Egypt, on account of their powers, which the Egyptians possessed by means of enchantment; that the reason for building the Pyramids was the following dream, which happened to Surid **three hundred years prevous to the Flood.** It appeared to him that the **earth was overthrown,** and that the inhabitants were laid prostrate upon it, that the stars wandered confusedly from their courses, and clashed to-

1. Cottrell, L., "The Mountains of Pharaoh", page 62, **Trinity Press 1956** London.

165

gether with tremendous noise. The king though greatly affected by this vision, did not disclose it to any person,[1] but was conscious that **some great event** was about to take place . . .".

If these above quotations carry any weight and since the carbon 14 dating (of the Pyramids) fall around 2815 - 2615 B.C., then our conclusion would be that the Noahic Flood took place after the above events. Therefore, the Biblical chronological date of 2356/5 B.C. for "Noah's Flood" is certain.

On page 162 we have the pre-Flood patriarchs listed according to the Biblical, Assyrian, Babylonian, and the Egyptian data. The sixth name of each list seem to have the same connotation which implies corruption. The number "six" also speaks of the "works of man" rather than God, and also declares that the "sixth generation" became defiled by their inter-marriage and corrupt worship (fertility cult) during the days of Jered according to Enoch.[2] The sixth generation is listed as follows:

	Biblical	Assyrian	Babylonian	Egyptian
6th Generation:	Jered	Dumuzi	Daonos	Set(Seth - his death)
	("go down")	(Tammuz)		(Inter-marriage with Sethites)

Historians believe that Dummuzi is closely related phonetically to the word "Tammuz", which was the name of one of the defiled objects of religious worship also of the Canaanites (fertility cult). During the post-Flood period it is known that Nimrod (meaning "Rebel") revived that same evil worship and went so far as to call one of his sons "Tammuz" according to tradition. Since the Babylonian name "Daonos" is listed as the sixth and seems to be closely related to the words Dumuzi or Tammuz, it may have the same connotation. Last of all "Jered" means "to go down", which indicates spiritual declension through the inter-marriage of the Sethites (Godly line) with the Cainites(ungodly line) during the days of Jered.

The sixth generation god of the Egyptians is listed as "Set" or "Seth". Perhaps the worship of this Egyptian god emphasizes the inter-mingling of the Cainite line with the Sethite line. Another strange coincidence is that Seth, according to our chronology, died in 2970 B.C. during the days of Jered (3552 - 2590 B.C.), and during the reign of King Perabsen (2980 - 2956 B.C.), and the ancestral worship of "Seth" was prevalent during Perabsen's reign (see page 212).

Among the Egyptian gods listed on pages 203 and 211, one finds that each of the god's names have a special meaning and correlates in some way with the first 10 characters in the Biblical genealogy. This is quite significant. This implies ancestral worship which is heathenistic.

In the providence of God, He from the very first, has hidden truth in His Word in symbolic language. An example of this is found in Malachi 4:2: "But .unto you that fear my name shall the **Sun** (Son) of Righteousness arise (appear) with healing in His wings" (for those who get under His atoning covering). Therefore, the "Sun" is an object of symbolism picturing the coming Messiah,"Son of Righteousness", who is Jesus Christ

1. Cottrell, L., "The Mountains of Pharaoh", page 62, **Trinity Press**,Lond.
2. "The Book of Enoch", page 34,35

our Lord and Saviour, and the Son of God (the Light of the world).

Therefore, because of these hidden meanings, the polytheistic heathen worshippers, because of lack of the revelation of the Spirit of the one true God and a lack of heart repentance, would misinterpret and would, literally, worship the "Sun" falsely. The first three character gods in the Egyptian genealogy also symbolize the one true triune Hebrew God of Abraham (Father), Isaac (offered son), and Jacob (Spirit blessed), as Ptah (Father - Adam); Ra or Re (Son - Sun - Seth the son of Adam); Shu (Spirit - Enos Spirit worship). Therefore, as "Seth" is a chosen son of the first Adam, he is a type of "Christ" the Son of Righteousness, who is, in turn, Biblically portrayed as the "Sun of Righteousness" (Malachi 4:2), therefore the Spiritual Light of the world. The heathen think to capitalize on this wrongly through ancestral worship as in the days of the Egyptian King Perabsen (Seth died - son of Adam - Sun worship); in the days of Rameses II (Joseph's, son of Seth, famine - deliverance brought by Seth the Sun god), and then later in the days of Seti I, II etc. (400 year Tanus Stella) as mentioned earlier.

In contro-distinction, a revival of true faith came during the days of Seth the chosen son of Adam. According to Gen. 4:26, the revival began after Enos was born to Seth in the year 3777 B.C.. The first three patriarchs (Adam, Seth, Enos) lived well beyond the years of the births of both Enoch and Methuselah, and the latter two had a spiritual revelation of the coming world judgment of the Noahic Flood. Thus, the spiritual instructions from Adam, who had walked and talked with God (Gen. 3:8), and the spiritual revelation were passed on by both Enoch and Methuselah.

Methuselah, the eighth generation, was already 243 years old, and Lamech, the ninth generation, was 56 years old when Adam died (3082 B.C.). Adam's death was well within the pre-Flood monarchal period of both the Babylonian and Egyptian Dynasties. Lamech, the father of Noah, died five years before the Noahic Flood. Noah was 84 years old when Enos died, during whose life the revival began. Therefore, Noah carried the first hand revelation of the knowledge of the one true God from the time of the third generation after Adam (time of Enos) on to 2006 B.C. (Noah died), or to within three years of the birth of Abraham (2003 B.C.). In other words there was no excuse for a lack of spiritual source of revelation of God during this whole period. Man was inexcusable for his downhill trend of corruption, which soon led to world destruction. This was a plain rejection by mankind in the face of a continual spiritual revelation and Godly warning.

In the midst of all this, God chose Enoch and gave him a personal revelation of the coming world judgment of the Noahic Flood. Enoch was born in the year 3390 B.C. in the 622nd year of Adam. When Enoch was 65 years old (687th year of Adam), God blessed him with the birth of a son whom he called "Methuselah". On the day of the birth of this son (3325 B.C.), Enoch had a revelation from God concerning God's coming judgment, which would fall on the whole corrupt world. God revealed to him that when his son, that he now held in his arms, died there would be a world wide destruction. Therefore, Enoch named his son "Methuselah", which means "man of the dart", "man of death", or "man of destruction". In other words, when Methuselah died destruction would come on the whole world. Chronologically, it was fulfilled to the letter, since he died the very year in which the Flood came (2356 B.C.). Adam was 687 years old when Methuselah was born, and Methuselah lived 969 years (Gen. 5:25-27). The total of 687 years plus 969 years equals 1656 years which spans the pre-Flood period perfectly.

"Enoch walked with God after he begat Methuselah three hundred years . . ." Gen. 5:22.

It is apparent from this scripture that the revelation he had put a Godly reverental fear in his heart, so that he walked with God 300 years following this tragic revelation.

Gen. 5:24 continues, "And Enoch walked with God: and he was not; for God took him" (Old Testament Rapture or a type of Christ's present Bride).

Enoch's walk was so close to the Lord so that God decided to translate him and take him to heaven alive. This clearly typifies the coming "Rapture" of the true church, or Christ's Bride, which will take place at the end of this present age (Dispensation of Grace - Rom. 11:25). As Enoch held his child in his arms he never knew what hour, day, or year he would die. Thus, the coming destruction was always "imminent" to Enoch, as the coming of the Lord Jesus Christ is to the true Bride of Christ today. For "by faith Enoch was translated that he should not see death; and was not found, because God had translated him: for before his translation he had this testimony that he pleased God" Heb. 11:5. Enoch's translation took place in 3025 B.C. just 57 years after Adam's death, and in the midst of the thriving monarchal rule and corrupt city life of the early dynasties which flourished in all parts of the rich Fertile Crescent from the Persian gulf into Egypt.

We should recall, at this point, that the Cainites (Proto-Irano) were re-infiltrating the Fertile Crescent at this time. As we stated before, they brought with them the tomtom music and the organ and the harp in the person of Jubal (Gen. 4:21); the knowledge of metallurgy (industry) and weapons of warfare in the person of Tubal-Cain, the artificer in brass and iron (Gen. 4:22); the beauty queen of the daughters of men in the person of Naamah (beautiful) for "they were fair" to look upon (Gen.4:22; 6:2); the fugitive and wandering spirit of Cain (Gen. 4:12,20); the Cain spirit of nakedness that was evident in Ham (Gen. 9:22); the spirit of polygamy, bloodshed, and debauchery in the person of Lamech (Gen. 4:19,23), and finally the spirit of the degrading life of the city (Gen. 4:17). All these influences were penetrating and permeating the quiet, Godly way of life of the Semitic peoples by inter-marriage, and the fruit of such a choice would soon be manifested in complete destruction, as it will during our last generation of our present age.

Enoch's walk with God, with the revelation of coming judgment, centred during the time of the life of Jered (See our New Bible Chronology Chart). He was a testimony for God in the face of the God forbidden inter-mingling of the races. The dramatic warning in Enoch's message lashed out at the corrupt generation: ". . . Behold the Lord cometh with ten thousands of his saints, to execute judgment upon all and to convince all that are ungodly among them of all their ungodly deeds which they have ungodly committed, and of all their hard speeches which ungodly sinners have spoken against Him . . ." Jude 14,15.

Enoch's message was heard by the early Sumerian civilization; by the Early Dynasty I of Babylon; and by the First Egyptian Dynasty.

Noah carried on the message of Salvation and coming judgment during the time of the Babylonian Early Dynasties II and III, and also during the time of the Second to the Seventh Egyptian Dynasties, during which time they began to build Pyramids as the means of protection of their dead bodies from coming destruction. Noah was the chosen vessel of God and a preacher of righteousness to those last generations:

"And (God) spared not the old world, but saved Noah the eighth person (in the Ark), a preacher of righteousness, bringing in the Flood upon the world of the ungodly . . . The Lord knoweth how to deliver the Godly out of temptations, and to reserve the unjust unto the day of judgment to be punished. But chiefly them that walk after the flesh in the lust of uncleanness, and despise government. Presumptuous are they, self willed, they are not afraid to speak evil of dignities" II Pet.2:5,9,10.

Genesis chapter six gives us a resume of the conditions that existed during that final phase just before the Flood. Rev. A.W. Rainsbury in his message "The Day of Noah", states concerning the pre-Flood wickedness:

"Whether they were Sethites or Cainites, believers or unbelievers they cared not 'they took them wives of all which they chose'.

The result was disastrous abandoned by restraining Spirit of God, (v. 3), their children became monsters of evil. Verse four tells us, 'The Nephilim were on the earth in those days, and also afterwards, when the sons of God came unto the daughters of men, and they bore children of them'. The verse concludes, 'They were heroes, the men of renown'. In other words they glamorised crime just as T.V.. The word 'Nephilim' means 'those who fell upon others; brigands, thugs, tyrants'. The verse concludes, 'They were the heroes, the men of renown'".

The Word of God continues in Gen. 6:7-14: "And the Lord said, I will destroy man whom I have created from the face of the earth; both man, and beast, and the creeping things, and the fowls of the air; for it repenteth me that I have made them. But Noah found grace in the eyes of the Lord. . . . Noah was a just man and perfect in his generations, and Noah walked with God. . . . **The earth also was corrupt before God,** and the earth **was filled with violence.** And God looked upon the earth, and, behold, **it was corrupt; for all flesh had corrupted his way upon the earth.** And God said unto Noah, **The end of all flesh is come before me; for the earth is filled with violence** through them; and, behold, I will destroy them with the earth. Make thee an Ark of gopher wood; rooms shalt thou make in the Ark, and shalt pitch it within and without with pitch" Gen. 6:7-14.

"And the Lord said, My Spirit shall not always strive with man, for that he also is flesh: yet his days shall be an **hundred and twenty years** (2476 - 2356 B.,C.)" Gen. 6:3 (See New Bible Chronology Chart).

1. Rainsbury, A.W., "The Day of Noah", Keswick Week Manual,1970,p.32

Thus, you can readily see why we accept the dates and chronology of the Massoretic text without qualification. It is exactly 1656 years from Adam's expulsion from the Garden of Eden to the Flood, and the Bible correlates with history perfectly.

Philip Mauro, in his book "The Chronology of the Bible", says, "Thus, if the figures given in the Hebrew text of the Old Testament are accepted as correct, there is no possibility of arriving at any other conclusion then that the period of time from the creation (expulsion - added) of Adam to the Flood (whereby his entire posterity then living, with the exception of the Family of Noah, were wiped out) is exactly 1656 years. As to this there is perfect agreement among all chronologists who accept as correct the Hebrew text of the Old Testament".[1]

CHAOS IN EGYPT HARMONIZES WITH PRE-FLOOD DAYS

Since we have no detailed extant historical information from the Babylonian and Assyrian civilizations dating during Noah's time, we must turn to the contemporary history of Egypt. Noah was born in 2956 B.C. and lived to a ripe old age of 950 years. Therefore, he died in 2006 B.C. only three years before the birth of Abraham (Gen. 5:32; 9:28,29; 7:6; 11:10). Noah was 600 years old when the Flood began (Gen. 7:11) and 601 years old when it ended (Gen. 8:13). Therefore, Noah was born 319 years after the first Egyptian Dynasty gained power under King Narmer (Menes - 3270 B.C.), and probably shortly after the days of King Perabsen during the Second Dynasty. The Carbon dating of the first Dynasty was found to be plus, minus 3250 B.C. (See Chart).

The Third Egyptian Dynasty, the first Pyramid builders, begins, according to the Biblical chronology, in the year 2818 B.C. for the accession of their first ruler, Sanekht. There were four rulers in the Third Egyptian Dynasty, and we have allowed a total of 60 years for that period. This allows an average of 15 years for each king. The first King Sanekht ruled 28 years and the following King Zoser reigned 19 years. Therefore, Zoser (Djoser) would have reigned from 2790 to 2771 B.C.. He was one of the first Pyramid builders and built the famous "Step-Pyramid". He also constructed ships 160 feet long in which they sailed to Lebanon in a flourishing Mediterranean trade. It is rather interesting that our Biblical chronological date for Zoser's reign includes the exact years of the possible "Sothic Cycle" (2785 - 2773 B.C.), which is, in turn, known, historically, to have been during Zoser's reign (Quote below). Once again this verifies our Biblical chronological data to the letter.

Quoting from the book "Ancient Egypt" page 137, 138, we have the following information:

"He noticed that the bright Star Sothis (Sirius, the Dog Star) appeared on the eastern horizon on this particular day at the same moment as the sun. . . . Modern research established that a Sothic cycle would have commenced between 1325 and 1322 B.C., during the 19th Dynasty. This fact is confirmed by ancient scribal records. Therefore, the previous cycle would have begun in the Pyramid Age between 2785 and 2782 B.C.

1. Mauro, Philip, "The Chronology of the Bible", page 22

(Gardiner has suggested as late as 2773 B.C.[1]) . . .There are also excellent reasons for thinking, however, that the calendar was not in fact
introduced until the reign of Zoser in the 3rd Dynasty, perhaps by
Imhotep . ."[2] (Our chronology correlates the two since Zoser reigns from
2790 to 2771 B.C.).

This is very important and coincidental having the "Sothic Cycle" fall
exactly in the reign of Zoser, since we have simply correlated the
average length of all the first seven Egyptian Dynasties from various
authors and then allowed the seventh Egyptian Dynasty to end with the
Biblical date of the Noahic Flood. Above and beyond this confirmation,
the carbon 14 dating of the Third Dynasty is plus, minus 2770 B.C.,
which also correlates perfectly. With these verifications, we confident-
ly proceed knowing that by following the Biblical chronology we are on
the right path.

The following Egyptian King, Snofru, no doubt reigned about the time of
the death of Cainan (2777 B.C.), when Noah was 180 years of age (2776
B.C.). Snofru imitated the previous King, Zoser, by building several
Pyramids but filled in the terraced stages and smoothing the slopes,
resulting in the first real Pyramid built at Meydum (tentatively ascrib-
ed to Huny). Another two were built at Dahshur, four miles south of
Sakkara. A black diorite tablet now in the museum at Palermo tells of
some of his exploits. In those days Lebanon was covered with dense cedar
to furnish the interior of their large stone buildings and for ships.
Egypt traded with Nubia where she obtained spices and gold; with ports
along the Palestine coast; with Byblos where she traded for leather and
dyes; with Crete for painted vases, and for the silver from Taurus. King
Snofru said, "We brought forty ships laden with cedar trunks: We built
ships of cedarwood: one of meru-wood, two ships 150 feet long. We made
the doors of the King's palace of cedarwood".[3]

Snofru had successful campaigns into both Nubia and Libya where he captured 7000 slaves (Nubian). They were used as slave labor for his energetic Pyramid building construction, while the 200,000 cattle taken were
needed to feed the scores of workers laboring on his extensive projects.

During the fourth Dynasty (2760 - 2640 B.C.), Cheops (Khufu), the second
King, followed in the foot steps of the Third Dynasty by excelling them
all in building the grandest monument of the ages known as the "Great
Pyramid" (Noah was about 240 years old - 2716 B.C.). The base covers 13
acres, and is 768 feet square (now 750), and 482 feet high (now 450).
It is estimated to have contained, 2,300,000 stones of an average thickness of three feet each, and an average weight of two and one half tons.
Its outer surface consisted of successive coats of roughhewn blocks of
limestone, of which the outer coat was made smooth by exquisitely carved
and close-fitting blocks of granite.[4]

The following two kings, Cephren (Khafra) and Mycerinus completed the

1. Gardiner, Sir Alan, "Egypt of the Pharaohs"p.65,249,Oxford Univ.Press
2. Kees, Hermann, "Ancient Egypt", page 137,138, Faber & Faber Ltd
3. Keller, Werner, "The Bible as History"p.73,74, Econ Verlagsgruppe,Ger.
4. Halley, "Pocket Bible Handbook", page 93, 1948

three architechural giants at Giza. At the Boston Museum we have a beautiful slate statue of Mycerinus and his queen (page 173). They reigned during the latter days of Jered's life (Noah was around 275 years old).

As the following Fifth Egyptian Dynasty came to a close, Noah was around 450 years of age and his God given revelation to build the Ark came in his 480th year (2476 B.C.), just 120 years before the Noahic Deluge. Noah's building the Ark inland for 120 years was a spectacular witness to the dying, corrupt world. Noah began his work on the Ark either near or during the days of Pepy II's reign in Egypt (6th Dynasty), during the latter part of the Sumerian period, and during the final phase of the Early Babylonian Dynasty III.

As Noah built the Ark for 120 years as a last testimony and witness to a judged generation, real evidence of world wide chaos and wickedness was clearly being manifested and revealed by a certain Egyptian writer in his "**Admonition of the Egyptian Sage**", who recalled or wrote down concerning the disastrous situation which took place in Egypt during the close of the Sixth (the reign of Pepy II) and Seventh Dynasties. This dramatic picture of the contemporary Egyptian chaos during the pre-Flood period is found in an extremely tattered papyrus in the Leyden Museum collection dated no earlier than the Dynasty XIX.[1] The papyrus relates a torrid story of conditions during the latter years of the reign of Pepy II. These conditions continued until they were ended by the World Wide Noahic Flood.

Even during Pepy I's reign interference in Egypt from Bedouin tribes from Palestine was a menace. The first inter-racial united nations force under the official "**Uni**" were gathered to bring them to heel. Mr. Uni reports, "His majesty made war on the desert peoples and his majesty gathered an army: in the south . . and among the Jertet-Mazoi - and Nubians. I was entrusted with the whole campaign.[2] Perhaps we could be a bit terse by interpolating and saying: "As it was in the days of 'Uni' so will it be in the days of the 'United Nations'". Consequently the following chaos will also follow in both situations:

EGYPTIAN PRE-FLOOD CHAOS
(Just before 2356 B.C.)
"ADMONITION OF THE EGYPTIAN SAGE"

"In the course of the early part of the 23rd century B.C. at the close of **Pepy II longest recorded reign of ninety years,** the power of the pharaohs was reduced by the internal weakness. Theban and Heracleopolitan rulers and governors fought for supremacy. Central administration finally collapsed completely so hereditary landowners ruled in their separate districts by means of miniature courts. Every man did that which was right in his own eyes. Local governors began to fight and usurp one another's possessions. People gathered to the most powerful man whether he was just or not. Temple worship stopped, respect for their Egyptian gods vanished, crafts ceased, and temples were robbed. Poverty and misery was rife everywhere. Soldiers began to plunder rather than defend. Therefore, Nubian, Palestinian, Libyan hordes sweep

1. Gardiner, Sir Alan, "Egypt of the Pharaohs" p. 109, Oxford Univ. Press
2. Keller, Werner, "The Bible as History", page 741, Econ Verlagsgruppe

172

MYCERINUS· AND HIS QUEEN
Slate statue in the Boston Museum

across the land and things began to topple. The popular stratum of people rose in a terrifying revolution. Ancient statues in the pyramid temples were overturned and broken. Graves were robbed and buildings were stripped of every precious thing. Nothing was holy anymore. Horrible and dreadful chaos followed". The Egyptian Sage continues (Remember this happened in Egypt while Noah was finishing the Ark):

"The land turns around as does a potter's wheel . . all female servants are free with their tongues. When their mistress speaks it is irksome to the servants . . . Great ladies, who were mistresses of goodly things give their children in exchange for beds. The children of princes are dashed against the walls . . . The poor man is full of joy. Every town says, 'Let us suppress the powerful among us'. . . . He who once possessed no property is now a man of wealth. The poor of the land have become rich. He who had no dependants is now a lord of serfs . . He who once possessed them looks at them, but they are not his . . . The desert-dwellers are throughout the land. The provinces are laid waste . . Behold the delta is in the hands of those who know it not . . Men sit behind the bushes until the benighted traveller comes in order to plunder his load. . . . Everyone thinks only of violence. Pestilence ravages the land. Blood flows every where. Masses of corpses are thrown into the river. The river has become the usual burial place. . . Behold the river is full of blood. When you seek to drink from it, you recoil. . . . The crocodiles in the river are sated with the dead which they have to eat. Willingly people cast themselves to them. . . Now gold and lapis lazuli, silver and malichite, carnelian, and precious jewels hang about the necks of slave girls. But distinguished ladies go dressed in rags. Mothers of households sigh: 'If we only had something to eat'. . . . One eats only vegetables and washes them down with water. One takes food from the mouths of swine. . . . High and low alike say: 'If only I might die'. Small children cry: 'Would that I had never been born'. . . . All warehouses are plundered and the guards are slaughtered. . . . The legal texts from the courts of justice are cast into the front hall. Men trample them in the streets. . . . Behold the flame leaps high. . . . The land weeps. Everything lies in ruins. . . .It is said of god Re, that he is the herdsman of all men: there is no evil in his heart. When his herds are few he passes the day to gather them together. . . But would that he had perceived men's nature in the first generation (1656 years before this - added), then he would have suppressed evil; he would have stretched forth his hand against it. But men desire to give birth, and so sadness grew up and needy people on every side. . . . There is no pilot (no god). Where is he today? Doth he sleep perchance? Behold, his might is not seen".

Thus, the **"Admonitions of an Egyptian Sage"** ends, and direct information from Egypt ceases completely.

Noah has now put the finishing touches to the great Ark built far from any ocean. People laughed him to scorn and joked about his ship so far from the sea. They berated his message of coming judgment and called him a fanatic, but the news of his building the ship for the past 120 years had traveled to the ends of the earth.

Even the animals showed more sense than that last generation, as they gathered to be selected to enter the Ark of safety. God spoke to Noah, "Come thou and all thy house into the Ark: for thee have I seen righteous before me in his generation. . . . For yet 'seven days', and I will cause it to rain upon the earth forty days and forty nights; and every living substance that I have made will I destroy from off the face of the earth. And Noah did according unto all that the Lord commanded him" (Gen. 7:1,4,5 - **Seven years** before the Battle of Armageddon God will call His Bride upward into the Ark of heavenly places - "**Rapture**").

Suddenly, it was reported that Methuselah had finally died after 969 years of God's extended grace (The oldest man in the world). This must be a tragic sign. It had been rumored that world judgment would come at the time of Methuselah's death! In fact, they had heard that this was why Methuselah's father, Enoch, had reverently feared God and had walked with God for 300 years, and suddenly he was not, because God took him (**Raptured**). Noah and his family were actually going into the Ark! Would it truly happen in **seven days**? Would the Lord Jesus Christ really return in **seven years** after the **seven year covenant**?

In Egypt the Sixth Dynasty had ended in complete chaos, and the Seventh Dynasty had just begun. Could someone solve the problems and bring peace? But alas, everyone seemed to have failed. The last report from Egypt stated that they had had **seventy rulers in as many days**. Then it happened! Noah was right after all.

". . . In the six hundredth year of Noah's life (the 969th year of Methuselah), in the second month, the seventeenth day of the month, the same day (2356 B.C.) were all the fountains of the great deep broken up, and the windows of heaven were opened. And the rain was upon the earth forty days and forty nights" Gen. 7:11,12.

During the late Sumerian and pre-Noahic Flood period, the cities of Ur, Kish, Umar, and Lagash were leading commercial centres with Uruk (Erech) as their capital. The last Sumerian King Lugal-Zaggisi began his reign from Uruk in 2381 B.C., ruling 25 years, and he died in 2356 B.C. during the Noahic Flood. History describes the disaster with these simple words: "Uruk (Erech) was smitten with weapons (God's weapons - Ezekiel 14:21); the Kingdom was carried to Agade". This reference to Agade refers to Sargon's (Ham's) tribal rule which began 40 years after the Noahic Flood (2315 B.C.).

The Chinese Emperor, the fourth and last pre-Flood King (Laou), began to reign in 2357 B.C., according to the Westminister Historical Atlas Map, 1945, and it is stated that in his days the "Flood" came. All races, the Sumerians along with all contemporary civilizations, completely disappeared in in one year (2356 B.C.). Therefore, historians state that for 200 years thereafter "History is almost silent". Monarchal rule disappeared and nomadic life soon re-appeared, and no wonder!? The Cambridge Ancient History, page 144, I, Part 2 states that Lugal-zaggisi's reign ". . with utmost distinctness ends an age".

How long dost thou halt between two opinions? Chose you this day whom ye shall serve, before it is everlasting too late.

"But **as the days of Noah were, so shall also the coming of the Son of man be**" Matthew 24:37.

"And **woe** unto them that are with child, and to them that give suck in those days. But pray ye that your flight be not in the winter, neither on the sabbath day: For then shall be 'Great Tribulation', such was not since the beginning of the world to this time, no, nor ever shall be, And except those days should be shortened (limited) there should no flesh be saved: but for the elect's sake (out of the Tribulation) those days shall be shortened" Matt. 24:19-22.

"For as in the days that were before the Flood they were **eating and drinking, marrying and giving in marriage,** until the day that Noah entered into the Ark, and knew not until the Flood came, and **took them all away; so shall also the coming of the Son of Man be. . .Watch therefore; for ye know not what hour your Lord doth come.** . . . Therefore, be ye also ready: for in such an hour as ye think not the Son of Man cometh. . . . But as the days of Noah were, so shall also the coming of the Son of Man be . . . There shall be weeping and gnashing of teeth" Matthew 24:38,39,42,44,51. (As Noah had seven days warning before the Noahic World Wide Flood came, so will mankind, in these present last days of this age, have seven years warning before the **"Great Tribulation"** ends. This gives them their last chance to repent before Jesus Christ returns to judge, personally by God's Word, all mankind at the Bema Seat of Christ Jesus. These last seven years will be introduced with a false seven year covenant, by a false Jewish Messiah, who introduces a period of seven years of false peace, which turns, deceptively, into **"Jacob's Troubles"** (Jer. 30:7), or seven years of the **"Great Tribulation"** (Rev. 7:14). Be ye also ready for the Son of Man cometh when ye think not. Repent, for the Kingdom of God is at hand! .

- - - - - - -

THE DATE OF THE BIRTH OF CHRIST

Christ's birth took place during Herod's reign (37 - 4 B.C. - Matthew 2:3). The question is, "What year in Herod's reign was Christ Jesus born"?

First, let us consider information gleaned from extant history clarifying the general attitude of the people as to their general expectations, thoughts, and aspirations concerning the expected Jewish Messiah during the early part of Herod's reign (34 B.C.).

"An interesting sidelight on the situation is afforded by a passage in the Slavonic version of Josephus's translation in the Loeb Classical Library (Josephus, Vol. II, pp. 636-8). Part of the passage, inserted in Book I of the Jewish War, reads as follows:

But Herod spent little (time) in Jerusalem, and marched against the Arabs. At that time the priests mourned and grieved one to another in secret. They durst not (do so openly for fear of) Herod and his friends.

For (one Jonathan) spake: 'The law bids us have no foreigner for king. Yet we wait for the **Anointed (Messiah)**, the meek one, of David's line. But of Herod we know that he is an Arabian, uncircumcised. The **Anointed (Messiah)** will be called meek, but this (is) he who has filled our whole land with blood. Under the **Anointed (Messiah)** it was ordained for the lame to walk, and the blind to see, (and) the poor to become rich. But under this man, the whole have become lame, the seeing are blinded, the rich have become beggars. What is this, or how? Have the prophets lied? The prophets have written that there shall not want a ruler from Judah, until He comes unto whom it is given up; for Him do the Gentiles hope. But is this man the hope for the Gentiles? For we hate his misdeeds. Will the Gentiles perchance set their hopes on him? **Woe unto us,** because God has forsaken us, and we are forgotten of him! And He will give us over to **desolation and to destruction.** Not as under Nebuchadnezzar and Antiochus (is it)? For then were the prophets teachers also of the people, and they made promises concerning the captivity and the return. And now neither is there any whom one could ask, nor any with whom one could find comfort'.

But **Ananus the priest** answered and spake to them: 'I know all books. When Herod fought beneath the city wall, I had never a thought that God would permit him to rule over us. But now I understand that **our desolation is nigh.** And bethink you **of the prophecy of Daniel;** for he writes that **after the return of the city of Jerusalem shall stand for seventy weeks of years, which are 490 years, and after these shall it be desolate'.** And when they had counted the years, (they) were thirty years and four (34 years). But Jonathan answered and spake: 'The numbers of the years are even as we have said, but the Holy of Holies, where is He? For this Herod he (according to the prophet) can not be called the Holy One - (him) the blood-thirsty and impure'".

The historical value of this passage is very questionable. Yet it does communicate something of the atmosphere of the period, and illustrates how that Daniel's prophecy was regarded in informed circles in the reign of Herod. The debate, as is pointed out in the notes on the passage, is supposed to have taken place in 32 B.C., and if the 490 years were then considered to have thirty-four years to run, the terminal date would have been 3 A.D.. At any rate we can not be very far out if we conclude that the end of the days was supposed to have begun about the turn of the century".[1]

At this point (1986 B.C.) it could be stated that we have noticed a tremendous advance in knowledge during the last twenty years concerning the dating of ancient historical events, because of late archeological discoveries. Therefore, only now are we properly prepared, with knowledge at hand, to begin to put these final pieces together. Only now do we have a definite foundation to calculate the Seventy Weeks of Daniel in both the "Historic" and "Prophetic" sense.

"It has been said that the only way in which the credit of the Science of Chronology can be restored is to adhere strictly to the actual statements of the original text, and to deal with these statements in accordence with the laws of the Science of History, which places the criterion of credibility and the test of truth in the testimony of **witnesses** at once honest, capable and contemporary. The identification of **the dates of the** dedication of Solomon's Temple and the birth of Christ **with the** years An. Hom. 3000 and 4000 respectively must be jealously scrutinized and the facts must not be warped in order to bring about the exhibition of this result.[2] As you have noticed in the reading of this book, the correlation of thousands of facts has been based on the scriptural and historical facts, and are scientifically interwoven into this plan with no ulterior motive of making it fit to any pattern. In fact, the pieces continually fell so wonderfully together that we could only marvel and praise God for the perfection of His Holy Word. We had been determined from the start to be honest and accurate with facts or we would have thrown the whole work in the trash long ago. Yet, after completing the manuscript, we found that the dedication of Solomon's Temple fell exactly on the 3000th year since the expulsion of Adam from Eden. The temple was literally built the last seven years of the first 3000 year period. Naturally, we would then wonder whether there might be a possibility that the Messiah might have arrived, or have been born, in the 4000th year? That year, in our chronology, fell in the year 12 B.C., which is a Hebrew Jubilee year. A very appropriate time for the Messiah to have arrived, and also implied by Luke 2:1-5.

Literally, it seemed utterly impossible that there would be any information found, that early, indicating the Messiah's birth. Yet soon, we were greatly surprised and encouraged to look comprehensively into the matter, and to say with the Psalmst, "I have considered the days of old, the years of ancient times" Psa. 77:4. At this time we also noticed that Herod's Temple (18 B.C.) was begun exactly 1000 years after Solomon's (1018 B.C.). Solomon's Temple was completed seven years later (1012

1. Schonfield, Hugh J., "Secrets of the Dead Sea Scrolls", p. 41. A Perpetua Book, A.S. Barnes & Co., San Diego, CA. Reprinted by permission.
2. Anstey, M., "The Romance of Bible Chronology", Vol 1, p. 87.

B.C.). Could it be possible that 1000 years later the physical temple of Messiah's body would make its appearance (12 B.C.)?

As we began to search for evidence of Christ's birth dating around 12 B.C., we immediately found that there were indications that there were **two periods of census** in Palestine some 2000 years ago, as the following quotations might imply:

"The solution to this vexing problem (census) is that **Quirinius apparently was twice associated with the government of the province of Syria.** Sir William Ramsay accepted the inscriptural evidence contained in Titules Tiburtinus, construing the words 'iterum Syrian', i.e. **'a second time Syria'** to refer to Quirinius".[1]

"The **census of Acts 5:37** which was marked by the insurrection led by Judas of Galilee, was held **in A.D. 6.** In that year Judaea was incorporated into the Roman provincial system, and a **census** was held in order to assess the amount of tribute which the new province should pay to the imperial exchequer. **The census was conducted by P. Sulpicius Quirinius,** at that time imperial legate of Syria. The suggestion that Israel should pay tribute to a pagan overlord was deemed intolerable by Judas, and by the party of the Zealots (q.v.), whose formation is to be dated from this time.

"**The census of Luke 2:1 ff., in the course of which Christ was born in Bethlehem, raises a number of problems.** It is, however, widely agreed: (1) that such a census as Luke describes could have taken place in Judaea **towards the end of Herod's reign** (37 - 4 B.C.); (2) that it could have formed part of **an empire-wide enrolment, as Luke 2:3 states** (According to our computation **12 B.C. was a 'Jubilee' year** when, after fifty years, all property was returned to the original Jewish owner. This could be the appropriate reason for the Romans taking their census at this time - added). . . . There is **evidence of census activity in various parts of the Roman Empire between 11 and 8 B.C..** . . . This suggests that in Luke 2:2 the enrolment made at the time of Jesus' birth has been confused with this later and better known enrolment. It is possible, however, **that Quirinius governed Syria from 12 B.C.** to the coming of Titius as its governor in 9 B.C. (So, e.g., Marsh, Founding of the Roman Empire, p. 246, n.1), and **that Augustus decided to make an enrolment after consultation with Herod when the latter visited him in 12 B.C..** . . Ramsay has adduced additional inscriptural evidence **that Quirinius commanded the Homanadensian campaign as legate of Syria between 12 and 6 B.C..** . . .Halley's comet seen in 12 B.C. (This date is incorrect. It has been now proven that it arrived **Aug. 25, 11 B.C., which coordinates with scriptural facts better** - added), was a brilliant spectacle well fitted to be the harbinger of Him who was to be the Light of the World. The Italian astronomer Argentieri's conclusion that this comet was the star of the Magi rests on two questionable assumptions, that Jesus was born on a Sunday and that He was born on the 25th of December".[2]

It does seem logical that the Romans might plan an enrolment during the

1. Unger, M.F., "Archeology and the Old Testament", p. 65, Zondervon P.
2. The New Bible Dictionary, page 203,223, InterVarsity Press, Leicester

179

year of the **Jewish Jubilee,** since there would have been extensive exchange of land at that time (**12 B.C.**). Therefore, **we could definitely say that Joseph and Mary were returning to the place of their forefathers to claim their possessions and also to be enrolled by the Romans in 12 B.C. (a Jubilee year).** Mary was also bringing in her womb the Messiah of Israel, Emmanuel, God with us, and the King of Kings and the Lord of Lords **who was appearing to claim His possession and Kingdom on this special Jubilee year.** No wonder ⌐the angels shouted in "Jubilee" and said, "Fear not; for behold, I bring you good tidings of great joy, which shall be to all people. For unto you is born this day in the city of David a Saviour, which is Christ the Lord" Luke 2:10,11. This "Jubilee" was exactly 4000 years after Adam's "Jubilee disaster" in Eden in the year 4012 B.C.(Herod's New Port at Caesarea dedicated in 13/12 BC).

The above statement, that Herod was in Rome during the year 12 B.C., would give the possible reason why he did not personally know the date of Christ's birth; therefore, being away at Rome, he questioned the Magis, after returning, about Christ's birth, and ". . . enquired of them diligently what time the star appeared (first in Persia 63 days before - added). And he sent them to Bethlehem, and said, Go and search diligently for the young child; and when ye have found Him, bring me word again, that I may come and worship Him also. When they had heard the king, they departed; and, lo, **the star** which they saw in the east, went before them (Halley's comet), till it came and stood over where the young child was. When they saw the star, they rejoiced with exceeding great joy. . . .Then Herod, when he saw that he was mocked of the wise men, was exceeding wroth, and sent forth, and slew all the children that were in Bethlehem, and in all the coasts thereof, **from two years old and under** (**The birth of Christ, no** doubt, **took place at least a year or more before the appearance of the star**, which was inferred by their considerations, and since the **wise men were following the star at the time Herod questioned them after he returned from Rome)** according to the time which he had diligently enquired of the wise men" Matthew 2:8b-10,16.

Therefore, since **Publius Sulpicius Quirinius was Consul at Rome from 12 B.C. to 9 B.C.,** and not long after his inauguration, conducted a campaign against the unruly Homanadensians of Central Asia Minor, and since Herod was in Rome in 12 B.C. considering Caesar Augustus' tax enrolment, we conclude that the first Caesar's Roman census, as the following scripture confirms, took place early during this period:

"In those days it occurred that a decree went out from Caesar Augustus that the whole Roman Empire should be registered. **This was the 'first' enrolment and it was made when Quirinius was (first) governor of Syria"** Luke 2:1,2 (Amplified Version).

Mr. D.I. Cole, former lecturer at the planetarium, in the monthly notes of the Astronomical Society of Southern Africa, Royal Observatory, Observatory Cape, adds to the testimony when he writes, "Halley's comet is due to appear again in 1986 A.D. (has appeared last year - added), for it keeps to a strict time schedule on its tremendous elliptical course through space. Below, we have an account by Chinese astronomers of its appearance **in 11 B.C.** with most precise details of its course. Their observations are recorded in the Wen-hien-thung-khao encyclopaedia of the Chinese scholar Ma Tuan-lin:

"In the first year of (the Emperor) Yen-yen, in the seventh month, on the day Sin-ouei **(25 August) a comet** was seen in the region of the sky known as Toung-tsing (beside the Mu of the Gemini). It passed over Ou-tschouiheou (Gemini), proceeded from the Ho-su (Castor and Pollux) in a northernly direction and then into the group of Thaiouei (tail of Leo) . . . On the 56th day it disappeared with the Blue Dragon (Scorpio). Altogether it was observed for 63 days".

Mr. Cole continues, "During the latter part of **August 11 B.C.** (as quoted above) Halley's comet appeared in the constellation of Gemini, a little north of Castor and Pollux. The latitude of Bethlehem is 31° 42' and it so happens that 2000 years ago Castor and Pollux were almost exactly 31° 42' north of what astronomers call the celestral equator. This means that Castor and Pollux in their daily journey across the night sky passed through the zenith of Bethlehem, and, since Halley's comet came very close to Castor and Pollux, it must also **at one time have stood directly over Bethlehem** - 'the place where the young child was'".[1]

It is said that the head of Halley's comet could have been seen at a certain time high in its zenith overhead, and its tail would have extended at times to the horizon (1986 appearance was very distant). This would have been a fantastic sight as a testimony of **the Star out of Jacob (Num. 24:17)**. The eleven sons of Jacob are identified as **eleven stars (Gen. 37:9,10)**, therefore, Joseph, who typified Christ, the Jewish Messiah, was the **12th Star.** Could it be possible that **Halley's comet which has just appeared in 1986 A.D., and also appeared after Christ's birth 1998 years ago, could be now heralding Christ's final coming, sometime after-wards, and soon completing a 6000 year period?**

Mr. Cole continues, "Now, although its seems unlikely then that Jesus was born around December, January, we can not say in which other month He was born. But since shepherds are more likely to be out with their flocks during the lambing season in spring, which extends over February, March, and April, these months seem more probable for the Nativity".

Therefore, if we chose March before Passover time, this would have made Christ's age approximately one and a half years when **Halley's Comet appeared on August 25, 11 B.C..** If we choose Jan. 6, 12 B.C., it would make Him a year and eight months old, and allow His conception by the Holy Ghost to have taken place 9 months before. This would put His conception before the Sabbatical year Passover 13 B.C.. If Christ was conceived before Passover 13 B.C., then during the Passover, after the end of the old Nisan year (Palm Sunday - April 6, 32/33 A.D.) plus 5 days later until the first day (11th) of the new Nisan year (Qurnran Calender's date of Passover was three days later on April 14, 33/34 A.D.), when Christ was crucified, Christ would have passed His 46th

1. Cole, D.I., Monthly notes of Astronomical Soc.,Observatory Cape, S.A.

Passover since conception. If Christ's death took place on his 46th Passover, then **"His resurrection"** falls during the first few months of His 46th year from conception.

These above details might well coordinate with John 2:20,21, "Then said the Jews **forty and six years was this temple in building,** and wilt thou rear it up in three days? But He spake **of the temple of His body"**. In 29 A.D., when this event happened, Herod's temple had been in the process of being built for **46 years (18 B.C. to 29 A.D. – one less year for the period 1 B.C. to 1 A.D. – equals 46 years),** but "Christ spake of the temple of His body". His body temple was raised again over three years later, and would not His body be also then 46 years old? Therefore, John 2:20,21 may, literally, in a hidden way, give His age at His death? At least since His conception.

During his final years of ministry the Jewish people said of Jesus in John 8:57, **"Thou art not yet 'fifty years' old,** and hast thou seen Abraham?" This clearly suggests, according to historians, that in the course of His ministry Jesus was in His forties. Also, according to Irenaeus, there was a tradition to that effect among the Asian elders".[1] Now, if Jesus was beginning His 46th year when he was resurrected, how about the difficulty found in Luke 2:23 where it says, "And Jesus Himself began to be **'about' thirty years of age,** being (as was supposed) the son of Joseph, which was the son of Heli?" This would then be interpreted as Christ being **"about"** 30 years **(29 years old plus, in 29 A.D.) old since His dedication in the temple at the age of 12 years (Luke 2:42) at Zero B.C.**. At the time of His dedication in the temple (Barmitzvah) He said to His parents, **"Wist ye not that I must be about my Father's business"** Luke 2:49. He had fulfilled this statement by "being about His Father's business" for "about 30 years" (Zero B.C. to 29 A.D.) since His Barmitzvah. He began to witness to the intellectuals in the temple at that point.

– – – – – – –

1. The New Bible Dictionary, page 223, InterVarsity Press, Leicester

If we accept the above statements as true concerning the birth of Christ Jesus, then we can apply a **"Historical Application"** (365 1/4 day year) to the 69 weeks of Daniel (Dan. 9:24,25). **Historically,** the command was given by Cyrus to rebuild the Jerusalem Temple in 539/8 B.C. (Ezra 1: 1,2; II Chron. 36:23). This was his accession year (his first **year** completed in 538/7 B.C.), therefore, the completion of the "second year" (537/6 B.C.), and the third year in 536/5 B.C., during which, they went to Jerusalem after three years of preparation under the Persian rule. The work on the Jerusalem Temple ceased, later (Ezra 4:24), for only a few months in the second year of the reign of Darius King of Persia (520 B.C.), but soon continued on until the Jerusalem Temple was completed in the sixth year of the reign of Darius the King (Ezra 4:24; 6:15). This event (Temple worship restored) would have providentially taken place in the year 516/5 B.C., or exactly 70 years since the destruction of Jerusalem.

There follows the above an interregnum period until the seventh year of Artaxerxes (Ezra 7:7,8 - 459/8 B.C.) when they began the period of beautification of the temple under Ezra until Artaxerxes' 20th year. This work continued for 13 years until Nehemiah came to rebuild the walls in 446/5 B.C. (Neh. 2:1). The following 12 years work on the walls began in the 20th year of Artaxerxes and continued until the city was dedicated. Nehemiah finished the work on the wall and the dedication of the City of Jerusalem in Artaxerxes' 32nd year, or after 25 years(inclusive would be 26 years). Thus, we have the following information (In the historic view, they are **not** consecutive years):

		B.C.	
1. Preparation for building of temple		(539/8 - 536/5)-	3 yrs
2. Building the temple (inclusive)		(535/4 - 516/5)-	20 yrs
3. Beautifying temple & building wall(inclusive)	(459/8 - 434/3)-		26 yrs

Jerusalem built in **"Troublous Times"** - 7 wks of Dan. 9:25 - Total 49 yrs

Now these above 49 years, though not consecutive, could represent the "seven weeks" of Daniel in the **Historical Application,** ". . . the street shall be built again, and the wall, even in troublous times" Dan. 9:25 (7 weeks of years times seven equal 49 years). This brings us to the 32nd year of Artaxerxes or to the year 434/3 B.C.. Now, after the **seven weeks,** we have **62 weeks** (Dan. 9:25 - 62 weeks of years times seven equal 434 years). Therefore, if we subtract 434 years from 434/3 B.C., when Nehemiah completed his work, we come to Zero B.C. when the **"Messiah the Prince"** (Dan. 9:25; Luke 2:42) was dedicated in the Jerusalem Temple at **12 years of age** (Zero B.C.). Is this "Historic Application" the only approach? No, there is also a "Prophetic Application" which also pinpoints the time when the **"Messiah the King"** was presented at Jerusalem riding a white donkey on Palm Sunday (Psalm 118:22-29; Dan. 9:26; Luke 19:28-46):

PROPHETIC APPLICATION TO THE 69 WEEKS OF DANIEL

The "**Prophetic Application**" to the 69 weeks of Daniel would be according to the late Dr. A.J. McClain's suggestion in his book, "Daniel's Prophecy of the Seventy Weeks", Zondervan Press (This book is a must in these last days).

In his book the 69 weeks start with the command to build the wall of Jerusalem (Neh. 2:1) which is dated from March 14, 445/4 B.C. extending on until April 6, 32/33 A.D. when Christ rode into Jerusalem on the donkey on Palm Sunday completing the 69 weeks of Daniel just five days before His crucifixion (Luke 19) on April the 11th. The Qumran Calender makes Passover fall 3 days later - 14th - falling on the first day of the new Nisan year 33/34 A.D..

The above "Prophetic Application" should be figured with a prophetic year of 360 days as the book of Revelation confirms. Thus, we highly recommend the reading of the above book. Therefore, we suggest both, a "Historical", and a "Prophetic" application to the 69 weeks of Daniel. The "Historical" pin-pointed the Barmitzvah of Christ as "Messiah their Prince" when He "began to be about His Father's business" (Luke 2:42-49 - 12 years old - Zero B.C.); while the "Prophetic" application applied to the presentation of Christ Jesus our Lord as "Messiah their King" on "Palm Sunday" (Luke 19:42,44 - the last few days of Nisan year 32/33 A.D. and the beginning of 33/34 A.D.).

Therefore, we conclude and suggest that Christ Jesus our Lord may have made His appearance as Messiah the Christ child (born of the virgin Mary - Isa. 7:14) on the 4000th year since Adam's expulsion from the Garden of Eden (4012 B.C.). That year of His birth (12 B.C.) was a "Jubilee Year", a year of release and repossession of all family property. Therefore, Joseph and Mary were returning to Bethlehem to repossess their family heritage as the descendants of David. Mary carried in her womb the "Messiah of Israel", who was coming to "possess His Kingdom" on that "Jubilee" year. Therefore, an appropriate time for the Roman enrolment, census, and for the possession of the "Roman World Empire tax" (Luke 2:1).

That year (12 B.C.) was also the 1000th anniversary since Solomon's temple was "dedicated" in 1012 B.C.. Thus, Christ's "bodily temple" made its appearance and was "dedicated" 1000 years later (12 B.C.) to be "wholly presented" to God's plan, service, and to His sacrificial will.

Christ's birth, on perhaps Jan. 6, 12 B.C., or September[1], was during Herod's visit to Rome; therefore, his lack of knowledge concerning that event. The above birth date would have made Christ either one year and eight months old, or around one year, when the Magis arrived following the Star, Halley's Comet, which appeared on Aug. 25, 11 B.C.. It appeared for 63 days giving sufficient time for the Magis from Persia to travel that distance to Jerusalem.

Mary and Joseph escaped with Jesus to Egypt during the latter months of 11 B.C., during Herod's slaying of all those children up to two years

1. If Christ was born in September (Day of Atonement) then His conception would still be in the same year December, 13 B.C.

184

old (he hoped to destroy the Christ child - Matt. 3:13; Rev. 12:4 - since Christ Jesus' age was approximately between one year and up to one year and eight months old). Joseph and his family lived in Egypt for seven years (11 B.C. to 4 B.C.) until Herod died. Then they returned to Israel in 4 B.C. (Christ was eight years old). Since "Archelaus did reign in Judaea in the room of his father" (Herod - Matt. 2:22), and being fearful of him, they traveled to Nazareth where they lived.

Then four years later, when Jesus was 12 years old (Luke 2:42-49 - Zero B.C.), Joseph and Mary went to Jerusalem to dedicate Jesus in the temple (as "Messiah the Prince") on His "Bar-mitzvah" day. Therefore, Jesus said concerning Himself, "Should I not be about my Father's business" (Luke 2:49). Even though extant history, and the Bible, gives very little information about the next 29 years.(about 30 years since His Bar-mitzvah - Luke 3:23), Christ fulfilled the above scripture by beginning to testify to the intellectuals in the temple.

Christ Jesus was later identified by John the Baptist as "the Lamb of God which taketh away the sins of the world"(John 1:29). At that time Christ was approximately 42½ years old since His conception (13 B.C.) by the Holy Ghost. He ministered for three and half years and was crucified on the first day of the new Nisan year 33/34 A.D. and on this 46th Passover since His conception as God's Passover Lamb. Christ died soon after He had completed His 45th year since His conception, and said also concerning His own body, "46 years was it in building. Destroy it and I will raise it up in three days"(This scripture applies both to Herod's "physical temple", and also to Christ's "spiritual body temple" risen over three and half years later). Thus, He, Christ Jesus our Lord and Saviour, fulfilled all scripture (John 2:19,20; 8:57,58; Luke 3:23; Isa. 53; Psa. 22; Isa. 7:14; 9:6; Micah 5:1-3, and many more). "And there are also many other things which Jesus did, the which, if they should be written everyone, I suppose that even the world itself could not contain the books that should be written" (John 21:25) Amen.

- - - - - - -

SOLOMON'S, HEROD'S, & THE TEMPLE OF CHRIST'S BODY
(B.C. 1 to A.D. 1 is one year)

Solomon's - I Kings 6:1

```
   4th yr  5th    6th    7th    8th    9th    10th   11th yr reign-Solomon
   |  1018BC 1017BC 1016BC 1015BC 1014BC 1013BC 1012BC - Nisan year
:2nd mo:      :      :      :      :      :      :      :
```
```
                                                     8th mo. beautified
 : *    : *    : *    : *    : *    : *    : *    : * / : I K.6:38a
 :1st yr:2nd yr:3rd yr:4th yr:5th yr:6th yr:7th yr:
                                            :_____:
```
I K.6:38b "So was he seven years in building it" - 12th mo. Nisan year
```
                                            :    the temple complete
                                            :
```
19 BC - Materials gathered.
```
:                                1000 yrs after Solomon's temple
Herod's temple            Christ conceived :      dedicated
begun in year         just before Passover : (Herod's Caesarea Port
   18 BC                           :-------: dedicated-13/12 BC)
   :      1st    2nd    3rd    4th    5th    6th yr of Herod's temple
19BC : 18BC   17BC   16BC   15BC  14BC   13BC   12BC -Herod to Rome
   :    :      :      :      :      :      :    : Jan. 6, 12 BC was
   :                                       ;-------Christ's birth-Sept.
 : *:   : *    : *    : *    : *     :/*   :/*
   :                                 1st    2nd Passover since concept.
   :                                   1 yr old since conception.
(Jubilee year)            John conceived 6 mos before Christ.
Roman census in
```
12 BC-Lk.2:1-6 - **Herod returns from Rome & contacts Magis 11 B.C.**
```
  :  7th   8th    9th   10th   11th   12th   13th year of Herod's T.
 12BC  11BC   10BC   9BC    8BC    7BC    6BC    5BC
  :     :      :      :      :      :      :      :
```
```
 :/*    :/*    :/*    :/*    :/*    :/*    :/*    :
 / :3rd / :4th / :5th / :6th / :7th / :8th / :9th - Passover
   2 yrs 3 yrs  4 yrs  5 yrs  6 yrs  7 yrs  8 yrs since conception
   :
```
Aug.25,11 BC Halley's comet (Wisemen). Christ & family to Egypt.
```
   :-Herod dies & Archelaus rules(Christ & family to Nazareth)
   :
  14th  15th   16th   17th   18th   19th   20th year of Herod's Temple.
 5BC   4BC    3BC    2BC    1BC    1AD    2AD    3AD(1 yr from 1BC to 1AD)
 :     :      :      :      :      :      :      :
```
```
:/*    :/*    :/*    :/*    :/*    :/*    :/*    :
/ :10th/ :11th/ :12th/ :13th/ :14th/ :15th/ :16th - Passover
9 yrs 10 yrs 11 yrs 12 yrs 13 yrs 14 yrs 15 yrs since conception.
                   :    1st    2nd yr since Bar-mitzvah.
        Christ's Bar-mitzvah
          at 12 years
```

```
 21st    22nd    23rd    24th    25th    26th    27th    28th year Herod's T.
3AD     4AD     5AD     6AD     7AD     8AD     9AD    10AD    11AD
:       :       :       :       :       :       :       :


:/*____ :/*____ :/*____ :/*____ :/*____ :/*____ :/*____ :
/ :17th/ :18th/ :19th/ :20th/ :21st/ :22nd/ :23rd/ :24th - Passover
16yrs   17yrs   18yrs   19yrs   20yrs   21yrs   22yrs   23yrs since conception
3rd     4th     5th     6th     7th     8th     9th    10th since Bar-mitzvah


                        Caesar Augustus dies
                            :       1st     2nd     3rd     4th year Tiberius C.
29th    30th    31st    32nd    33rd    34th    35th    36th year Herod's T.
11AD   12AD   13AD   14AD    15AD   16AD    17AD    18AD   19AD
:       :       :       :       :       :       :       :


:/*____ :/*____ :/*____ :/*____ :/*____ :/*____ :/*____ :
/ :25th/ :26th/ :27th/ :28th/ :29th/ :30th/ :31st/ :32nd - Passover
24yrs   25yrs . 26yrs   27yrs   28yrs   29yrs   30yrs   31yrs since conception
11th    12th    13th    14th    15th    16th    17th    18th since Bar-mitzvah


5th     6th     7th     8th     9th    10th    11th    12th year Tiberius C.
37th    38th    39th    40th    41st    42nd    43rd    44th yr of Herod's T.
19AD 20AD   21AD    22AD    23AD    24AD   25AD    26AD    27AD
:       :       :       :       :       :       :       :


:/*____ :/*____ :/*____ :/*____ :/*____ :/*____ :/*____ :
/ :33rd/ :34th/ :35th/ :36th/ :37th/ :38th/ :39th/ :40th - Passover
32yrs   33yrs   34yrs   35yrs   36yrs   37yrs   38yrs   39yrs since conception
19th    20th    21st    22nd    23rd    24th    25th    26th since Bar-mitzvah


Herod's Temple 46 years in building - John 2:20

13th    14th | 15th    16th    17th    18th year of Tiberius Caesar.
45th    46th | 47th    48th    49th    50th year of Herod's Temple.
27AD  28AD   29AD   30AD    31AD   32AD    33AD    /-Crucified-Ap.11,32/33
:       :       :       :       :       : Christ  / A.D.-Qumran calender 3
                                          died   // later on Ap.14,33/34 AD.
:/*____ :/*____ :/*____ :/*____ :/*____ :/* - -/
/ :41st/ :42nd/ :43rd/ :44th/ :45th/ :46th Passover since conception.
40yrs   41yrs |42yrs   43yrs   44yrs   45yrs since conception - beginning
27th    28th |29th Bar-mitzvah(about 30)(of 46th year-seen 40 days +3

    (46th yr of Herod's Temple)    (46th year of Christ's body temple)
```

"Jesus answered and said unto them, Destroy this temple, and in three days I will raise it up. Then said the Jews, Forty and six years was this temple in building, and wilt thou rear it up in three days? But He spake of the temple of His body" John 2:19-21.

"Then said the Jews unto Him, Thou art not yet fifty years old, and has thou seen Abraham" John 8:57.

APPENDIX B

THE SABBATIC AND JUBILEE YEARS

We know that the Sabbatic and Jubilee years were sealed by a firm covenant between the Nation of Israel and their God; disobedience (illustrated in Jer. 34:8-11) brought God's four sore judgments (Jer. 34:17; Ezek. 14:21; Rev. 6:8), and also final destruction of Jerusalem by Babylon (Jer. 34:21,22).

This Sabbatic covenant in the year 591/0 B.C., in Zedekiah's day (Jer. 34:8) was confirmed by a ceremony when the men of Israel walked between the parts of a slain animal to verify their covenant with God (Jer.34: 19). God's reverse covenant concerning Abraham's Seed and the Promised Land, took place when God passed through the parts of the sacrifice as a Lamp during His covenant with Abraham and is recorded in Gen. 15:17,18.

Leviticus 25:8-11 and Deut. 15 gives us complete instructions concerning the Sabbatic and Jubilee years. It is quite clear that the year of Jubilee immediately followed the seventh Sabbatic year. There were seven Sabbatic years in a period of 49 years, and the 50th year was the Jubilee. Since the Jubilee year was every fifty years, there were exactly two Jubilee years in every century. Therefore, when the year of Jubilee was once established and designated, it remained on the same two dates (or years of) each century down to the time of Christ, and was easily remembered.

The next problem is to find from the Bible and extant history when the Jubilee year may have fallen. William Whiston in his translation of Flavius Josephus lists several in Dissertation five, pages 979-983. He mentioned a Jubilee in the 43/44 year of Nebuchadnezzar's reign which, in turn, would be year 562/1 B.C.. This date agrees exactly with our chronology (see chart on page 47 - Jubilee year). The following year Evil-Merodach, King of Babylon, began to rule after the death of Nebuchadnezzar. It was also the same year that he released Jehoiachin in the 37th year of his captivity (II Kings 25:27).

That Jubilee year, 562 B.C., correlates with, and is confirmed by the Jubilee mentioned in II Kings 19:29 dated in 712 B.C.. The latter being the third Jubilee, or exactly 150 years, previous. Let us consider that Jubilee in Hezekiah's day.

Hezekiah began to reign in year 727/6 B.C.. In his fourteenth year (II Kings 18:13), Sennacherib attacked Judah (713/12 B.C.). This very same year was also the year in which Hezekiah prayed, after being sick, and God extended his life another fifteen years (II Kings 20:6 - he ruled 29 years). It was also the same year that Hezekiah prayed over the letter from the Assyrians (II Kings 19:14). That year (713 B.C.) was a Sabbatic year, and the following was a Jubilee year (712 B.C.), and in 711 B.C., during the following third year, they could freely eat of the land. Thus, our dating is perfectly confirmed and established by II Kings 19:29 where the prophet Isaiah, in answer to

188

Hezekiah's prayer, brings him a message from the Lord: "Ye shall eat this year (Sabbatic - 713 B.C.) such things as grow of themselves, and in the second year (712 B.C. - Jubilee year) that which springeth of the same, and in the third year (711 B.C.) sow ye, and reap, and plant vineyards, and eat the fruits thereof".

It is during Hezekiah's reign that we add the extra 9 months prorated in behalf of Jehoiachin's three months reign, which puts Hezekiah's death over into a portion of his 30th year. This works well with events involved around the Sabbatic and Jubilee years. Thus, by establishing these two Jubilee dates (712 B.C., 562 B.C.), we establish by repetition the same dates which fell in each consecutive century from then on. The year position for each century for each of the two Jubilee years never changed. In other words, the first Jubilee of each century always fell on years such as 862, 762, and 662 B.C. etc., while the latter Jubilee of each century always fell on years 812, 712, and 612 B.C. etc. (Jubilee's origin may also date from Adam' explusion - 4012 B.C. - a Jubilee disaster).

The Sabbatic year referred to in Zedekiah's day (Jer. 34:8-17), when correlated with the above Jubilee years, fell during the 6th year of Zedekiah's reign in the year 591/0 B.C.. He reigned eleven years.

The suggested Sabbatic year mentioned in Ezra 10:9-17 would have taken place in the 19th year of Artaxerxes (455 B.C.), while the one mentioned in Nehemiah would have fallen either in the Sabbatic year 441 B.C. (Artaxerxes 24th year), or in 434 B.C. (Artaxerxes 32nd year - inclusive).

Another Sabbatic year mentioned in Josephus' Dissertation V and found on page 982 states: "The sixth Sabbatic year that I find was that in the days of good Judas the Maccabee, when Antiochus Euportor (Epiphanes) besieged him in the temple B.C. 163. (This would have been a Sabbatic year followed by a Jubilee year in 162 B.C.). This year was observed as a rest for the land, I Macc. VI. 49,53; Antiq. B. XII, Ch. IX, sect. 5, page 371".

The next instance is found in Josephus' Book XIV, Chapter XVI section, page 442, where a Sabbatic year is mentioned. Book I, Chapter XVII, section eight, page 635, identifies the same attack on Jerusalem as taking place in Herod's third year. Since Herod began to rule in 37 B.C., the third year would date the attack on Jerusalem in 34 B.C.. This truly was a Sabbatic year and proves that we have found the perfect correlation of dates in the Jewish Sabbatical System.

By correlating the above dates, we also know that 13 B.C. was a Sabbatic year in which Christ was conceived by the Holy Ghost, and the following year was the year of His birth, which was a year of Jubilee (12 B.C.). The following Sabbatic years would have fallen on these Nisan years: 5 B.C., 3 A.D. (3 A.D. rather than 2 A.D. - one year less for period one B.C. to one A.D.), 10 A.D., 17 A.D., 24 A.D., 31/32 A.D., 38 A.D., (39/40 A.D. - Jubilee year) 46 A.D., 53 A.D., 60 A.D., 67 A.D., 74 A.D., 81 A.D., 88 A.D. (89/90 A.D. - Jubilee yr - 1989/90 AD is 6000 yrs since Adam's expulsion), 96 A.D.. They continue on the same dates each century since then.

Following the above Sabbatic year 31/32 A.D., there was a death and resurrection year (32/33 A.D.) which Christ Jesus fulfilled. Therefore, Christ Jesus the Jewish Messiah, the Son of God and the Prince of Peace, died and rose again at the close of the Nisan Resurrection Year (32/33 A.D.) following the Sabbatic year rest, so that we might also enter by faith into His Sabbatical rest for our lives (Hebrew 4: 1-11) and be filled with the hope of resurrection life (John 5:27-29).

Since we realize that the Sabbatic year and the year of Jubilee were not introduced until the law was given by Moses (Lev. 25); therefore, according our chronology, the Exodus fell in the year 1498 B.C. when Israel came out of Egypt. By correlating dates back to that period, based on our conclusion on the Sabbatic and Jubilee years, we find that our Exodus rightly took place on a "Sabbatic year" according to the scripture (Lev. 25 - 1498 B.C.).

If we extend our figures on back to Adam's expulsion from Eden, we find that the Edenic Jubilee year disaster fell exactly on 4012 B.C.. Adam lost his Jubilee rest by his sin. Providentially, the Jubilee year, immediately previous to the Exodus, fell in the year 1512 B.C., which was exactly 2500 years after 4012 B.C.. Therefore, that year was the 50th Jubilee period, or, in other words, it was a "Jubilee of Jubilees". A very appropriate time for Israel to be freed from Egyptian bondage. Only during the following years "after" that Jubilee did God bring Israel out of Egypt during the second Sabbatic year following that Jubilee. Praise God!

Thutmose II, the final Egyptian King of Israel's bondage was born in 1528 B.C., just ten years after Moses fled to the wilderness (1538 B.C.). He was made king in 1510 B.C. just two years after the 50th Jubilee (1512 B.C.). Thus, Thutmose II was absolutely the first 18th Dynasty Egyptian King that <u>"knew not Moses"</u> fulfilling Exodus 4:19: <u>"And the Lord said unto Moses in Midian, Go, return into Egypt: for all the men are dead which sought thy life"</u>.

One interesting Jubilee year fell in 1062 B.C. when David, who typified Christ, the Jewish Messiah, began to reign. Another important Jubilee year fell, providentially, in the year that Solomon dedicated his temple (1012 B.C.). Thus, as they dedicated the temple all slaves were released and all land was returned. It was truly a new beginning. The temple foundation was laid just seven years before in 1018 B.C. (inclusive). Just one thousand years later (see charts pages 186,187), the foundation of Herod's temple was laid in 18 B.C.. Then, seven years later (12 B.C. - inclusive), the temple of Christ's body made its "Jubilee" appearance in the world as the Messiah of Israel. He was introducing a new beginning, a new age of the New Covenant of the New Testament promised by Jeremiah the prophet in Jer. 31:31-35. This New Covenant was sealed by Christ's own blood, the Lamb of God, rather than by the blood of bulls and goats. Are you beginning to see Him high and lifted up? Have your eyes been opened to see His Glory? Christ Jesus, our Lord and promised Jewish Messiah, will gladly dwell in your heart as your hope of Glory, if you will allow Him. May He come in now as you repent and let Him fill your temple with all His fulness so that you may receive your Sabbatical rest and your Jubilee release today. Amen.

THE SIGNIFICANCE OF THE 6000 YEARS

The following scriptures are quoted since they have a significant relationship to the six or seven thousand year period:

"And God saw every thing that He had made, and behold, it was very good. And the evening and the morning were the <u>sixth day</u>. Thus, the heavens and the earth were finished, and all the hosts of them. And on the <u>seventh day</u> God ended His work which He had made; and <u>He rested on the seventh day</u>, from all His work which He had made. And God <u>blessed the seventh day</u>, and <u>sanctified it</u>; because that in it He had rested from all His work which God created and made" Gen. 1:31 - 2:1-3.

"For a <u>thousand years</u> in thy sight are but as yesterday when it is past, and as a <u>watch</u> in the night" Psa. 90:4.

"But, beloved, be not ignorant of this one thing, that <u>one day</u> is with the Lord <u>as a thousand years</u>, and a <u>thousand years as one day</u>" II Peter 3:8.

"I (Messiah) will go and return to my place (heaven), till they (Jewish people) acknowledge their offense, and seek my face (Messiah's face): in their (Jewish people's) affliction (Jacob's Troubles - Jer. 30:7) they will seek me (Messiah) early" Hosea 5:15.

"Come (Israel says), and let us return unto the Lord: for He hath torn (afflicted), and He will heal us (Israel); He hath smitten, and He will bind us up. After two days (2000 years) will He revive us (Israel), in the third day (3rd millennium) He will raise us up, and we shall live in His sight (Nationally & spiritually); then shall we (Israel) know the Lord: His going forth is prepared as the morning; and He (Messiah) shall come unto us as the rain (spiritual revival), as the latter rain (after the past 1900 years) and the former rain (before the past 1900 years during the book of Acts) unto the earth" Hosea 6:1-3.

Whether the six days of Creation were six literal days, or a day of a thousand or six thousand years, or perhaps a multiple of that number, we do not know for certain. We tend toward the view of six literal days of recreation. But what we do know from the scripture is that during the seventh day (or period?) God rested (Gen. 2:3). But since the Adamic fall and during this period of restoration and recreation of all things, has God rested, or is He still in the process of perfecting His fallen creation, and does His Sabbatical rest still lie ahead during a 1000 year period sometime in the future? These are questions which interest us as we consider the Creation Story in scripture.

"Dr. Silver has told us that the rabbis generally believe, on the basis of the Biblical creation week, that the world would continue 6000 years. In other words, the six days of reconstruction, not creation, typified the 6000 years of human history. This supposition, to their minds, was confirmed by the statement of Moses in Psalm 90:4"[1] (Quoted above).

Isaiah refers to the Messianic Age as follows:

1. Cooper, David L. "Messiah: His First Coming Scheduled," page 512

"And it shall come to pass in the last days, that the mountain of the Lord's house (temple) shall be established in the top of the mountains (on the seven hills of Jerusalem), and shall be exalted above the hills; and all nations shall flow unto it. And many people shall go and say, come ye, and let us go up to the mountain of the Lord, to the house of the God of Jacob; and He (Messiah) will teach us His ways, and we will walk in His paths: for out of Zion shall go forth the law, and the Word of the Lord from Jerusalem. And He (Messiah), the son of David, shall judge among the nations, and shall rebuke many people: and they shall beat their swords into plowshares, and their spears into prunning-hooks: nations shall not lift up sword against nation, neither shall they learn war any more (1000 years)" Isa. 2:2-4.

The reason why we shall have peace is because the Messiah (Christ) the Prince of Peace, rules (Rev. 20:4b), and satan our adversary, will have been put in the pit for a 1000 years:

"And I saw an angel come down from heaven, having the key of the bottom-less pit (see Rev. 1:18) and a great chain in his hand. And he laid hold on the dragon, that old serpent, which is the devil, and satan, and bound him (satan) a 1000 years and cast him into the bottomless pit, and shut him up, and set a seal upon him (satan), that he should deceive the nations no more till the 1000 years should be fulfilled: and after that he must be loosed a little season. And when the 1000 years are expired, satan shall be loosed out of his prison, and shall go out to deceive the nations which are in the four quarters of the earth, Gog and Magog, to gather them together to battle: the number of whom is as the sands of the sea" Rev. 20:1-3,7,8 (satan is released to inspire and identify the wicked who have never been subjected to the Messiah from their hearts. The loosed satanic spirit identifies them as they gather to oppose the Messiah, and they are destroyed to bring final purity to the world after seven thousand years).

This complete period since the expulsion of Adam simply represents 7000 years of recreation and purification. No wonder David said, "Create in me a clean heart O God . . . restore unto me the joy of thy Salvation" Psa. 51:10a,12a.

Recreation and restoration are uppermost in His thoughts and blueprint of our God during these 7000 years in which we now live. Also, recreation and restoration comes only through faith in the Lord Jesus Christ, the Messiah of Israel and the whole world.

"'Why does the Messiah tarry? When will He come?' These were questions which continually agitated the young Rabbi's (Leopold Cohen, D.D.) mind. One day, while poring over a volume of the Talmud, he came upon the following citation: 'The world (age) will stand 6000 years. There will be 2000 years of confusion (Adam to Abraham), 2000 years under the law (Abraham to Messiah), and 2000 years of the time of the Messiah'".[1]

Thus, the 6000 year age of which we are completing now is clearly divided into three two thousand year periods by the Talmud. Anstey also indicated this in his interesting book on Bible Chronlogy when he states:

1. Good News Magazine, page 15, Vol. 17, 1967

"The only way in which the credit of the Science of Chronology can be restored is to adhere strictly to the actual statements of the original text, and to deal with these statements in accordance with the laws of the Science of History, which places the criterion of credibility and the test of truth in the testimony of witnesses at once honest, capable, and contemporary. The identification of **the date of the dedication of Solomon's Temple and the birth of Christ with the years An. Hom. 3000 and 4000** respectively must be jealously scrutinized, and the facts must not be warped in order to bring about the exhibition of this result".[1]

At this point we might add this following information and suggest that Adam's expulsion from Eden was not necessarily at the beginning of the sixth day of Creation (Gen. 1:24-31 - the animals were first created). Thus, such consideration may add another dimension to the sixth day. If it has been almost 6000 years since Adam's expulsion, and another 1000 years until God's perfect age (God's day of rest - His seventh day), then how long was the first portion of the sixth day of creation until Adam's expulsion? Could it have been another 7000 years, or seven times seven thousand years, or even, seventy times seven? We can easily see that there are infinite possibilities, even as God is infinite. Then before this, we have the previous five days of Creation. Therefore, one can easily see that there are infinite possibilities of expansion. Thus, we wisely limit our exposition to this age, the present seven thousand years in which we are living, and in which the Bible is expecially concerned.

Therefore, the six days of "recreation" since the expulsion of Adam from the Garden of Eden are almost up, when we equate one day as a thousand years (II Pet. 3:8; Psa. 90:4). And according to the Talmud, the final Messianic Age (2000 years) started after 4 days, or after 4000 years (our chronology establishes the 4000 years as from 4012 to 12 B.C.).

When comparing our chronological date of the expulsion of Adam (4012 B.C.) with Bishop Ussher's date (4004 B.C.), we have an eight year variancy. Dropping down 4000 years later, we have 12 and 4 B.C. respectively for the birth dates of Christ. There is real proof that Christ was born during this very limited period, and probably before 6 or 7 B.C.. Thus, the Science of Chronology pinpoints, and limits, the birth of Christ to a concise 5 or 6 year period 4000 years since Adam's expulsion (Between 12 and 7 B.C.).

Now, let us refer back to our quotation from Hosea 5:15. The Messiah arrived after 4000 years, but was rejected (Isa. 53). As a result of this the Messiah prophetically states through Hosea 5:15:

"I will go and return to my place (heaven) till they (the Jewish people) acknowledge their offence".

What offence? The offence of Israel was the rejection of the true Jewish Messiah by delivering Him (Matthew 27:19-25) to be crucified by the Romans. Yet, the scripture states, "I will go and return to my place". This prophetically proves that He overcame death, arose again, and ascended into heaven (His original place or eternal abode) and is there waiting until His enemies be made His footstool (I Cor. 15:25),

1. Anstey, M., "Romance of Bible Chronology", page 87

193

and until Israel acknowledges their mistake of rejecting Him.

The Scripture also states, "That at the name of Jesus every knee should bow, of things in heaven, and things in earth, and things under the earth; and that every tongue should confess that Jesus Christ is Lord, to the Glory of God the Father" Philippians 2:10,11. They must bow before Him, their Jewish Messiah, and seek His face. Where can they see His face? II Corinthians 4:6 states that "God who commanded the Light to shine out of darkness, hath shined in our hearts, to give the Light of the knowledge of the Glory of God in the 'Face' of Jesus Christ".

Now, after 2000 years, Israel has providentially returned to her own promised land (Gen. 15:17-18), and Hosea has already prophesied what Israel will say in these last days, "Come, and let us return unto the Lord: for He (Messiah) hath torn, and He will heal us; He hath smitten, and He will bind us up. After two days (the last 2000 years since 12 B.C.) will He revive us: in the third day (millennium) He will raise us up, and we shall live in His sight" Hosea 6:1,2.

Israel will say, "Come let us return unto the Lord: for He hath torn . . . .". Why has the Lord allowed the enemies of Israel to tear her for the last two thousand years? Simply because the Lord has justly filfilled His promised Word (Deut. 4:25-40) as a result of their rejection of the Messiah. He also prophecied, "Smite the Shepherd (reject the Messiah) and the sheep (Israel) shall be scattered" Zechariah 13:7. When sheep are scattered without a proper, true shepherd, they are often torn by the wild beasts (Gentile nations). Thus, the beastly Gentile nations have been allowed by God to providentially tear Israel, to chasten her in His love, to bring her back to her true Jewish Messiah and Shepherd, the Lord Jesus Christ.

"He hath torn, and He will heal us; He hath smitten, and He will bind us up. After two days (2000 years) will He revive us". Therefore, the Jewish people have had 2000 years until Abraham; 2000 years from Abraham unto Christ Jesus the Messiah, and since then, they have almost fulfilled the final two days (2000 years). Then God will spiritually revive all those that remain (Nationally and spiritually). This does not nullify the possibility of the revival of many Jewish believers in Jesus Christ the true Messiah also during the end of our present Church Age (Dispensation of Grace - Eph. 3:1-10); nor does it nullify the revival to faith in Jesus Christ (Rev. 19:10) of the 144,000 Jewish people through God's dealing nationally with Israel during the end of the 70th week of Daniel (Tribulation Period - Dan. 9:27), but it does limit the full "National repentance of Israel" to just before, or during the first few transitional years of the millennium. Since He says, "After the 2nd day I will revive you"(or after the last two of six days).

Thus, National spiritual revival of Israel may carry over into the first few years of the seventh millennium, and that peace will continue through the 7th one thousand years, except for the two transitional periods, one at the beginning and the other during the culminating period when the last remnant of sin and rebellion is annihilated. Thus, the true Messiah, Jesus Christ of Nazarath, will be reigning "in peace" during the Jewish "Golden Age"; therefore, His personal return to reign will be the introductory event which will completely resolve the chaos which is in the world today.

We also find that even in the early centuries of the Christian Era this view was prominent among both Christian and Jewish people:

"According to Ephraim Syrus (325 - 378 A.D.), a Syrian theologian, there was a general universal tradition among the Jewish people that the world would last 7000 years, and as man was made on the sixth day, and fell by sin, so the Messiah would come to redeem the world in the sixth millennium".[1]

The great evangelist of the early part of this century, Charles H. Spurgeon adds his agreement to the matter when commenting on the portion of Scripture, "They shall reign with Him a thousand years" Rev. 20:4. He said, "Here is another point upon which there has been a long vigorous contention. It was believed in the early church, that the seventh thousand years of the world's history would be a Sabbath, that, as there were six days of toil in the week, and the seventh was a day of rest, so the world would have six thousand years of toil and sorrow, and the seventh thousand would be a thousand years of rest".

"Also a very early Christian Chronologer, Theophilus, Bishop of Antioch (176 - 186 A.D.) states that the leading writers of the Christian church were dominated with the idea of six millenary ages of the world, which they regarded as equally divided into two periods of 3000 years".[2]

The well known christian author, Clarence Larkin in his book "The Second Coming of Christ", page 42, adds his comment:

"Now we know that the length of the Millennium is 1000 years (Rev. 20: 1-9), and if it corresponds with the "Seventh Day" of the "Creative Week", why should not the remaining six days be of the same length? If so, and those days correspond with the past of human history, then from the date of the "Creative Week" (Expulsion - added) up to the beginning of the Millennium should be 6000 years of human history".[3]

Thus, we conclude and agree with these eminent scholars concerning their theory that the "Seven Days" of the "Recreative Week" are typical of seven "One Thousand Year Periods", and that this theory is warranted by the testimonies of these scholars, and may also be warranted by the Holy Scripture.

Now, what other signs are evident today to prove that we are near the end of six thousand years, and near the battle of Armageddon (Rev. 16: 16), which battle will climax the period of this Age, Jacob's Troubles (Jer. 30:7), and introduces the Millennium reign of Christ Jesus, the Jewish Messiah?

1. Anstey, M., "The Romance of Bible Chronology", page 216
2. Fuller, David Ottis, "Spurgeon's Sermons on the Second Coming" p. 17
3. Larkin, C., "The Second Coming of Christ" p.42, permission Larkin Est

195

1. Chronologically, the six thousand years should end before the end of this century.

2. As Jerusalem was taken away from the Jewish people (70 A.D.) shortly after the Roman's first attack in 66/67 A.D., so Jerusalem has been given back in our day exactly 1900 years later (1967 A.D. - non-inclusive) to the Jewish people as a sign that we are in the last phase of this age.

3. The Jewish people were in the land 1900 years ago, and now they are returning to the same land 1900 years later.

4. Israel's independence, in 1948 A.D., typified by the fig tree (Jer. 24:1-8; Luke 21:29,39), and the independence of the scores of other countries portrayed by the phrase "all the trees" (Luke 21:29), are scriptural signs of the end just before Messiah, Jesus, returns (Luke 21:28). There have been 50 African nations, minus one, who have received their independance since 1948 A.D..

5. "And there shall be signs in the sun, and in the moon, and in the stars; and upon the earth distress of nations, with perplexity, the sea and the waves (of evil humanity) roaring; men's hearts failing them for fear, and for looking after those things which are coming on the earth: for the powers of heaven shall be shaken and then shall they see the Son of man coming in a cloud with power and glory" Luke 21:25-27. The Greek word for "sign" (simeon) can mean "sign", "mark", or "signal". Marks and signals have already been placed on several of the planets by man, and as to a "sign", the sun and moon will soon turn to the color of blood during the "Tribulation" of these last days (Matt. 24:29).

7. Increased world corruption and sinfulness are signs of the end of this age. "For that day shall not come, except there come a _falling away first_ (into sin), and that man of sin (Anti-Christ or false Christ) be revealed" I Thess. 2:3b (See also Dan. 8:23).

8. The amalgamation of many religious groups under one head will fulfill the scriptures concerning a religious harlot (Rev. 17 - tares - Matt. 13:30),who will be destroyed in the last days (Rev. 17:16).

The amalgamation of the true christians into the Bride and Body of Christ (the wheat - Matt. 13:24-30) will be soon taken to heaven (Luke 21:36; II Thess.2:7)an is in the process of being completed.

9. The political aspect of the revived Roman Empire is beginning to be manifest through the ten Common Market countries. They will soon enter into a world sphere of control in the end time (ten horns, ten toes) as a united ten world confedercies (Rev. 17:12,13; 13:16,17).

10.The army of the north (Russia - Ezek. 38,39; Joel 2:12-26) is challenging Israel's existence today, and must be destroyed on the mountains of Israel in the last days (Ezek. 38:8).

11.The kings of the East must marshal an army of 200,000,000 men to fulfill the last phase of this age (Rev. 9:14-16; 16:12), and this is a possibility now (China - 1 billion in 1982 A.D.).

12. Halley's Comet appeared for 63 days and stood over Bethlehem on Aug. 25, 11 B.C. approximately one year and a half after Christ's birth, which announced the Messiah, the "Star" out of Jacob as a sign to the wisemen from the East. Could Halley's Comet, about 1997 years later (appeared last year in 1986 A.D.), announce the "soon" return of the true Jewish Messiah on that great and terrible day of the Lord.

Are 6000 years significant in Bible Chronology? It seems that many believe so. Of course, this is a debatable question among many, and we certainly cannot be dogmatic. We may pass the close of the 6000 year period near the end of the century with no evidence of the Battle of Armageddon, yet again, evidence may multiply during this decade indicating that the end is at hand. God's Word implies that the final sprint in the last phase of this age will be short:

"For He will finish the work, and cut it short in righteousness: Because a short work will the Lord make upon the earth" Romans 9:28. Whatever be the turn of events, we must warn people of the present serious signs, and it also behoves us not to be among those who will say, "I wish we had all been ready" (See Luke 21:36).

– – – – – – –

GENEALOGICAL INSERTION OF CAINAN'S NAME
Luke 3:36

GENEALOGICAL CHART

4012 B.C. - - - ADAM'S explusion from Eden

           SETH   "CAIN" - "And the Lord set a "mark"
upon Cain, lest any finding
". . And they called – ENOS   him should kill him" G.4:15
his name Enos: then   ("Mark" of separation)
began man to call upon "CAIANAN"   ENOCH
the name of the Lord"
Gen. 4:26.   MAHALALAD
                         "That the sons of God saw
                         the daughters of men that
Jered or Irad – means – JERED - - - - IRAD – they were fair; and they
to "go down" or "decend".   took them wives of all they
which refers to corrupt-   which they chose" Gen. 6:2.
ion which set in by in-
ter-marriage.   MEHUJAEL
Walked with God - - - - ENOCH
    Gen. 5:24
        METHUSELAH
                  METHUSAEL
       LAMECH
                  LAMECH
      NOAH

      SHEM   NAAMAH(Beautiful) Possible wife of
                             Ham???

FLOOD – 2356–2355 B.C.

2353 B.C. - - - - born ARPHAXAD - - - Wife (Daughter of "Cain" line)?
two years after Flood
Gen. 11:10   (Cain's line) – The two names were
(Inserted to identify – (Cainan's line). euphonic in sound and
  line - Luke 3:36)   chosen to identify the
                         genealogical line. Luke
                                3:36
        SALAH

2288–1824 B.C. – EBER – (Hebrew) – Father of the Hebrew
                      race since he lived four years
                      longer than Abraham. (Abraham the
                      Father of our faith)

The above genealogical chart could be a simple reason why "Cainan's" name was inserted the second time in the genealogical line after Arphaxad (Luke 3:36). In memorizing or writing the genealogy, the post-Diluvian student had to know which genealogical line to follow back to

Adam. Arphaxad (born two years after the Noahic Flood - Gen. 11:10) probably married a daughter of Cain's line through Ham's family (Ham's son "Cush" was black - the Hebrew language word "Cush" means "black"). Since Noah had all these racial groups in his loins, it would imply that either Noah or his fore-fathers, after the days of Jered, had inter-married with the "Cain" line; thereby inter-mingling the "mark" of separation into their line. Ham may have also married a wife from the line of Cain; therefore, the name "Cainan" was inserted before Arphaxad to designate the Messianic line, or "Cainan's line". If a student wanted to follow the line through "Cain", as some desired, through Arphaxad's wife, he would have inserted "Cain's" name in the genealogy to designate "Cain's" line. The name was inserted before Arphaxad (Luke 3:36) since the genealogical line split at that point both leading to Adam. This took place just after the Flood when the "Cainan" and "Cain" lines became one in the Noahic family. The names of "Cainan" and "Cain" were chosen to designate the two genealogical lines since they were euphonic in sound and similar in construction.

Furthermore the Hebrew and the Samaritan texts do not include Cainan's name among the Post-Diluvian patriarchs. Only the Septuagint makes the second insertion of Cainan's name and thus it is passed on to the New Testament text perhaps for the above reason. Neither do the early Christian fathers, Theophilus, Africanus, Eusebius, nor the Jewish Historian, Josephus, include the "second Cainan" in their lists. Martin Anstey remarks, "it was natural that the translators of the LXX (Septu-agint) should augment the chronology of the period by the centenary additions, and by the insertion of the second Cainan, in order to carry back the epoch of the Creation (explusion - added) and the Flood to a respectable antiquity, so that it might compare more favourably with that claimed for Babylon and Egypt. . . Possibly the desire to form a second list of ten Patriarchs from the Flood to Abraham, corresponding with the list of ten Patriarchs from Adam to Noah, may account for the insertion of the extra name (Cainan). In that case it would seem to have escaped the notice of the inventor of the extra name that the list of Patriarchs from the Flood to Noah, as given in the Hebrew Text of Gen. 11:10-26, already contains ten names and can only be reckoned as nine when the name of Shem is omitted from the list".

Anstey continues, "Many other arguments may be adduced to prove the spurious character of the addition of the second Cainan:

1. It is omitted from the Hebrew Massoretic Text, and also from the Samaritan, as well as from all the ancient versions and Targums of Gen. 11:12.

2. It is omitted from the Hebrew Text of the two passages I Chronicles 1:18-24, and also from many copies of the LXX version of that pass-age, though 21 copies collated by Dr. Parsons have it, in verse 18, and 6 copies have it in verse 24.

3. Josephus omits Cainan in his list of the Post-Diluvian Patriarchs and so does Philo by implicaton, for he reckons ten generations before the Flood from Adam to Noah, and ten generations after the Flood from Shem to Abraham, which leaves no room for Cainan in the second group.

4. Berosus (284 B.C.) and Eupolemus (174 B.C) represent Abraham as living in the 10th generation after the Flood, whereas if the name of Cainan had been included Abraham would have been living in the 11th generation after the Flood.

5. Origen marks the name of Cainan with an obelisk in his copy of the LXX., to mark his rejection of it as not genuine.

6. Eusebius excludes him by reckoning only 942 years from the Flood to Abraham, and in this he is followed by Epiphanius and Jerome.

7. The name is evidently a late invention of the Chiliasts, who reckoned up their Chronology by periods of a thousand years, and where the facts were stubborn they invented others, and thus retained their theory.

The fact that the name of the second Cainan occurs in the genealogy of Mary, the mother of our Lord, in Luke 3:36 is easily explained. The Bible, as it was held in the hands of the common people, in the time of our Lord, was the LXX. The LXX was to them what our Authorized Version is to us. Scholars like Paul, and students of the Word like our Lord and His Apostles, had access to the Hebrew Text also, but Luke, the only writer of any book contained in the New Testament who was not a Jew (Gentile - Col. 4:10-14) and the one writer whose Gospel was specifically addressed to a Greek reader (Luke 1:3), would naturally use and quote from the Greek version in common use, and if the copy of the LXX which he used contained the spurious addition of the name of the 'second Cainan', the error would of course be reproduced in his Gospel, just in the same way as any error of translation in the A.V. would be reproduced by any layman occupying a modern pulpit, and acquainted only with the Scriptures in the Authorized Version.

It is just possible, of course, that Luke never wrote the word Cainan in Luke 3:36, for it is omitted in the Codex Bezae, the great Cambridge Uncials, except the codes Bezae D, though it is spelt Cainam instead of Cainan in some of them".[1]

Thus, we conclude that the name "Cainan" should not be included in the Hebrew genealogy the second time (Luke 3:36) and creates no problem in our Chronology.

- - - - - - - -

1. Anstey, Martin, "The Romance of Bible Chronology", p. 82,84,86

IMPORTANT OLD TESTAMENT DATES DURING 6000 YEARS

(Jubilee year the 12th & 62nd year of each century B.C.)
(Sabbatic year the 13th & 63rd year and every seven years)

```
: - Adam's sin - Paradise lost - Jubilee year disaster   4012 B.C. - - - - - - - - - :
:   Date of Adam's expulsion from paradise to his death   4012-3882     :          ↑
:   The FLOOD    (NOAHIC WORLD WIDE FLOOD)                2356-2355   Ageing        :
:   Abraham (First call at Ur - 1935/4 B.C. - Acts 7:4)   2003-1828   period        :
:   Abraham's second call at Haran - Gen.12:1-4;Acts 7:4  1928 B.C.   begins        :
2300 Isaac(Sodom & Gomorrha destroyed in 1904 BC)yr before 1903-1723     :          :
:   Jacob's Life                                          1843-1696   2300          :
:   Jacob to Haran for "twenty years" at 77 years of age  1766-1746     :           |
:   Joseph - sold to Egypt - 1735 B.C. at 17 years old -  1752-1642   years         :
:   Jacob's return to Canaan (HYKSOS INVADE EGYPT) - - -  1746 B.C.     :           :
: - - Jacob to Egypt (Joseph revealed-age 39)Sabbatic year 1713/12 - JUBILEE YEAR - | - :
: : Moses(Expulsion of Hyksos over a period 1582-1580 BC) 1578-1458                  :
: : EXODUS(Thutmose II dies - Red Sea - Hatshepsut reigns)1498 B.C. - Sabbatic year |
: : JOSHUA INTO Canaan (Moses dies - see Jude 1:9)         1458 B.C.                 :
: : Israel in Canaan 300 yrs to Jephtah's rule-Jud.11:26   1458-1158                :
: : Joshua dies after 25 years in Canaan - Josephus        1434/3 BC                :
:700 Joshua dies-to end Nehemiahs ministry(1000yrs)Neh8:17 1434/3-434/3          |3000
:yrs Samuel's Life                                         1188-1065              | yrs
: : Saul as King                                           1102-1062               :
: : David as King              Jubilee year - 1062-1022                            :
: : Solomon as King (Solomon's temple begun in 1018 B.C.)  1022- 982              :
: - - Solomon's temple built last 7 yrs of 3000 yrs (incl.) 1018-1012 JUBILEE YEAR - | - :
:   The Divided Kingdom for 260 years                     982- 722                 |
:   Israel's Captivity to Assyria                          721 B.C.        4000 yrs |
:   Babylonian Captivity of 1st group(Submission-605 B.C.) 606- 536     (70 years) |
:   "Jehoiachin's Captivity"- 2nd group(Mar.16th-Nisan yr- 597/6 BC)               |
:   BABYLONIAN CAPTIVITY OF FINAL GROUP (JERUSALEM FELL)   586- 516     (70 years) |
:   First return from Babylonian Captivity(Built Temple)   536 B.C.                |
:   Cyrus' command & preparation of return to Jerusalem    539- 536     ( 3 years) |
:   Zerrubabel constructs & dedicates temple - Jerusalem   536- 516 B.C.(20 years) |
:   Period of no repair in Jerusalem (on city or wall)     516- 459      Not       |
:   Esther became Queen of Persia and saves Jewish nation  478 B.C.   consecutive  |
:   Ezra's beautification of the Jerusalem Temple          459- 445     (14 years) |
:   Nehemiah rebuilt wall & rededicates city (inclusive) - 445- 434     (12 years) |
:   7 wks-Dan.9:25(7x7=49 Jerusalem,str.,walls troublous times - - -    49 years)  |
:   62 wks-Dan.9:25(62x7=434BC to Messiah Prince dedicated 434-0 BC - - in Temple) |
:   CHRIST BORN(Angelic announcement & Herod in Rome-Jan.6,12 BC or Sept. or Oct.)─┘
:   Wisemen visit(Astronomical announcement-Halley Comet-Aug.25,11 B.C.- Herod returned)
3700 Ante-Diluvian World                          (1656 yrs) 4012-2356 B.C.
:   First year Flood to call of Abraham(Haran) (428  yrs) 2356-1928
:   Israel's sojourn in Egypt including Canaan  (430  yrs) 1928-1498
:   Abraham's cov. with Abimelech(Amenemhet II)(400  yrs) 1898-1498
:   Exodus to 4th yr of Solomon (Temple begun) (480  yrs) 1498-1018
:   Call of Abraham(Haran)to 4th yr of Solomon (910  yrs) 1928-1018
:   Begin Solomon's T. to Herod's Temple begun (1000 yrs) 1018-  18 B.C.
:   Dedicate Solomon's T. to T. Christ's birth (1000 yrs) 1012-  12 B.C. - JUBILEES
:   Adam's expulsion to dedication Solomon's T.(3000 yrs) 4012-1012 B.C. - JUBILEES
:   Adam's expulsion until the birth of Christ (4000 yrs) 4012-  12 B.C. - JUBILEES
:   Herod's T. begun(18 BC)to Great Snagogue T.(2000 yrs) BC18-1982 A.D. - Aug. 4th
LAST JUBILEE IST C.(12 BC)TO LAST JUBILEE THIS C.(2000 yrs) BC12-1988/89 A.D. (6000 YEARS)
```

PALAEOLITHIC AGE - before 4012 B.C. time was not calculated and man did
  not age. Adam & Eve had perfect fellowship with God (Heaven on earth).
4012 B.C. - NEOLITHIC AGE - Adam & Eve sinned and began to age with ex-
: pulsion from Eden during the 6th day of recreation (Psa. 104:30).

: - 1656 years to Noah's Flood - Gen. 5:3-31;7:6;9:28,29;11:10,11.

:   2476 B.C. - Noah began to build the Ark for 120 year testimony.

:- - - -2356 B.C. - Noahic Flood (Most logical historical time).

  :   "First Intermediate Period" - "History Almost Silent"

: - 428 years to Abram's call (427 years plus 1 year for Flood).

  :   2003 B.C. - The birth of Abraham.

:- - - - - 1928 B.C. -Call of Abram (Left Haran - Gen. 12:1-2 - - -:
                                                   :
       :   4 generations-Abram,Isaac,Jacob and Levi - 215 yrs :

Gal.3:17:- 430 years (Jacob to Egypt - Half of Egyptian Captivity) - - :
1713 B.C
     :   4 generations-Kohath,Amram,Moses & Gershon-215 yrs :
Ex.12:40,
41;14:4;:- 1498 B.C. (EXODUS) Thutmose II dies, Hatshepsut reigns- - - :
18:30;15:19. :(Deut.2:14;Num.1:1;20:1 - 2nd yr,1st Mo.Kadesh)-1496 BC-:
                                                   :

Gen. 15:16

  • :   "And it came to pass in the (Canaan - 1458 B.C.-:
              476 480th year after the children   Deut. 2:7;8:2
480 years -.- : -‾‾ - of Israel were come out of the
             yrs land of Egypt, in the 4th year
      • :   of Solomon's reign over Israel"- I Kings 6:1
Solomon's
1st year - . -: - - 1022 B.C. - The first year of Solomon's reign,
& 4th yr-1018 BC :

                  "And thou shall bear iniquity of
Evil-I K.11:6,9-12: ‾40‾ the house of Judah 40 days: I have
              yrs appointed thee each day a year"- Ezek. 4:6
          :

         - - 982 B.C. - Rehoboam & Jeroboam reigns.
        :   "For I have laid upon thee the years
               of their iniquity, according to the
390 yrs - : - - number of the days, three hundred and
               ninety so shalt thou bear the iniquity
        :   of the house of Israel" Ezekiel 4:5
Ezek. 1:1,2;4:5 -*Fifth year Jehiachin's "Captivity" - July 5, 592/1 BC
               : - 592 years from 0 B.C. to the 5th year
                    of Jehiachin's Captivity.
             : - - 0 B.C.

* The fifth year of Jehoiachin's "reign" would be 593/2 B.C.

APPENDIX F

CHRONOLOGICAL TABLE I (ANTE–DILUVIAN)

*(Adam began to age at the expulsion from Eden – Time began)

In the beginning God created the heaven and the earth not a waste – Gen. 1:1 (not in vain–
Isa. 45:18 – Creation perfect)

And the earth was without form and void – Gen. 1:2 (Imperfect, waste – Lucifer and 1/3 of
his angels fell into sin – Isa. 14; Ezek. 28; Luke 10:18; Rev. 12:7-10

1st Day of Recreation   (Darkness and Light) – Gen. 1:3-5
2nd Day of Recreation   (Firmament in the midst of the waters) – Gen. 1:6-8
3rd Day of Recreation   (Earth appears from the waters and green) – Gen. 1:9-13
4th Day of Recreation   (Sun, Moon, and Stars appear) – Gen. 1:14-19
5th Day of Recreation   (Living creatures in sea and air) – Gen. 1:20-23
6th Day of Recreation   (Cattle & creeping things on earth – man created) – Gen. 1:24-31

Cities	gods of Assyrian	Characters Biblical	gods of Babylonia	gods & kings of Egypt
Eridu	Alulim (Father)	(Adam)	Aloroa (Father)	Ptah-Creator,Opener; All in Himself
– – Man's expulsion from Eden – – *4012-3082 BC				
Eridu	Alalmar	(Son)(Seth) 3882-2970 BC	Aloparus (Son)	Ra – Sun god Re or Atum – later
Badgurgurru "Called on God" Gen. 4:26	Emenluanna	(Spirit-worship-Enos) 3777-2870 BC	Almelon(Spirit)	Shu – Spirit god, or Su – Air god
Larsa	Kickunna	(Cainan) 3687-2777 BC	Ammenon (Geb)	Seb – Earth god
Badgurgurru	Enmengalanna	(Mahalalad) 3617-2722 BC	Amegalarus	Hosiri – Ruler of the dead – Osiris
Badgurgurru	Dumuzi(Tammuz)	(Jered) 3552-2590 BC	Daonos	Seth-Typhon god of Famine (Carbon 14)
Larak	Sibzianna	(Enoch) 3390-3025 BC = Rapture – Sky God – Horus(3250 BC)	Eudorachus	Horus (date)
Sippar(Akkad) Emenduranna (Menes 1900 yrs-Akhneten's Jubilee)		(Methuselah) 3325-2356 BC	Amempsinos(Moon)	Thut (:) Dyn I Menes 3270 B.C.
Shuruppak(Turah)Uburratum		(Lamech) 3138-2361 BC	Otiartes	Ma-at – Truth Dyn II Hotep-shani 3055 BC
– – Adam died in 3082 B.C.				
Shuruppak(Turah)Zinsuddu		(Noah) 2956-2006 BC	Xisuthros	Horus – Sky god
Seth dies – 2970 B.C..		28 year reign – – – – –	(Seth or Seti worship – King Persbsen) Dyn III Sanekht 2818-2790 BC
(Carbon 14 date 2770 B.C.)		Pyramid Kings – – – – –	Djoser(Zoser)(19) 2790-2771 BC
(Day Star – Sothic Cycle)		: (Step Pyramid at Saqqara)	Sneferu
(Calender begun by Imhotep)		:	Neberkhet
(between 2785-2773 B.C. in)		:	Dyn IV Shaaru 2753-2729 BC
(reign of Djozer (Zoser).)		:	(Khufu) Cleops (23) 2729-2706 BC
		:	(Khafra) Chephren (23) 2706-2683 BC
Methuselah lived		:Pyramid Kings (Men-kau-re)	Mycerinus 2683-2655 BC
969 years & died just before Noah's Flood			Dodef-Re 2655-2649 BC
		Sinia mines –	Shepses-kaf 2649-2642 BC
		(Sebek-ka-Re)	Thamphis 2642-2633 BC
Shuruppak – Listed 700 gods around 2600 B.C.			Dyn V Userkaf 2633 BC
Shem born – 2453-1853 B.C. –Shem saved-Noah's Ark.			Dyn VI Teti(Tety) 2500 BC
3rd Dyn Uruk-Urukagina 2388-2381 BC (7)-The last 2 Kings of the Sumarian Civilizat.			
Lugal-zaggisi 2381-2356 BC(25) "With utmost distinctness it ends an age" p.			
144, I Part 2, Cambridge Ancient History			Dyn VII Neith-aqert 2361 BC
– Noahic Flood – 2356-2355 BC – Seventy rulers in seventy days – Noahic Flood Chaos			

Adam lived 930 years

APPENDIX G TABLE II
EGYPTIAN POST-DILUVIAN CHRONOLOGY

Noahic Flood 2356-2355 B.C. - "History Almost Silent" - Chaotic Dark Age

"FIRST INTER-MEDIATE PERIOD"   130 years

DYN VIII		Neferkauhor I	B.C. 2225-2197/
		Neferkauhor II	2197-2175
DYN IX		Kheti	2175-2154
DYN X		Yntef I	2154-2120
		Yntef II	2120-2111
DYN XI	(Gen. 21:32; 26:1)	Mentuhetep I	2111-2062
	(Proto-Philistine)	Mentuhetep II	2062-2054
	(Hamitic Race)	Mentuhetep III	2054-2003

Abram's Cov.)-- born 2003 B.C.   (v)Mentuhetep IV   2003-1991
with 3 gens.) DYN XII Amenemhet I = (Ammenemes) = Abimelech I   1991-1961
Gen. 21:23)(Abram to Canaan 1928 B.C.) Gen. 12:10-20 - Sesostris I (Bible)   1961-1926
-x- - - 1898 B.C. Gen. 21:32 Amenemhet II = (Ammenemes) = Abimelech II (Bible)   1926-1894
:- - - - Forbidden cov. - Ex. 23:32; 34:12,15; Deut. 7:2 Sesostris II(On Tomb)   1894-1879
:-3rd gener. DYN XIII(7th year - Sirius Star) Gen. 21:23 Sesostris III   1879-1841
:   Cov. broken - Gen. 26:20 Amenemhet III = Ammenemes = Abimelech III (Bible)   1841-1793
:   Amenemhet IV = (Ammenemes) = Abimelech IV   1793-1784
Gen.15:13 (Proto-Philistine migration to N. Africa & Sebknefrure to Crete)   1784-1781
Bondage   DYN XIV HYKSOS penetrate the Middle East - confusion - - - - - -   1781-1746
400 years DYN XV Hyksos Army Gen. rules over Egypt - Rameses I (invade)   1746-1730
until Tanus Stella-1713 BC-Joseph-Jacob to Egypt Land - Rameses II (Salitos)   1730-1711
sins   : HYKSOS PERIOD   Apepi I   1711-1705
of   :   Apephis (61 years)   1705-1644
Amorites :   Palestine ruler-(Khayan co-regent 10 yrs   1654-1644)
are   : DYNS XV & XVI in opposition - - - - - - - Khayan (30) Khian   1644-1614
full :   (1642 B.C. - Joseph dies)/ Aa-seh-re   1614-1611
:   : DYN XVII (HYKSOS PERIOD IN EGYPT ENDS) Apepi II   1611-1580
:   : DYN XVIII "Knew not Joseph" Moses born - - Ahmoses I -in 1578 BC   1580-1557
: (400 year Tanus) Hebrew bondage-(Pharaohs of - (Moses-Midian-Amenophis I -1538 BC.)1557-1536
: Stella   the Hebrews'-   Thutmose I   1536-1510
:v   :   bondage)-   Thutmose II (EXODUS)   1510-1498
B.C. : (Thutmose II subject 3 years to the Queen - Hatshepsut coup d'etat1501-1498)
1498-EXODUS Thutmose II dies in the Red Sea - - - - - - Hatshepsut (EXODUS)   1498-1477
?   : (1476 B.C. - Battle at Megiddo)-(Hornet)- - Thutmose III (alone)   1477-1445
:   :- - - (1441 B.C. Temple of Bast built in 4th year-Amenophis II   1445-1421
: : :   :   Thutmose IV   1421-1413
: : : 500 years   Amenophis III   1413-1376
IK.6:1 : : later   (Akhnaten) - - - - - Amenophis IV   1376-1359
480 : : Temple of   Smenkhkare   1359-1356
years : : Bast rebuilt   (King Tut) - - - - - - Tutankhamun   1356-1347
to 4th : : in 941 BC in   Ay   1347-1343
year of: : 21st year of Zerah (Osorkon I)   Haremhab   1343-1316
Solomon: v DYN XIX   Rameses I (20 mos)   1316-1315
: : :- - - 1313 B.C. - 400 year Stella of Tanus - - Seti I (in his 2nd yr)1315-1294
: : ?   Rameses II   1294-1228
: 500 : Crete-Philistines attack 5th year - Merneptah   1228-1218
: yrs : Abimelech of Judges (Ephemeral King - - - - Ammenemes = Abimelech 1225-1222)
1018 : v- -1213 B.C - 500th year of Tanus Stella - - - Sethos II   1218-1212
B.C. : ?   (Rameses) Siptah   1212-1211
v   v v(600th)   (Merneptah) Siptah   1211-1205

			Continued	Twsre (Irus)	B.C. 1205-1200

```
 ?   ?   ?  DYN XX                                       Setnakhte                              1200-1198
 |   |   | Philistines(Crete) attack 8th year -Rameses   III                                    1198-1166
 |   |   |                                     Rameses   IV                                     1166-1160
480  |   | 600th year of the Tanus Stella      Rameses   V                                      1160-1156
yrs  |   |                                     Rameses   VI                                     1156-1148
 |   |   |                                     Rameses   VII                                    1148-1147
 |  500  |                                     Rameses   VIII                                   1147-1140
 |  yrs  | Philistines mentioned latter years- Rameses   IX                                     1140-1135
 |   |   |                                     Rameses   X                                      1135-1132
 |   |   |-1113 BC - 600th year Tanus Stella-19th year- Rameses   XI                            1132-1105
 |   |    DYN XXI (David born 1092 ruled 1062-1022BC) Smendes (K. Saul 1102 BC)1105-1080
 |   |  Edom - Hadad in court of King Pseusennes-.- - -Pseusennes I (Q.Tahpenes)1080-1040
 |   |  Hadad married Tahpenes' sister    (Amenemnisu) Neferkheres                              1040-1036
 |   |  (David contemporary of Q.Tahpenes -I K.11:20,21)Amenofthis                             1036-1028
 |   |  1022 BC - Hadad to Edom at David's death - - - -Osochor                                1028-1022
 |- -  Solomon's 4th year - 1018 BC -Temple begun - - -Psinachas (King Solomon)-1022-1013
 |     Solomon married a daughter of Pharaoh 2 C.8:11- Pseusennes  II-1002 BC 1013-1000
 |     K ing Solomon was contemporary of the Egyptian- Siamun                                  1000- 983
 |     DYN XXI I Kings 11:40-Solomon contp. of - Sheshank I (Shishak)                          983- 962
 |- - 941 BC - 500 year of T. Bast in 21st year of- Osorkon I(Zerah)Ethiopian 962- 927
```

				Takeloth I	927- 920
				Osorkon II	920- 891
				Takeloth II	891- 872
				Sheshank III	872- 833
				Pamay	833- 827
				Sheshank IV	827- 791

```
             DYN XXIII(Uzziah's 36th 1st Olympiad- Psdibulat- in 776 BC)  791- 768
                                                   Osorkon III            768- 760
                                                   Psammus                760- 750
             DYN XXIV                              Piankhy                750- 734
                                                   Bocchoris              734- 731
             DYN XXV    (Ethiopian - Shabake =So or King-Shabalaka)(General 3 yrs) 731- 718
Egypt fell in 711 B.C. (Sibahki a General in 720 B.C.)-Shebitku /(Gen. 4 yrs)     718- 711
Sabbatic attack 713/12-Jubilee-Isa. 37:9,30;IIK.19:9 - Tirhakak /(Gen.714 BC)      711- 672
                                                   Tanutamon              672- 669
                                                   Necho I                669- 664
             DYN XXVI                 (Psamtik I) Psammetiches I          664- 610
                                                   Necho II               610- 596
                                     (Psamtik II) Psammetiches II         596- 589
Jer.43:8,9;44:30 - 586 BC - Israel to city of Tahpanhes Hophra (Apries)   589- 569
                                                   Amasis                 569- 526
                                    (Psamtik III) Psammetiches III        526- 525
             DYN XXVII                   (Persian) Campysis               525- 521
                                                   Smerdis                521- 521
                                       (Hytaspes) Darius I                521- 486
483 BC Vasti deposed & Esther Queen in 478 B.C. - - - -Xerxes I (Ahasueres) 486- 465
     (Purim - 473 B.C.)                            Artaxerxes I           465- 423
                                                   Darius II (Nothus)     423- 404
                                     (Muemon) Artaxerxes II               404- 359
                                      (Ochus) Artaxerxes III              359- 338
* (Josephus) Babylon built by Nimrod (2234 BC) & 1903) Arses              338- 336
years before Alexander the Great took Babylon in 331 BC-Darius III   B.C. 336-*331
```

2400 BC-Post-Flooder's acquired Sumer cuneiform writings,lists, & grammatical texts.
Carried by Ark   Noahic Flood - 2356-2355 B.C. - Dark Age - "History almost Silent"
& excavation. 1ST INTERMEDIATE PERIOD-Village Kish 2334BC-SargonI=Ham grandfather of Nimrod

2305 BC	Igris-Halam(ELBA DYN 2305-2205)Sargon I 1ST DYN AGADE(Akkad)B.C.2315-2258		
2275 BC	Ar-Ennum	- contp. of	" Tribal Life (Eber born - 2288*)
2260 BC	Ebrum(City)	- contp. of	Rimush (City Life)(Eber 30 yrs)2258-2249
(Son Sura-Damu)	Ebrum(took Mari)	- contp. of	Man-ishtu-shu 2249-2334
2230 BC	Ibbi-Sipis	- contp. of	Nimrod (Naram-Sin) **2234-2197
2220 BC	Dubuhu-Ada	- contp. of	" (City State - Babel - Gen. 10:10)
2210 BC	Irkab-Damu	- contp. of	Nimrod the grandson of Ham - Gen.10:6-8

2205 BC Elba fell to Nimrod 150 years after Flood-1600 years before Nebuchadnezzer-605 BC

1ST DYN LAGASH - - - - - Ur-bau (Governors)	"	2204 B.C.
*Eber-father of "Hebrew")Nam-makhni	Sharkali-sharr	2197-2172
race lived 285 years be-)Ur-gar	Iqiq In three years	2172-
fore Abram and lived 4)Dar-azag	Nanum four kings	
years longer. Gen.10:21;)Lu-Bau	Imi rule	
Luke 3:35 K.J.V.)Lu-Gula	Elulu Babel (Babylon) fell in BC -2169	
2ND DYN LAGASH - Gudean invasion - - - - - -	GUDEAN-(Dudu 2169-2148) Babel fell-	2169-2120
Ur-Nimgirshu - Gudea - 2140-2121 BC	(Su-Turul 2148-2133)	
Ur-lama - Gudeans defeated by - - - - - - -	-Utuhegal - 5TH DYNASTY ERECH	2120-2112
	LAW CODE - Ur-Nammu - 3RD DYNASTY OF UR	2112-2095
Nammakhni	Shugli	2095-2047
LARSA DYNASTY	Amar-Suen(Sin)	2047-2038
B.C.	Shu -Suen(Sin)	2038-2029
2025 Naplanum (Amurru invasion - 2022 B.C.)	Ishbi-Suen(Sin)-yr 4 Amurru attack-	2029-2005
	ISIN DYNASTY - - - - Ishbi-Erra	- 2017-2005
2004 Emisum (ABRAHAM BORN AT Ur - 2003 B.C.)	Invasion of the Elamites(Amurru)	- 2005-1997
	ISIN DYNASTY - - - - Ishbi-Erra	- 1997-1984
1976 Samium - Abram 27 years old - called at Ur in 1935/4)	Shu-ihishu	- 1984-1974
(B.C. & went to Haran & after death of Terah)	Iddin Dagon	- 1974-1953
1941 Zabaja (in 1928 BC Abram to Canaan-Gen.12:4;Acts 7:4)	Ishme-Dagon	- 1953-1935
1932 Gungunan (Chedorlaomer invasion 1934 BC-Elam)LAW CODE -	Lipit-Ishtar	- 1935-1923
1905 Abi-sare(Abram smites Elam-1920BC)Ilushuma smites Elam-1908 BC.Ur-Ninurta		- 1923-1895
1894 Sumu-ilu(1ST DYN BABYLON-Sumu-ilum 1894 BC	Bur-Sin	- 1895-1873
1865 Nur-Adad Sumulailum 1880 BC	Lipit-Enlil	- 1873-1868
1849 Sin-idinnam(Sodom fell 1904 BC-Isaac born next yr 1903)	Irra-imitti	- 1868-1861
1843 Sin-aribam-contemporary- Zabum 1844 BC	Enlil-bani	- 1861-1837
1841 Sin-iguisham " "(Jacob's Life 1843-1696)Zambyu		- 1837-1833
1836 Silli-Adad " "	Istar-pisa	- 1833-1830
1834 Warad-Sin Apilsin 1830 BC	Urdukuga	- 1830-1827
1822-1763 BC Rim-Sin - DYN ENDS	Sin-magir	- 1827-1816
1812 Ilumael(60)(SEALAND) Sinmuballit 1812 BC	Damin-ilisu	- 1816-1794
1752 Itti-ili-nibi(56) Hammurabi 1792 BC - LAW CODE(Joseph's Life 1752-1642 B.C.)		
1721-Revolt in Sealand Samsuiluma 1749 BC - 28th year of Samsuiluma revolt- -1721		
1696 Damiq-ilisu(36) Abi-eshub 1712 BC		
1660 Iskibal(15) Ammiditana 1684 BC	KASSITE DYNASTY	
1645 Ussi(29) Ammizadge 1646 BC	Candas -	1611-1595
1616 Gukisar(55) Shamash-ditanna 1626-1595 BC(Mursilis defeats Babel)Agum I 1595-1574		

**Josephus dates Nimrod's Babylon 1903 years before Alexander took it in 331 B.C.

1561 BC Perguldaramas(50)	Kashtiliash I	B.C. 1574–1553	(22)
(1st Dyn Sealand continues)	Ussi	1553–1546	(8)
: – Aberatash	1546–1545	(2)
: Kashtiliash II	1545–1543	(3)
(Tazzigarrumash) : Nazi–Maruttash I	1543–1540	(4)
These contemporary of Shamshi Adad III – – – : Harba–Shipak	1540–1540 mos?
: Tiptakzi	1540–1540 mos?
: Agum II	1540–1540 mos?
: – Meli–shipak I	1540–1522	(?)
1511 Adara–Kalamma(28) <u>Israel Exodus–1498 B.C.</u> Burnaburiash I	1522–1496	(?)
Kashtiliash III	1496–1493	(?)
1483 Akurulanna(26)(Puzur Ashur III contemporary– Ulamburiash)	1493–1445	(?)
1457 Melamkurkurra(7) Israel entered Canaan in the year – – – – – – – 1458 B.C.
1450–1441 Ea–gamel(9)–contemporary of Ulamburiash.Agum III	1445–1429	(17)?
Kara–indash(Reversed by	1429–1412	(18)?
Kadashman–Harbe I: some)1412–1410	(2)
<u>Mesopotamia in Israel(1418–1410BC) called on – – Kurigalzu I</u>	1410–1396	(15)
Kadashman–enlil I	1396–1394	(?)
Burnaburiash II	1394–1366	(29)
Kara–hardas?	1366–1366 mos?
Nuzi–bugash	1366–1364	(2)
Kurigalzu II	1364–1342	(23)
Enlil–Nasir III defeats Babylon – – – – – – – Nazi–maruttash II	1342–1318	25/26
Kadashman–turqu	1318–1302	17/18
These contemporary of Shalmaneser I – – – – – : – Kadashman–enlil II	1302–1288	15/7
: – Kudur–enlil(coreg.7 yrs)	1288–1288 mos.
689/8 BC Sennacherib took seal 600 years later – Shagarakti–shagiash(Seal)1288–1274	(13)
↑Kassite Dyn weakened–Elamite power in 1286 BC Kashtiliash IV	1275–1267	(8)
: – –1st–Interregnum	1267–1259	8/7
: Enlil–nadin–shum	1259–1257	1½
Tukulti–Ninurta I was + – – – – – – :Elam – –+ Kadeshman–harbe II	1257–1255	1½
contemporary of these :helps Adad–shum–iddin	1255–1249	(6)
:–v̇ – – – Adad–shum–nasir(usur)	1249–1219	(30)
Meli–shipak II	1219–1203	(15)
Merodach–baladan I – – Marduk–apal–iddin	1203–1193	(13)
Ashur–dan I attacked by Ilbaba–shum–iddin	1193–1192	1/6
Enlil–nadin–ahhe	1192–1190	(3)
Marduk–shapik–seri	1190–1174	(17)
Enurta–nadin–shum	1174–1169	(6)
1153 BC Elam power ends(132½ yrs) in 16th year– Nebuchadnezzar I	1169–1130	40/22
Enlil–nadin–apli	1130–1126	5/30
Sennacherib took Babylon 418 year later – 1107 Marduk–nadin–ahhe	1126–1105	22/15
(Famine in his 18th year↑in 1108 B.C. – – /)Itti–marduk–balatu	1105–1097	8/1
Contemporary–Ashur–bel–kala – – – – – – – – Marduk–shapik–ser–mati	1097–1086	(12)
Contemporary Ashur–bel–kala & Shamshi Adad IV – Adad–apal–iddin	1086–1065	(22)
Marduk–ahhe–eriba	1065–1063	(1½)
Marduk – Merodach–sum–adin	1063–1052	(12)
2ND DYN SEALAND Nabu–shum–libur	1052–1045	(8)
Eulmas–sakin–sumi (18) Shimmash–shipak	1045–1028	(18)
Ninurta–kudurra usur I (5 months) Ea–mukin–zer (Bel)	1028–1027	5mos
Kassu–nadin–ahhe (3) – – – – – – – – – – Kashshu–nadin–ahhe	1027–1025	(3)
600 yrs ↓ Buzi Dynasty – – Eulmash–shakin–sum	1025–1009	(17)
418 yrs Enurta–kudur–usur I	1009–1007	(3)

```
                                    Shuriglum-shugamuna       1007-1007 (3mos)
                        Elamite Dyn- Marbiti-apal-usur        1007-1002 (6)
            Dynasty of E - - - - - - -Nabu-mukin-apli         1002- 967 (36)
                                    Enurta-kudur-usur II       967- 966 (8mos)
                                    Marbiti-ah-iddin           966- 965 (8mos)
            418 years               Shamash-mudamiqiabt        965- 933 (32)?
                                    Nabu-shum-ukim I           933- 932 (2)
600 years      921 BC-Assyria against-Nabu-apal-iddim         932- 899 (34)
                                    Marduk-bel-usate           899- 896 (4)
                    | - - - - Marduk-zakir-shum I             896- 875 (22)
                    |          Marduk-nadin-shum               875- 865 (11)
                    |          Marduk-balatsu-uqbi             865- 864 (1)?
                    |          Bau-arh-iddin                   864- 856 ?
                        No. II ?   (          )               856- 840 ?
                                   (          )               840- 832 ?
                                    Adad-shum-ibin             832- 823 ?
                                    Shamas-Iva                 823- 810 ?
                                    Iva Lush                   810- 781 ?
                                    Ninurta apla-x             781- 781 ?
                                    Marduk-bel-zeri            781- 771 ?
                                    Marduk-apal-usur           771- 766 ?
                                    Iribu-marduk               766- 761 ?
                                    Nabu-shum-ukin II          761- 747 (15)
                                    Nabonassar(Nisan) Feb.27,747- 734 (15)
                                    Nabu-nadin-zer             734- 732 (3)
                                    Nabu-shum-ukin III         732- 732 (1mo.)
                                    Nabu-ukin-zeri             732- 729 (4)
                    Tiglath Pileser III - Pulu (Pul of Assyria) 729- 727 (3)
                                    Ulula                      727- 722 (6)
                    - - - - - - - - - - - - Merodach-baladan II 722- 709 (14)
                    |   Defeats Babylon - Sargon II of Assyria 709- 704 (6)
                    Same           2nd Interregnum - - - - - - 704- 703 (1)
                    | - - - - - - - - - - Merodach-baladan II  703- 701 (3)
                    Rebellion - - Bel-ibni                     701- 700 (1)
                                    Assur-nadin-sum            700- 694 (7)
                                    Nergal-ushezib             694- 693 (2)
                                    Mushezib-marduk            693- 689 (5)
- - Sennacherib took Babylon & Seal - - 3rd Interregnum - - - - - - - 689- 681 (9)
                                    Assur-akh-iddin            681- 672 (10)
                    Treaty with Esarhaddon - Shammash-shum-ukin 672- 650 (23)
                                    Ashurbanipal               650- 647 (14)
                                    (Civil War - - - - - - - - 649- 648)(2)
(1600 years - Nimrod destroyed Elba in 2205 BC). Kaudalann   647- 639 (9)
                                    Ashur-stillu-ili           639- 626 (14)
                        4th Interregnum - - - - - - - 626- 626 (mos.)
                                    Nabopalasser               626- 605 (22)
1st of Four World Empires begin - - - - - - - - - Nebuchadnezzer II  605- 562 (44)
            Jer. 27:7 - Amel-Marduk - Evil Merodack (Son)     562- 560 (3)
            Nergal-shar-usur - Neriglazzer(Brother-in-L)560- 556 (5)
            Jer. 27:7 - (Son) Nabonidus - Labashi-marduk      556- 553 (4)
                                    Belshazzer(co-reg.w/above553- 539 (15)
(Darius the Median) - Cyrus destroys Babylon - - - - - - - - - - - - Oct. 13,539 BC Dan.5:31
2nd World Empire - - - - - - - - - - - - - - - -Cyrus reigns     538 B.C.
```

Noahic Flood - 2356 - 2355 B.C. "HISTORY ALMOST SILENT" - THE DARK AGE

FIRST INTERMEDIATE PERIOD   80 years   B.C.

2260 BC Ibrum (Ebla Dynasty) Contemporary to Tudija(Tent Dwell)2275-2258

	:	Adamu	
	16	Kittamu	Jangi
	:	Mandaru	Harnaru
	Post-Flood	Harsu	Imsu
	:	Hanu	Didanu
	Tent Dwellers	Nuabu	Zuabu
	:	Belu	Abaza
	:	Uspia	Azarah
Monarchal rule restored - - -	City Dwellers-	APIASAL	2080 B.C.
	:- Ila-kabkabi - - -	Aminu	
	: Jakmesu	Jazkur-ilu	
Reversed in one chronlogy -	: Hajanu	Ilu-mer	
	:-Hale	Bamanu	
	Zariku(Kikkia)	Sulili	19 -1935 B.C.

(1920 BC-G.14:1 Elam defeat)-	Akia(Eri-Aku or Arioch)	1935-1920
(King Ellasar)	Puzur-Ashur I	1920-
	Shallam-ahi	
Ur freed in 1908 B.C. by -	Ilushuma	1910-1902
	Erishum I	1902-1862
	Ikunn	1862-1858
	Sargon I	1858-
Ila-kabkabu? -	Puzur-Ashur II	
	Naram-sin	
	Erishum II	
641 yrs-1814 BC-Temple of Anu-	Shamshi-Adad I	1814-1781 (33)
	Ishma-Dagon I	1781-1731 (50)
Rimus? -	Mut-Askur	1731-
	Asinum	Ruled about
	Puzur Sin	40 years
	Ashur-Dugal	1691-1686 (6)
6 rulers - Assur Apla-idi;	Nasir-sin	1686-
in less Sin Namir;	Ibzi-Istar	(6)
than six Adad-salulu;	Adasi	
years - - - - - - - - -	Belu-bani	1681-1672 (10)
	Libaja	1672-1656 (17)
	Sarma-Adad I	1656-1645 (12)
Ashur-dan I	Ip-tar-sin	1645-1634 (12)
relaid foundation	Bazaja	1634-1607 (28)
of Temple Anu	Lullaja	1607-1602 (6)
641 years later	Kidin-ninna	1602-1589 (14)
	Sarma-Adad II	1589-1587 (3)
Isrishum -	Erishum III	1587-1575 (13)
Contemporary of Agum I -	Shamshi Adad II	1575-1570 (6)
-forefather next two kings-	Ishme Dagan II (Father)	1570-1555 (16)
Aberatash to Meli-shipak -	Shamshi-Adad III (Son)	1555-1540 (16)
(Contemporary)	Ashur Nirari I	1540-1515 (26)
Ulamburiash contemporary -	Puzur-Ashur III (EXODUS)	1515-1492 (24)
1500 B.C. - - - - - - - -	Enlil Nasir I	1492-1480 (13)
	Nur-lil	1480-1469 (12)
↓ ↓	Ashur-Shaduni	1469-1469(1mo)

209

Annotation	King	Dates	(Years)
↑ ↑	Ashur-Rabi I	1469-1468	(?)
1500 (Israel Canaan-1458 BC) Walls neglected –	Ashur-nadin-ahhe I	1468-1450	(19)
	Ulamburiash	1450-1438	(13)
Mitannian	Enlil-Nasir II	1438-1433	(6)
domination	Ashur-Nirari II	1433-1427	(7)
	Ashur-bel-nisheshu	1427-1419	(9)
Mesopotamia-1418 BC-Chushan-rishathaim – –	Ashur-rim-nisheshu	1419-1412	(8)
641	Ashur-nakin-ahhe II	1412-1403	(10)
yrs	Eriba-Adad I	1403-1377	27/29
-1360 BC and rule ends 1275 B.C. – – – – –	Ashur-Uballit I	1377-1342	(36)
Babylon defeated by – – – – – – – –	Enlil-Nasir III (Bel)	1342-1333	(10)
	Arik-den-ili	1333-1322	(12)
	Adad-Nirari I	1322-1289	(33)
Temple 1261 BC(Kadashman-enlil & Kudur-enlil contp.-	Shalmaneser I(Library)	1289-1260	(30)
↑ Contemporary interregnum-	Tukulti-Ninurta I	1260-1224	(37)
	Ashur-nadin-apil	1224-1224	1/4
	Ashur-Nirari III	1224-1219	(6)
1219 B.C. Adad Shum Nasir contemporary of – –	Enlil-kudur-usur	1219-1215	(5)
	Ninurta-apal-ekur	1215-1203	13/3
– Temple foundation of Anu relaid in 1173 B.C.-	Ashur-dan I	1203-1158	(46)
	Ninurta-tukulti-Ashur	1158-1157	(2)?
Cambridge Anc. History, Part II – 60 yrs later	Mutakkil-nusku	1157-1153	(5)?
Temple complete	Ashur-resh-ishi I	1153-1136	(18)
1107 BC Marduk-nadin-ahhe defeat Assyria(1113 BC)	Tiglath Pileser I	1136-1098	(39)
↑Itti-Marduk-balatu & Marduk-shapik-seri contp.┐	Ashurid-apal-ekur II	1098-1097	(2)
Extend Shalmanser I's Library – /:-Ashur-bel-kala		1097-1080	(18)
Adad-apal-iddin contemporary of these Kings-:	Eriba-Adad II	1080-1079	(2)
	:-Shamshi-Adad IV	1079-1076	(4)
580	Ashur-Nasir-pal I	1076-1058	(19)
yrs	Shalmaneser II	1058-1047	(12)
	Ashur-Nirari IV	1047-1042	(6)
418 yrs later	Ashur-Rabi II	1042-1002	(41)
Sennacherib attacked	Ashur-resh-ishi II	1002- 998	(5)
Babylon	Tiglath-Pileser II	998- 977	22/32
	Ashur-dan II	977- 954	(24)
Assyria defeats the Amorites – – – –	Adad-Nirari II	954- 932	(23)
	Tukulti-Ninurta II	932- 926	(7)
	Ashur-Nasir-pal II	926- 904	(23)
Battle Karkar 902 BC.,K. Baalazar 909-903 contp-Shalmaneser III		904- 868	(37)
	Shamshi-Adad V	868- 855	(14)
Urartu (Armenia) begins attacks on Assyria – – Adad-Nirari III		855- 827	(29)
Jehoash contemporary of Assyrian King – – – – Shalmaneser IV		827- 817	(11)
(Contp. Uzziah)809 BC Jun.13 eclipse 8th yr of- Ashur-dan III		817- 799	(19)
Urartu destroys Royal family of King - Ashur-Nirari V		799- 791	(9)
772 B.C. Pul retakes Ramath area - URARTU RULES ASSYRIANS 46 YEARS - -		791- 745	(47)
Assyria defeats Urartu by Pul or as - Tiglath Pileser III		745- 727	(19)
	Shalmaneser V	727- 722	(6)
(Took Babylon - Sargon II - 709-704 BC)		722- 704	(19)
After 418 yrs Sennacherib destroys Babylon 689 BC-Sennacherib		704- 681	(24)
↓-Shalmaneser's Temple complete 580 yrs before King Ezarhaddon in 681 BC.	-681- 668	(14)
	Ashurbanipal	668- 633	(36)
(Smith Sidney, Early History of Assyria p. 355)	Ashuruballit I	633- 629	(5)
(NINEVEH FELL UNDER ASHUR-UBALLIT II 612 BC)	Ashur-stillu-ili	629- 612	(18)

APPENDIX J TABLE V
ALL BIBLICAL AND HISTORICAL CHRONOLOGY CORRELATED

The "Jubilee" year may have been set by the year of Adam's expulsion (4012 B.C.), since all "Jubilee Years" follow on the 12th and the 62nd years of each century.
(Adam's fellowship with the one true God - "Palaeolithic Age"- no sin-time not calculated-ageless)

BIBLICAL (Father Adam)	(Neolithic Age) EGYPTIAN	ASSYRIAN	BABYLONIAN
ADAM'S EDEN EXPULSION JUBILEE - 4012 B.C.	Ptah-Father god	Alulim	Aloros Adam ages-Sin
CAIN'S BIRTH? 4000 Cain?			Time begins
ABEL'S BIRTH?			
Jubilee - 3962 :			
3950 : - "Mark upon Cain" - to the land of Nod - East-Gen. 4:16			
Jubilee - 3912 :	Jarmo Culture		
3900		Hassuno	Eridu
SETH'S BIRTH (Adam's chosen son)- 3882 :	Ra - Sun god(Son)	Alalmar	Alaparos
Jubilee - 3862 :	Badarian	Samarra	
Jubilee - 3850 :		(Tepe Gawra VIII)	
ENOS' BIRTH (Spirit Blessed) - 3812	CHALCOLITHIC - Copper		
3777 :	Shu - spirit god	Emenluanna	Amelon
(Called on God - Gen. 4:26) 3762 - Jubilee			
Jubilee - 3712 :	Early Amratian		
3700	Enoch?		
CAINAN'S BIRTH 3687 :	Seb - earth god	Kichunna	Ammenon
Jubilee - 3662	(Gerzean)	From Iran-Ubaid	
MAHALALAD'S BIRTH 3617	Hosiri-death god	Enmengalanna Megalaros	
(FIRST WRITING) Jubilee - 3612 :	(Proto-Sumerian Period)		
Jubilee - 3562 :	Metal, scales, cylinder seals - Uruk North Ubaid		
JERED BIRTH means "descend"- sin -3552	Irad(Tammus-Fertility god)Dumuzi	Daonos	
The 6th Generation - Jubilee - 3512 :	Godly intermarried with Cainites (Mark) corruption		
3500	(Sumarian Period)	Early Proto-literate	
Jubilee - 3462 Mehujael?	(Cuneiform Writing)		
3450	(Gerzean - late)		
ENOCH'S BIRTH (Resurrected) 3412	(Tepe Gawra VII)	Early Dynasty I	
3390 :	Horus-sky god(Rapture)Sibzianna	Eudorachus	
Jubilee - 3362 :	Femdet Nasr	Late Proto-literate	
METHUSELAH'S BIRTH 3325	Thut-moon god	Emenduranna	Amenpzines
(Monarchal Rule) Jubilee - 3312 :	EGYPTIAN OLD KINGDOM		
3300 Methusael?	"Jawa" - Trans-Jordon Fortress		

211

BIBLICAL   EGYPTIAN   ASSYRIAN   BABYLONIAN

Carbon 14 date 3250 B.C. - - B.C. 3270 DYN I 1. Narmer (Menes) -1900 to Akhnaten's Jubilee 6th yr-1370 BC
                                 Early Literate

BIBLICAL	B.C.	EGYPTIAN	ASSYRIAN	BABYLONIAN
Carbon 14 date 3250 B.C. - - B.C.	3270 DYN I	1. Narmer (Menes)		
Jubilee -	3262	2. Athothis (Zer)		
	3250	3. Merit-nit		
Jubilee -	3212	4. Uenephes (Uadji)		
	3200	5. Usaphaidos		
	3162	6. Enezib		
LAMECH'S BIRTH	3138 :	Maat-truth god	Uburratum	Otiartos
Calender Mesoamerican Astec Olmec-3114 - long count (Aug. 13)				
Jubilee -	3112 Lamech?	7. Semempses		
ADAM DIED	3082	8. Bienaches (Ka'a)		Early Dynasty II
Jubilee -	3062	9. Ubienthes(Tepe Gawra VI)		
	3055 DYN II	1. Hotep-shani		
ENOCH RAPTURED - - -	3025	2. Ra-neb		
Jubilee -	3012 Jabel?	3. Neteri-nu (Neteren)		
	2980	4. Perabsen		
SETH DIED (SETH WORSHIP)- -	2970	- Seth worship by Perabsen and revived in 1713 B.C.(Sun god-Son of Adam)		
Jubilee -	2962	5. Khasekhem (Sendji)	Zinsuddu	(Flood)Xisuthros
NOAH'S BIRTH(Flood his 600th yr)-	2956	6. Ka-ra (Neferkara)		
Jubilee -	2912			
	2872			Late Literate
ENOS DIED	2862			
	2845			
Carbon 14 date 2770 B.C. - - -	2818 DYN III	7. Khasekhemui(27)		
Jubilee -	2812 Jubal?	1. Sanakhte(28)(Nebka 19) - - -		THE (9 yr co-reg.)?
:-Day Star Calender in his years-	2790 - Dates	(Tepe Gawra V)		EGYPTIAN
:CAINAN DIED	2777 correct	2. Zoser(Djoser-19 yrs)		PYRAMID
:-Day Star between 2785-2773 B.C.-	2771 -w/Day S.	3. Snofru(18)(13) Step Pyramid		KINGS
(Camb. Anc. H. I, Part) Jubilee -	2762 :	Bend Pyramid		OF Dates
(II, P. 170.	2753 DYN IV	1. Shaaru (24) Great		THE very
MALALALAD DIED	2722/2729	2. Cheops (Khufu 23 yrs)-Pyramid		THIRD accurate
Jubilee -	2712/2706	3. Chephren(Khfra 23 yrs) Giza		AND
	2683	4. Mycerimus (28) Pyramids		FOURTH
Jubilee -	2662/2655	5. Dadef-Re (6)		DYNASTIES
(Correct-Bible Arch. Revi.-1984)-	2649 - - -	6. Shepses-kaf(7) Sinai Mines		OF
	2642 :	7. Thamphis (9) - - - -		EGYPT
				LAND - - -

212

BIBLICAL	B.C.	EGYPTIAN	ASSYRIAN	BABYLONIAN
	2633 DYN V	1. USERKAF		
Jubilee – 2612		2. Sephres		
JERED DIED				
	2590 Tubal-Cain	3. Sisires (Shepses-ka-Re)		
Jubilee – 2562		4. Cheres (Neferf-Re)		
	2550	5. Rathures(Ni-uses)		
	2535	6. Mencheres(Men-kau-Hor)		
	2525	7. Djedkam (Ysesi)		
Jubilee – 2512		8. Unis (Wenis)		
	2500 DYN VI	1. Teti (Tety)		
"Spirit would not	2482	2. User-ka-Re		
always strive with man"	Jubilee – 2476 Ark begun	3. Pepy I (Noah's Ark begun 120 yrs before the Flood)		
	Jubilee – 2462	4. Meren-Re I (Menthesuphis)		
SHEM'S BIRTH	2453	5. Pepy II (90 year reign)		
	2450 Naamah? – Cainite with "mark" – perhaps the wife of Ham – in Ark			
	Jubilee – 2412		(Akkad or Accad)	
			Sumarian King Urukagina – 2388-2381	
	2388		Sumarian King Lugal-zaggisi – 2381-2356	
	2381	6. Menti-en-saf	(↑"With utmost distinctness it ends	
	2363	7. Neter-ka-Re	(an age" p.144,I Pt.2;Camb. A. History	
	Jubilee – 2362	1. Queen Neith-aqert – A woman rules-chaos before Flood		
LAMECH DIED"Darkest period 2361 DYN VII CHAOS – 70 rulers in 70 days – anarchy just before the Noahic Flood				
METHUSELAH DIED(Egypt H.) 2356 "From the 6th to the 10th Egyptian Dynasty history is almost silent"				
THE FLOOD BEGAN	2355 B.C.	"Splendour that was Egypt"– M.A. Murray		
THE FLOOD ENDS				
Sargon is Ham and carried-2354-over the Flood in tarred vessel – tradition story as was Moses'				
ARPHAXAD'S BIRTH-2 years -2353- after the Flood in 100th year of Shem – Gen. 11:10				
	2334 – Post-diluvian village of Kish			
SALAH'S BIRTH	2318	FIRST INTERMEDIATE PERIOD		
	2315 B.C. – Tribal village rule – – – – –		1ST DYNASTY OF AKKAD	
	Jubilee – 2312		Sargon (57) (Sharrun-kin)	
	2305 – – – – Tribal village of Ebla – – – – –		2305-2205 – –	
EBER'S BIRTH(Hebrews)		Nimrod's birth?-Naram-sin ="REBEL"		
	2288		Tudija 2275-2288	100
	2275 – 1st two of 16 Assyrian tent dwellers		2288-	yrs
	Jubilee – 2262	Assyrian tent dweller –Adamu		
	2258		Rimush (9)	

213

BIBLICAL   B.C.   EGYPTIAN   ASSYRIAN   BABYLONIAN

PELEG'S BIRTH-In his days-2254-was the earth divided"(Reinhabited - Gen. 10:25,32) 1ST DYN AKKAD
(1903 yrs from year of) 2249 (Salim-Jerusalem-referred to at Ebla) Man-ishtu-shu(15)
(Babylon attack-331 B.C.-2234 BC- Erech,Calneh,Accad,& Babel - Gen. 10:10 - Nimrod (37) 100
(by Alexander the Great) 2225 DYN VIII Neferkauhor I(28) (City State - Babel) yrs
REU'S BIRTH-Name means - 2224 - "FRIENDSHIP" - (THE TOWER OF BABYLON BUILT)Ebla
Jubilee - 2212
Nimrod destroys Ebla and-2205-in full power - 1600 years until Nebuchadnezzar II reigns-605 BC
2197 Neferkauhor II(22) Sharkali-sharri(25)
SERUG'S BIRTH-Name means-2192 - "FIRMNESS" (Infiltration by Guteans)
Heroic Age 2175 DYN IX Kheti I(11) 1600
Sumarian Legendary 2172 Iqiqi,Namum,Imi,Eluln(3) yrs
Period 2169 (Herakleopolis) Gutians(Lagash)destroy Babylon - FELL
NAHOR'S BIRTH-Name means-2162 - "SLAYER" - Jubilee - Language confounded - Gen.11:7
National Migration Period2154 DYN X Yntef I(34) Gutian persecution and migration
TERAH'S BIRTH-Name means-2133 - "WANDERING" (Herakleopolis) 5TH DYN OF ERECH
Yntef II(9) Gutians defeated by Utuhegel - Ur-Nammu
2120 3RD DYN UR-TOWER BUILT-LAW CODE - Ur-Nammu
2113
Jubilee - 2112/11 DYN XI Mentuhetep I(49) Shulgi
2095 Assyrian tent dwellers until Apiasal-monarchal
Jubilee - 2062 Mentuhetep II(8)
2054 Mentuhetep III(51) Tradition-Melchisedek built Jerusalem
2052 - Captured Herakleopolis
2047 - Amar-Suen
2038 Shu-Suen
(In Peleg's days there
was a migration period - 2032 B.C.-the present Korean Dynasties began)
2029 - (Invasion of Elamites-"Amurru" in 2025 & 2022 BC) - Ishbi-Suen-
2017 - ISIN DYNASTY - Ishbi-Erra
2015 - "In his days was the earth divided" - Gen. 10:5,25,32.
(AMURRU - ELAMITE PENETRATION)
-PELEG DIES - - - - Jubilee - 2012 Gen. 8:13 Elam
2011 Gen. 7:11.Influence
NAHOR DIES Ishbi-Erra
NOAH DIES-Lived 950 yrs- 2006 B.C. - Noah 600 years old at the Noahic Flood - Gen. 7:11.Influence
2005 B.C. - Elamites destroyed Ur and subjected them 8 yrs - Ishbi-Erra
B.C. Ur rebuilt by - Ishbi-Erra
2003 - ABRAHAM BORN-FATHER OF A MULTITUDE. Mentuhetep IV(12)
1997
1993 - SARAH BORN

214

BIBLICAL   ASSYRIAN   HAMITIC PROTO-PHILISTIM DYN XII   EGYPTIAN   ISIN DYN   BABYLONIAN

1991 B.C. HAMITIC PROTO-PHILISTIM DYN XII   Amenemhet I(30) = Abimelech I = Ammenemes   (AMURRU)
1985 - REU DIES (239 years old)   ELAMITE
1984 -   INFLUENCE
1974 - (Sinuhe to North Palestine)   IN THE
1971 - Shu-ihishu / Iddin-Dagon   Sesostris co-regent   GREAT
1962 - Jubilee SERUG DIES (230 yrs old)   FERTILE
1961 - Gen. 12:10-20 - Sesostries I(35)   CRESCENT
1953 -
1935 - Akia(Arioch)   Ishme-Dagon / Lipit-Ishtar
1934 - Abraham, Sarah, Terah, and Lot to Haran (Sinuhe to Egypt)ELAMITE CONTROL & TAKE NIPPUR
       Gen.14:1ff., Tidal, Amraphel of Shinar, Ellasar. Elam takes Ur and Palestine - -
1929 - Amenemhet co-regent
1928 - TERAH DIES. Abraham and Lot to Canaan - Famine - to Egypt Pharaoh Sesostris I   ELAMITES
1927 - Sarah taken to Pharaoh Sesostris's house - Plague - Gen. 12:10-20   SUBJECT
1926 - Gen. 20:32 - Abimelech II = Amenemhet II(32)   PALESTINE
1923 - 1920 BC - Puzur-Ashur I   Ur-Ninurta
1920 - Abraham rescues Lot & defeats Elamites in 14th yr as they attack Palestine & Hamites
1918 - Abraham takes Hagar (Egyptian) in 10th year in Canaan - Gen. 16:3
1917 - Ishmael born of Hagar
1915 - ARPHAZAD DIES at 438 years - born two years after the Noahic Flood - Gen. 11:10
1912 - Jubilee. 1910 BC-Ilushuma - in 1908 B.C. defeats the Elamites (Amurru) - ELAMITE DEFEAT---
1904 - ISAAC PROMISED; Sodom destroyed; Ishmael 13 years (Gen. 17) Covenant of circumcision
1903 - ISAAC BORN.1902 BC-Irishum I Lot begat Ben-ammi (Benjamites-wild Bedouin Tribe)
1901 - Abraham to Gerar. Abimelech (Amenemhet II) took Sarah and was baren - Gen. 20:18-
1900 - Isaac weaned 3rd year? Ishmael mocking at 17 years - Gen. 21:9.
1899 - Abraham and Abimelech II herdsman have trouble over a well - Gen. 21:25.   (32;34;12,15.
1898 - Abraham's covenant with Abimelech (Amenemhet II) 400 yr bondage-1898-1498 BC-Deut:7:2:Ex.23:
1897 - Sesostris co-reg.-Abram visit painted on his tomb   Ham
1895 - Bur-Sin(1ST DYN OF BABYLON)
1894 - Sumu-ilum (14) Canaan
1888 - Abraham's 40th yr in Canaan-Isaac offered as a lad 15 years of age - Gen. 22:2
1885 - SALAH DIES
1880 - Gen.10:15 - - Heth
         Sumulailum(36)
1879 - "Thrust into central Palestine(Shechem)Sesostris III(38) - Gen. 21:23
       DYNASTY XIII   Lipit-Enlil(5)
1873 - Zohar

215

BIBLICAL   ASSYRIAN   EGYPTIAN   ISIN DYN   BABYLONIAN   HITTITE

1872 - Day Star Sesostris 7th yr (Sirius Star) p. 84, C. Aldred, "The Egyptians"
    Irra-imitti(7)
1868 B.C.

1866 - SARAH DIES(127 yrs)-Cave of Machpelah bought from Hittites - 1866 B.C. - - - - - Ephron
    - - - Gen.23:7-10
1863 - Sabbatic - Isaac married in 40th year - Gen. 25:20 Hethites = Hittite-; - - - Gen.25:9,10
    Enlil-bani(24)
1862 - Jubilee year   Egypt "reclaimed the desert"
    Ikunn (4)
1861 -   Sargon I(44)
1858 -
1853 - SHEM DIES(Last Ante-Diluvian Patriarch)- lived 600 yrs(100 yrs 2 yrs after the Flood)
    Zabum(14)
1844 -
1843 - JACOB & ESAU BORN - Isaac 60 years old - Gen. 25:26
    Son's Heth,Zohar,Ephron -PRO-HITTITE K.
1841 -   Abimelech = Amenemhet III(38)   Gen.49:29,30;50:13
1837 - 3000 rooms-"Largest inhabitable building-Egypt"
    Zambyn(4)
    Istar-pisa(3)
1833 - Isaac 70 years old
    Urdukuga(3)
    Apilsin(18)
1830 -
1828 - ABRAHAM DIES(175 yrs) - Gen. 25:7
    Sin-magir(11)
1827/26 Jacob & Esau 17 years old - Gen. 25:27
1825 - Isaac's famine to Gerar - Abimelech III
1824 - EBER DIES(464 yrs)LAST PATRIARCH-Father of the "HEBREWS" - Gen.10:21;Lk.3:35 K.J.V.
1820 - Isaac's Beersheba well - Gen.26:25,32,33. Proto-Hittite Empire established - - Pithana (20)
1816 -   Damiq-ilisu
1814 -   Shamshi Adad I(33) - Built Temple of Anu
1812 - Jubilee year(Ilumael-60 yrs-Sealand Dynasty?)   Sinmuballit(20)
1803 - Esau takes a wife at 40 years of age - Gen. 26:34   1800 B.C. - Anita (20)
1793 -   Proto-Philistine - Amenemhet IV(9)=Abimelech IV=Ammenemes   Hammurabi(42) -|
1792 -   SECOND INTERMEDIATE PERIOD Code of Law - - -   10 yrs
1784 -   Sebknefrure(3) - Chaos
1782 -   Shamshi-Adad I disposed by Zimri-Lin(Mari)*Oath-legal doc.-10th -| Pirwa
1781/0 - Ishmael dies at 137 years (PERIOD OF CONFUSION - 1781-1746 B.C.)
1781/0 -   Ishme-Dagon I(49)
1775 -   (Area of Turkey)-HYKSOS DEFEAT THE KINGDOM OF THE PROTO-HITTITE
    DYN XIV Early Hyksos penetration into Palestine
1766 - JACOB TO HARAN(77 yrs)   Rim-Sin destroyed by Hammurabi ruled Babylonia.
1763 - Sabbatic yr ESAU TO EDOM   Hittite - HITTITE EMPIRE REESTABLISHED-MIGRATION - HITTITE E.
1762 - Jubilee HETHITES AGAIN TO TURKEY
1760(HIVITES-ISRAEL)HYKSOS TO PALESTINE - EGYPTIAN TO AFRICA & CRETE Hammurabi defeat Zimri-Lin
1759 - Jacob(84 yrs)takes Leah(hated)begets Reuben - Rachel also taken-7th yr-Gen.29:20,30,32

216

BIBLICAL   EGYPTIAN   ASSYRIAN   BABYLONIAN   HITTITE

1758 BC - Leah conceives Simeon - Gen. 29:33

1757 - Rachel baren - Gen.30:1;29:31   (PERIOD OF CONFUSION - 1786-1746 B.C.)

1756 - Leah conceives Levi - Gen. 29:35;Ex.6:16;Bilhah conceives Dan - Gen. 30:6

1755 - Leah conceives Judah - Gen. 29:35   MIGRATION

1754 - Leah left bearing - Gen. 30:9; Belhah conceives Naphtali - Gen. 30:8

1753 - Zilpah conceives Gad - Gen. 30:11   SEALAND DYNASTY - began in 1812 B.C.

1752 - Rachel conceives Joseph-14th yr-Gen.30:25:31:38;Zilpah conceives Asher-Gen.30:13

1751 - Zilpah begat Issachar-Gen.30:18;Joseph's 1st year of life

1750 - Leah begat Zebulun - Gen.30:20   Joseph's 2nd year

1749 - Leah begat Dinah - Gen. 30:21   Joseph's 3rd year   Samsuiluma(38)

1748 - Dinah's 1st yr. (HAMITIC EGYPTIAN PROTO-PHILISTINES to N. Africa, Crete, Caphtor).

1747 - Dinah's 2nd. (MIGRATION)   DYN XV Joseph's 5th year   (CONFUSION ENDS)

1746 - JACOB(97 yrs)CANAAN.RAMESES I(Army)Joseph's 6th year.HYKSOS INVADE EGYPT LAND-1746 BC.

1745 - Dinah's 4th yr. Jacob buys land at Shechem from Hivites-JACOB'S WELL-Gen.33:19.

1744 - Dinah's 5th yr.   Joseph's 8th year.

1743 - Dinah's 6th yr.   Joseph's 9th year.

1742 - Dinah's 7th yr.   Joseph's 10th year.

1741 - Dinah's 8th yr.   Joseph's 11th year.

1740 - Dinah's 9th yr. 21 years since Hethite to Turkey - The NEO-HITTITE EMPIRE-Tudhaliyas I(29)

1739 - Dinah's 10th year.   Joseph's 13th year.

1738 - Dinah's 11th year.   (Rameses I consolidating country under his army rule)

1737 - Dinah's 12th year.   Joseph's 15th year.

1736 - Dinah's defiled 13th yr - Jacob to Hebron & Benjamin born as Rachel dies - Gen.34,35.

1735 - JOSEPH TO SHECHEM & DOTHAM-sold in Dothan's 17th year to Ishmaelites - Gen. 37:2.

1734 - JACOB BUILT HEBRON(Arbah)7 YEARS BEFORE ZOAN - Num. 13:22; Gen. 35:27.

1733 - Servant in Egypt during Joseph's 19th yr.   Hebron built

1732 - Servant in Egypt during Joseph's 20th yr.   7 years be-

1731 - Servant in Egypt during Joseph's 21st   Mut-Askur   fore Zoan.

1730 - Ex. 1:11 - SETHOS-SALITOS AS RAMESES II HYKSOS KING - Num. 33:5; Gen. 47:11.

1729 - Servant in Egypt during Joseph's 23rd yr.   Num. 33:22;Gen. 35:27.

1728 - Servant in Egypt during Joseph's 24th yr.

1727 - Cities of Raameses & Pithom(Zoan) built by Rameses II 7 years after Hebron

1726 - Servant in Egypt during Joseph's 26th yr.

1725 - Servant in Egypt during Joseph's 27th yr.

1724 B.C. Put in prison in Egypt during Joseph's 28th year-Gen.41:1,46(11th yr in Egypt)

217

| | EGYPTIAN | ASSYRIAN | BABYLONIAN | HITTITE |

BIBLICAL
1723 BC - ISAAC DIES - 180 yrs. Joseph in prison in Egypt Rimus ?
1722 - JOSEPH RELEASED & 2nd to Rameses II(7 yrs plenty)Add graneries to city Zoan -|- Gen.41:46
1721 - Joseph married-Pithom(Zoan) & Rameses store cities. Sealand revolt 28th year -|-Samsuiluma
1720 - Manasseh son of Joseph born to him in Egypt (Tanus) Zoan built over 14
1719 - Joseph's 33rd year of age & the 3rd year of plenty   years and dedicated
1718 - Joseph's 34th year of age & the 4th year of plenty   in 1713 B.C.
1717 - Ephraim son of Joseph born to him in Egypt
1716 - Joseph's 36th year of age & the 6th year of plenty
1715 - Joseph's 37th year of age & the 7th year of plenty ends.
1714 - FAMINE - 38th year Joseph's famine begins.
1713 - Sabbatic-39th year of Joseph REVEALED-Rebirth god Seth-Typhon famine god. Jacob(130)-Egypt
1713 - Rameses II's greatest power over cattle,lands,people-400 yr Tanus Stella-dedicated-1713-1313
1712 - Jubilee Famine's 3rd year.   Asinum ?   Abi-Eshub(28)   B.C.
1711 -   Famine's 4th year.   Apepi I(6)   Pu-sarrumas(31)
1710 -   Famine's 5th year.
1709 -   Famine's 6th year.
1708 -   Famine's 7th year ends in Egypt and Palestine
1707/6 - Kohath born (133 yrs) Ex. 6:18 ?   Puzur-Sin ?
1705 -   Apephis(61) - Ruled over more than Egypt (Palestine)
1697 - Jacob blesses Ephraim & Manasseh along with his own sons - Gen. 48,49.
1696 - JACOB DIES IN EGYPT (147 yrs) - Gen. 47:28;49:33.
1691 -   Ashur-Dugul(6)
1686 -   -- Ashur Apla-idi ?
1685 -   Nasir-Sin ?   Ammiditana(38)
1684 -   Six rulers in -- --   Sin Namir ?
1683 -   less than 6 years   Ibzi-Istar ?
1682 -   Adad-Salulu?
1681 -   -Adasi;Belu-bani(10)
1680 -   Labarnas I(30)
1672 -   Libaja(17)
1663 - Sabbatic year
1662 - Jubilee year
1656 -   Sarma-Adad I(12)
1654 - Khayan co-regency(10) with Apephis in Palestine.
1650 -   Hittite cuneiform -Labarnas II?= Hattusillis II(30)

BIBLICAL   EGYPT(HYKSOS)   ASSYRIAN   BABYLONIAN   HITTITE

1646/5 - Dynasties XV & XVI in opposition   Ip-tar-sin(12) Ammizaduge(20)

1644 B.C. Palestine chief reigns as Khayan(Khian)(30) - ruled over more than Egypt - Palestine

1643 - On tomb of Baba of El-Kab: -- "I collected corn as a friend of the harvest God,

1642 - JOSEPH DIES   I was watchful at the time of sowing, When a

1641 - ARAM BORN(137 yrs) Ex. 6:20   famine arose, lasting many years, I distributed

1640 - Khayan destroys Thebes   corn to the city each year of the famine" Halley's Hand B.

1634 -   page 105

1626 -   Bazaja(28)   Shamash-ditana(31)

1620 -   (26)Mursilis I

1619 - LEVI DIES(137 yrs) - Ex. 6:16   Aa-Seh-re(3)

1614 -

1613 - Sabbatic - 100th year of the Repetition of the Birth of god Seth - 1713 to 1613 B.C.
                                          KASSITE DYN

1612 - Jubilee year   Gandas(16)

1611 -   DYN XVII - Apepi II(31)

1607 -

1602 -   Lullaja(6)

1595 -   Kidin-ninna(14)

1594 -   Agum I(22) - attacked by "

1589 -   Hantiles I(34)

1587 -   Sarma-Adad II(3)

1582 -   Erishum III(13)

1581 - AARON'S BIRTH - Numbers 33:38,39; Ex. 7:7

1580 - "Knew not Joseph" - - - Ahmose I(23) - Nefretiri(wife)

DYN XVIII - HYKSOS EXPELLED FROM EGYPT - PITHOM & RAAMSES DESTROYED

1579 -   (brother and sister)

1578 - MOSES BORN - saved by princess Set-Kamose -|-(Tethmois or Thermuthis)

1575 -

1574 - KOHATH DIES(133 yrs) - EX. 6:18.

1570 - Shamshi Adad I a forefather of 2 following Kings -Ishme Dagon II(16)   Shamshi Adad II(6)-contp. Agum I.

                                      Kashtiliash I(22)

1563 - Sabbatic year

1562 - Jubilee year

1560 -   Merit-Amon   Ahotep

1559 -   Zidantas II(9)

1557 - (Concubine) Senisonb Amenophis I(21)Set-Kamose(Tethmose or Thermuthis)

1555 -   (brother and sister)   Shamshi Adad III(16)

1553 -   Ussi(8)

	HITTITE Ammunas

BIBLICAL B.C.	EGYPTIAN	ASSYRIAN	BABYLON	HITTITE
1550 B.C.				
1546				
1545 – JOSHUA BORN	(Hebrew Slavery)		Aberatash(2) Kashtiliash II(3) Nazi-Maruttash I(4)	
1543 – REBUILT ZOAN(Pithom) & RAAMSES– Ex.1:11		Ashur Nirari I(26)	Harba-Shipak;Tiptakzi;	
1540 – MOSES TO MIDIAN	(40) Acts 7:23-29 –	PITHOM,RAAMSES REBUILT	Agum II;Meli-Shipak I	
1538 –				
1537 – CALEB'S BIRTH				
1536 – Mutnofret THUTMOSE I(26)	Ashmes (Ahmose)			
1535 – (Sister & brother)	(Half sister) – – – – Hatshepsut born			
1530 – Gershon born?				Huzziyas I
1528 – Thutmose II born				
1525 –				Telipinus
1522 – (JEWS BORN AFTER PASSOVER – SINS NOT IMPUTED FROM YEAR			Burnaburiash I ?	
1518 – (1518 BC TO EXODUS – Num. 32:11)–	– That Generation died 100 yrs later–1418 BC-Longivity			
1515 – Thutmose I's campaign to Narharia	Puzur-Ashur III(24)-restored Ashur-Ishtar Temple			
1513 – Sabbatic Repetition of the birth of Seth	*Frontier Treaty_ *			
1512 – 50 JUBILEES – – – since expulsion	– 2500 years ago. Exodus two sabbatic years later.			
1510 – (Conc.)Isis THUTMOSE II(12) Hatshepsut	– – – – at age 25			Tuhurwaili
1509 – Age 19				age 26
1508 – Age 20	1st born son			age 27
1507 – Age 21	Daughters			age 28
1506 – Age 22 Nefrure	Ra'nafru			age 29
1505 – Sabbatic Born–ThutmoseIIIAge 23				age 30
1504 – Thutmose III 1 yr old Age 24				age 31
1504/3 –AMRAN DIES(137) yrs old Age 25				age 32
1502 – 3 yrs old Age 26				age 33
1501 – MITTANIAN 4 yrs old Age 27				age 34
1500 – Parattarna 5 yrs old Age 28	Son dies(10 yrs) at Exodus(1498 BC)age 35			
1499 – MOSES RETURNS(Acts 7:30) Age 29				age 36
1498 – SABBATIC EXODUS 7 yrs old	Age 30–dies in Red Sea(Passover)Hatshepsut reigns–son dies.			
1497 – Thutmose III 2nd yr co-reign 8 yrs old	Hatshepsut's 2nd yr reign age 38			
1496 –MOSES AT KADESH3rd yr co-reign	Hatshepsut rebuilt temple(Sinai)	Kashtiliash III		
1495 – Thutmose III 4th yr co-reign 10 yrs old	Hatshepsut's 4th yr reign age 40			
1494 – Thutmose III 5th yr co-reign 11 yrs old	Hatshepsut's 5th yr reign age 41			

40 yrs

BIBLICAL	EGYPTIAN	MITTANIAN	ASSYRIAN	BABYLONIAN	HITTITE
1493 B.C.	Thutmose III	6th yr co-reign	12 yrs old	Hatshepsut's 6th yr reign Ulamburiash age 43	↑
1492 -	Thutmose III	7th yr co-reign	13 yrs Enlil Nasir I(13)	Hatshepsut's 7th yr reign age 44	40 yrs
1491 - SABBATIC	" III	8th yr co-reign	14 yrs old	Hatshepsut's 8th yr reign age 45	
1490 -	Thutmose III	9th yr co-reign	15 yrs old	Hatshepsut's to Pwene age 45	
1489 -	Thutmose III	10th yr co-reign	16 yrs old	Hatshepsut's 10th yr reign age 46	
1488 -	Thutmose III	11th yr co-reign	17 yrs old	Hatshepsut's 11th yr reign age 47	
1487 -	Thutmose III	12th yr co-reign	18 yrs old	Hatshepsut's 12th yr reign age 48	
1486 -	Thutmose III	13th yr co-reign	19 yrs old	Hatshepsut's 13th yr reign age 49	
1485 -	Thutmose III	14th yr co-reign	20 yrs old	Hatshepsut's 14th yr reign age 50	
1484 - SABBATIC	" III	15th yr co-reign	21 yrs old	Hatshepsut's 15th yr reign age 51	in
1483 -	Thutmose III	16th yr	Temple Thebes begun -	Hatshepsut's 16th yr reign age 52	
1482 -	Thutmose III	17th yr co-reign	23 yrs old	Hatshepsut's 17th yr reign age 53	
1481 -	Thutmose III	18th yr co-reign	24 yrs old	Hatshepsut's 18th yr reign age 54	
1480 -	Thutmose III	19th yr co-reign	Nur-lil (12)	Hatshepsut's 19th yr reign	Tudhaliya II
1479 -	Thutmose III	20th yr co-reign	26 yrs	Full power till 20th yr reign age 56	
1478 -	ISRAEL CONTINUES AT KADESH		27 yrs old	Hatshepsut's 21st yr reign age 57	
1477 - SABBATIC	" III	REIGNS IN HIS -	28TH YEAR.	HATSHEPSUT DIES AT THE -- AGE 58	the
1476 -	Thutmose III	"Hornet" EX.23:28;Deut.7:20		1st of 14 campaigns-Palestine-MEGIDDO	
1475 -	Thutmose III	24th Saussatar	30 yrs old	in his 2nd year reign alone	
1474 -	Thutmose III	25th yr reign as	31 yrs old	in his 3rd year reign alone	
1473 -	Thutmose III	26th yr reign as	32 yrs old	in his 4th year reign alone	
1472 -	Thutmose III	27th yr reign as	33 yrs old	in his 5th year reign alone	
1471 -	Thutmose III	28th yr reign as	34 yrs old	in his 6th year reign alone	Wilderness
1470 - SABBATIC	" III	29th yr reign as	35 yrs old	in his 7th year reign alone	Huzziyas II
1469 -	Thutmose III	's 6th campaign	Ashur-Shaduni(mo)Ashur-Rabi I		
1468 -	Thutmose III	31st yr reign as	37 yrs old Ashur-nadin-ahhe I(19)		
1467 -	Thutmose III	32nd yr reign as	38 yrs old	in his 10th yr reign alone	
1466 -	Thutmose III	's 8th campaign &	39 yrs old	. Destroys Carchemish,Aleppo	in 1st year
1465 -	Thutmose III	34th yr reign as	40 yrs old	in his 12th year reign alone	2nd year
1464 -	Thutmose III	35th yr reign as	41 yrs old	in his 13th year reign alone	3rd year
1463 - SABBATIC	" III	36th yr reign as	42 yrs old	in his 14th year reign alone	4th year
1462 - JUBILEE	" III	37th yr reign as	43 yrs old	in his 15th year reign alone	5th year
1461 -	Thutmose III	38th yr reign as	44 yrs old	in his 16th year reign alone	6th year
1460 -	Thutmose III	39th yr reign as	45 yrs old	in his 17th year reign alone	Tudhaliyas II
1459/8	MOSES DIES - ISRAEL IN WILDERNESS 40 YEARS - Acts 7:36			JOSHUA-Canaan-1458 BC	8th year ↓

BIBLICAL	EGYPTIAN	MITTANIAN	ASSYRIAN	BABYLONIAN	HITTITE

1458 BC JOSHUA TO CANAAN Thut. III took Syria,Palestine - 19th year (HORNET) in 9th year-40th yr.

1457 - Thutmose III 42nd stayed on shafalah in his 20th & Kadesh taken on the Orontes

1456 - Thutmose III 43rd yr reign & 49 yrs old. His 21st yr stayed in Egypt in 1st year 7

1455 - SABBATIC " III 44th yr reign & 50 yrs old. His 22nd yr alone & at home in 2nd year yrs

1454 - Thutmose III 45th yr reign & 51 yrs old. His 23rd yr alone & at home in 3rd year(take

1453 - Thutmose III 46th yr reign & 52 yrs old. His 24th yr alone & at home in 4th (Canaan

1452 - Thutmose III 47th yr reign & 53 yrs old. His 25th yr alone & at home in 5th year

1451 - Thutmose III 48th yr(Canaan land conquered by Joshua after 7 yrs - Josh.14:7-14) -:

1450 - Bethshen restored by Egypt - See Jud.1:27;Josh.17:12-16 yr alone &at home in 7th year

1449 - Thutmose III's Nubian campaign. His 28th yr alone & at home in 8th year

1448 - SABBATIC " III 51st yr reign & 57 yrs old. His 29th yr alone & at home in 9th year

1447 - Thutmose III 52nd yr reign & 58 yrs old. His 30th yr alone & at home - 10th year

1446 - Thutmose III 53rd yr reign & 59 yrs old. His 31st yr alone & at home - 11th year

1445 B.C. Thutmose III dies in Egypt & 60 yrs old. His 32nd yr Agum III NO BATTLES FOR 12 YEARS

1444 - Amenophis II(23)1st year

1443 - Amenophis II's 2nd Artatama

1442 - Amenophis II's 3rd year and has a Syrian campaign. SEALAND DYNASTY ENDS - 1441BC

1441 - SABBATIC " II's 4th year built Temple of Bast-500 years to 21st yr of Zerah-OsorkomII

1440 - Amenophis II's 5th year Arnuwandas I

1439 - Amenophis II's 6th year

1438 - Amenophis II's 7th year attacked- Enlil-Nasir II(6) - Shemash-Edom 3600 Hebrew taken

1437 - Amenophis II's 8th year

1436 - Amenophis II's 9th year campaign to Galilee(coast & Megiddo)-Israel on hill country

1435 - Amenophis II's 10th year

1434/3 JOSHUA DIES(110 YRS) 25 years Canaan - 1000 years to Nehemiah's work ended-434/3 BC Neh8:17

1433 - Amenophis II's 12th year Ashur-Nirari II(6)

1432 - Amenophis II's 13th year

1429 - Judges 1:26 - City of Luz built in the land of the Hittites. Kara-Indash(18) City of Luz

1427 - SABBATIC " II's 18th year Ashur-bel-nisheshu(9)

1422 - Amenophis II's 23rd year dies. * Covenant *

1421 - Thutmose IV(8) - wife Mutemuya daughter of Artatama (Alliance) - reigned 1443 BC

1420 - SABBATIC * Alliance * Hattusillis II

1419/8-A Generation had died-Num.32:11 -100 yrs Ashur-rim-nisheshu(8) - 1518 to 1418 BC-Jud. 2:10

1418 - Mesopotamia(8) - Cushan-rishathaim (100 years from 1518 B.C.- New Generation - Jud. 2:10)

BIBLICAL B.C.	EGYPTIAN	MITTANIAN	ASSYRIAN	BABYLONIAN	HITTITE
1414 B.C.					
1413 - SABBATIC	Amenophis III(37)	Shattarna I(2) - daughter Gilu-khepa			
	Year of Repetition of Birth of Seth - 300 yrs - 1713-1413 BC				
1412 - JUBILEE				Kadashman-harbe I(2)	
1412 - Mesopotamian prisoners in Israel-Egypt.			Ashur-nadin-ahhe II(10)		
1410 - OTHNIAL(40) Mesopotamia (Canaan) seeks			help from - - - King Kurigalzu I(15)		
1403			Eriba-Adad I(27/29)		
1400					Tudhaliyas III
1396				Kadashman-enlil I?	
1394				Burnaburiash II(29)	
1377					ArnuwandasI
			Ashur-uballit I(26)		Suppiluliman I(37)
1376 -(Akhnaten)Amenophis IV(17)		Tushratta-Artatama(25) Letter			
1375		Tushratta-Artatama			
1370 - EGLON(18) Moabites rule Israel: - Akhnaten occupied Tel El Amarna-6th yr Jubilee-1900th yr of Menes					
				Karahardas(mos)	
1366	A CALL FOR				
1366	EGYPTIAN HELP			Nuzi-bugash(2)	
1364	IN THE EL			Kurigalzu II(23)	
1363 - SABBATIC YR	AMARNA LETTERS	- Letter No. 68 from Shuwardata, Prince of Hebron			
1362 - JUBILEE YEAR					
1359	Smenkhkare(3)				
1356	Tutankhamun(9)		* CONTEMPORARY KINGS		
1352 - EHUD(20) - - - - -:80 yrs: - Judges 3:30 - The land had rest from the Moabite to Midians					
1350	the	:Artatama II			
1347	Ay(4)				
1343	Haremhab(27)				
1342	land	:Mattiwaza	Enlil-Nasir III-defeats-Nazi-maruttash II(26)		
1339					Arnuwanda II(1/5)
1338					Mursillis II(28)
1333	had		Arik-den-ili(12)		
1332 - JABIN(20) - border - :- incursions by Mursillis II who ruled Galilee by Jabin(Hazor)					
1328 B.C. - Eclipse in the- - - - - 10th year of Mursillis.					
1322	rest	:Shattuara I	Adad-Nirari I(33)		
1320			* Border agreement *		
1318				Kadashman-turqu(17)	

223

BIBLICAL	EGYPTIAN	MITTANIAN	ASSYRIAN	BABYLONIAN	HITTITE
1316 B.C.(20mos)	Rameses I	rest :		*Contemporary Kings	
1315 -	Seti I(20)	:-Rowton's Chron.-Seti attacked Murillis in Galilee unto Beirut		*	
1313 - SABBATIC YEAR		:- in 2nd year. Repetition birth of Seth(400th)since Jacob-Egypt.			
1312 - JUBILEE BARAK(40)		in :			Muwatallis(24)
1310 -		..			
1302 -		..	* Letter	Kadashman-enlil II(15/7)	
1294 - (66)	Rameses II	the :-wives Neferteri(ManeFrure)-			
1289/8 B.C.		..	Shalmaneser I(30)	Kudur-enlil(mos/7)	
1289/8 - Shagarakti's seal -		-taken 600 yrs later in 689 BC -	Shagarakti-shuriash(13)		
1286 -		land :- (Elamite infiltration)		Urhi-Teshub-Musillis III(7)	
1279 - *		- - - - :- -*-		Hattusilis III(26)	
1276 -		80 :Shattuara II(I)-			
1275 -		:Kartasura - conquered by-*		Kashtiliash IV(8)	
1274 - ELI BORN		:years:			
1273/2 - Rameses' 21st year:		:-a treaty with Hattusilis III			
1272 - MIDIANITES RULE ISRAEL (7)		- "Reduce stronghold of rebellion in Canaan" - Rameses II.		Interregnum(8)	
1266 - Midianite influence		..	Library collected		
1265 - GIDEON(40)					
1263 - SABBATIC YEAR					
1262 - JUBILEE YEAR					
1261 - 580 yrs before Esarhaddon(681 BC) King Shalmaneser I finished Ashur's Temple.					
1260 -			Tukulti-Ninurta I(37) - - - *took 28000 Hittites		
1259 -			Library collected	Enlil-nadin-shum(1½)	
1257 -				Kadeshman-harbe II(1½)	
1255 - (Amurru - Elamite influence)				Adad-shum-iddin(6)	
1253 -		Hurrians battle with - - - - -			Tudhaliya IV(30)
1249 -				Adad-shum-nasir(usur)(30)	
1240 - (Elam) Brother-in-law of King of Amurru, Istarmuwa makes a treaty with - - - - *					
1232 - In his 28th year & 7 years before his death Tukulti-Ninurta I conquered Babylon.					
1228 B.C.	Merneptah(10)				
1225 - ABIMELECH(3)-Amenmesse ephemeral Egypt King - "Israel is desolate, its seed is not"					
1224 - Egyptian Influence			Ashur-nadin-apli(1/4)		

BIBLICAL   EGYPTIAN   ASSYRIAN   BABYLONIAN   HITTITE

ASSYRIAN: Ashur-Nirari III(6)
HITTITE: Arnuwanda III(11)

B.C.	BIBLICAL	EGYPTIAN	ASSYRIAN	BABYLONIAN	HITTITE
1224					
1223			Enlil-kudur-usur(5) / Melī-šipak II(15)		
1222	TOLLA(23)		*	*	
1219		Sethos II(6)			
1218	(Eli's 40 year ministry begins)				
1216			Ninurta-apal-ekur(13) - - - -*	- - - -*	
1215					
1214					
1213	SABBATIC – Repetition of birth of Seth(500th year)–story told like unto Joseph's.				Tudhaliya V(17)
1212	JUBILEE YEAR	Siptah(Rameses)			
1211		Siptah(Merenptah)(6)	(Merodach Baladan I)Marduk-apal-iddin(13)		
1205		Twsre(Trus)(5)	Ashur-dan I(46)	*	
1203			*		
1202		DYNASTY XX			
1200		Setnakhte(2)			
1199	JAIR(22)				
1198		Rameses III(32)			
1195					Arnuwandas IV(5)
1193		Rameses III attacked by Amurru(Elam) *- - -	Ilbaba-shum-iddim(1/6)		BC
1193/2		– Libyans attacked Egypt. (Elam weakened)	Enlil-nadin-ahhe(3)SuppiluliumaII–Fell–1190		
1189		Philistines(Crete)attacked Rameses – 9th year & 13 years later controlled Israel–1176 BC			
1188	SAMUEL BORN		300 years back to Canaan entrance –		
1177	Period of Sin(1)	Philistines attack west & Ammon attacked Israel east of Jordon.			
1176	Philistine(20) & Ammon(18)	–rules Israel. Eli 98 years dies & Samuel 12 years old.			
1174			Enurta-nadin-shum(6)	*	
1173		Temple foundation of Anu relaid by Ashur-dan I 641 yrs after laid by Shamshi-Adad I.			
1169			Nebuchadnezzar I(40/22)		
1166		Rameses IV(6)			
1163	SABBATIC YEAR		16th year		
1162	JUBILEE YEAR				
1160		Rameses V(4)			
1158	JEPHTAH(6)	Ammon's defeat after 18 yrs & 300 years from the Canaan entry – Judges 11:26	Ninurta-tukulti-Ashur(2)		
1158					

225

BIBLICAL   EGYPTIAN   ASSYRIAN   BABYLONIAN

Mutakkil-musku(5)

Rameses: VI(6)-Philistine defeat – I Sam.7:1:41. | Peace with Amorites
in Nebuchadnezzer's 16th.   Ashur-resh-ishi I(18)   16th of Nebuchadnezzer

Rameses   VII(1)
Rameses   VIII(7)

No peace in Assyria since Babylon took it 90 years before.

Rameses   IX(5)

Tiglath Pileser I(39)

Rameses   X (3)
Rameses   XI(27)

Enlil-nadin-apli(5/30)

Famine in 18th year—Marduk-nadin-ahhe(22/15)
T. Pileser I finish Temple Anu 60 yrs after Ashur-dan relaid foundation.
Repetition birth(7th)600th year in 19th year Rameses XI – Seth Worship.
(Tiglath Pileser I defeated by Marduk-nadin-ahhe in
1107/6 B.C. 418 years before Sennacherib took Babylon in 689/8 B.C.)
Smendes(25)DYN XXI   Itti-marduk-balatu(8/1)
H. Rassam date of T. Pileser rebuilt temple. See our date 1113 B.C.
– I Sam. 9:16,17;11:11 – Slew Ammon.
– Israel fights Philistines – I Sam. 13:1;14:31
     Ashurid-apal-ekur II(2)   (12)
     Ashur-bel-kala(17)   Marduk-shapik-seri-mati

     Adad-apal-iddin(22)

– oppression – I Sam. 14:47,48,52.

Eriba-Adad(2)
Shamshi-Adad IV(4)
Temple Ishtar restored in Nineveh
Ashur-Nasir-pal I(19)
Prayer to Ishtar   Marduk-ahhe-ariba(1½)

BIBLICAL		
1157 B.C.		
1156 –	ARK RELEASED	
1153 –	Elam's power ends	
1152 –	IBZON(7)	
1148 –		
1147 –		
1142 –	ILON(10)	
1140 –		
1136 –		
1135 –	ABDON(8)	
1132 –		
1130 –		
1127 –	SAMSON(20)	
1126 –		40
1113 –	SABBATIC YR	
1113 –	JUBILEE YR	
1107/6	NO RULER(5)	
1105 –		
1103 B.C. –		yrs
1102 –	SAUL(40)	
1101 –	2nd year –	
1098 –		
1097 –	oppression(Philistine)	
1092 –	DAVID BORN	
1087 –	End 40 yrs–	
1086 –		
1084/3 –	Saul's 18th year Samuel dies – Josephus?	
	Pseusennes I(40/46)	
1080 –		
1079 –		
1078 –		
1076 –		
1075 –	David 17 years old kills Goliath – I Sam. 17.	
1065 –	Samuel dies 3 years before David's reign (123 yrs)	

226

	BIBLICAL	EGYPTIAN	ASSYRIAN	BABYLONIAN
1063	SABBATIC David's accession-477 years to Jerusalem's fall-586 BC.-			Marduk-sum-adin(12) --|
1062	JUBILEE DAVID(40) - - - - -		436 years back to Egyptian Captivity - 1498 B.C.	Josephus - 477 years
1058			Shalmaneser II(12)	to Babylon's
1056			*	
1052/1			Shalmaneser's 8th year attacked -	Nabu-shum-libur(8)
1047			Ashur-Nirari IV(6)	
1045/4			*	Shimmash-shipak(18)
1042			Ashur-rabi II(41)	Captivity
1042	Central year of David's reign -		-456 years to both Egyptian & Babylonian Captivities.	586 B.C.
1040		Neferkhares(4)		
1036		Amenofthis(8)		
1028 B.C.		Oschor(6)? -	- Menetho may have borrowed name from Dynasty XXII?	480 years
1028/7			(Bel)	Ea-mukin-zer(6 mos.)
1027				Kashshu-nadin-ahhe(3)
1025				Eubash-shakin-sum(17)
1022	SOLOMON(40)	Psinachas(9)	-436 years until Babylonian Captivity of Israel - 586 B.C.	
1018 B.C.	Solomon's Temple begun 4th year of Solomon-I Kings 6:1		-480 yrs to Egyptian Captivity.-|	
1013	SABBATIC YR	PseusennesII(13)		
1012	JUBILEE YEAR - Solomon's Temple completed in the 7th year (inclusive).			
1009				Enurta-kudur-usur I(3)
1007				Shiriglum-Shugamuna(3mos)
1007 B.C.	Elamite Dynasty - - - -+		*	Marbiti-apal-usur(6)
1002	(p. 344, Sidney Smith) - contemporary		- Ashur-resh-ishi II(5)- - - -	*
1002			*	Nabu-mukin-apli(36)
1000		Siamun(17)		
998		DYNASTY XXII		
983		Shishak I(21)	Tiglath-Pileser II(22-Cambridge H.)(32)	
982	REHOBOAM(17)JEROBOAM(22)			
978	5th year Jerusalem attacked by Shishak I - I Kings 14:25			
977			Ashur-dan II(24)	
967				Enurta-kudur-usur II(8mos)
966				Marbiti-ah-iddin(mos)
965	ABIJAM(3)			Shamash-mudamiqiabt?

227

	JUDAH	ISRAEL	EGYPTIAN	ASSYRIAN	BABYLONIAN
963 –	SABBATIC YEAR		Osorkon I(35)(Zerah)		
962 –	JUBILEE ASA(41)				
961 –		NABAT(2)			
959 –		BAASHA(24)			
954 –				Adad-Nirari II(23)	
951 –	Asa smote Zerah(Osorkon I) – II Chr. 14:13;16:8			*- - - - - - - - -*	
941* –	21st/22nd year Zerah rebuilt Temple of Bast-500th year from 4th of Amenophis II – 1441 B.C.				
	(SYNCHRONOUS HISTORY)				
935 –	OMRI & TIBNI(4)		(Inclusive) Babylon defeated by Adad-Nirari II – in 22nd-Treaty-Nabu-shum-ukin I(2)		
933 –				Tukulti-Ninurta II(7)	Nabu-apal-iddin(34)
932 –					
931 –	OMRI(8)				
929 B.C.	Omri's Capitol moved to Samaria				
			Takeloth I(7) Omri subjects Moab for 40 years until 887 B.C.		
927 –				Ashur-nasir-pal II(24)	
926 –					
923 –	AHAB(22)			Calah built – Capitol moved from Nineveh	
921 –	JEHOSHAPAT(26)			Calah as modern Nimrud dedicated. His 6th year against Babylon –*	
920 –			(23 alone)Osorkon II(29)	His 7th year take Khabar,M. Tigrus, Euphrates	
919 –				His 8th year marched to the Mediterranean	
914 –	(Beth-aden or Bit-Adini; Capitol Til-barsip) Assyria took Tyre & attacked Urartu				
913 –	SABBATIC YEAR				
912 –	JUBILEE YEAR				
908 –			ROYAL EPONYN LIST – – –	Shalmaneser's accession (co-regent)	
907 –				Shalmaneser's ascended as co-regent –	1st yr
906 –				Shalmaneser's as co-regent in his – –	2nd yr
905/4	SABBATIC YEAR		Contemporary of Osorkon II – –	Shalmaneser III reigns co-regent in his –	3rd yr
	Contp. Baalazar 909-903 BC				
904/3	Jehoshaphat's 17th Ahaziah co-regent			Sharru-baltu-nishe(to Ramanu) reign in his 1st –	4th yr
903/2	Jehoshaphat sick Joram pro-regent			Shulman-asharid(to Bit-Adini) reign 2nd –	5th yr
902/1 –	Ahab dies-Karkar;Ahaziah sick(Jehoram)			Assur-bunaia-usur(to " ")Battle Karkar-6th yr	
901/0 –	AHAZIAH(2) – sick (Jehoram)			Assur-bunaia-usur(to " ")	4th yr reign
900/9 –	Jehoram co-regent – – sick (Jehoram)			Abu-ina-ekalli-lilbur.Shalmaneser's	5th yr reign
899/8 –	Jehoram co-reg. & Ahaziah dies(Jehoram)			Daian-Assur(to Hatte)Marduk-bel-usate	6th reign
898/7 –	SABBATIC JEHORAM(12)			Shamash-abua(Til-abni) Civil War	7th yr reign
897/6 –	Elisha's famine(Osorkon and Takeloth)			Shamash-bel-usur(against Babylonia)	8th yr reign
896/5 –	Famine(" co-regency ")			Bel-bunaia(against Babylonia)Marduk-zakir-shum I	

228

B.C.	JUDAH	ISRAEL	EGYPTIAN	ASSYRIAN	BABYLONIAN
895/4	–	JEHORAM(8)	Famine(Osorkon and Takeloth)	Badi-lipushu (to Carachemish)	10th yr reign
894/3	–		Famine(" co-regency ")	Nergal-alik-pani(Hatte)Shalmaneser's	11th yr reign
893/2	–		Famine(" co-regency ")	Bir-Ranana(against Pakarhubuna)	12th yr reign
892/1	–		Famine(" co-regency ")	Urta-Mukin-nishe (against Iaeti)	13th yr reign
891/0	–	SABBATIC	Famine ends Takeloth II (19)	Urta-nadin-shum(Hatte) Hazael rules–	14th yr reign
890/9	–			Assur-bunua (against Namri)	15th yr reign
889/8	–			Tab-Urta (against Namri)	16th yr reign
888/7	–			Taklak-ana-sharri (against Hamanu)	17th yr reign
887/6	–	AHAZIAH(1)Jehu's tribute(Moab Free)		Adad-rimani(Damascus)Hazael defeats–	18th yr reign
886/5	–	ATHALIAH(6) JEHU(28)		Bel-abua (to Que-Cilicia) Jehu's 1st year	
885/4	–			Shulmu-bel-lumur(against Kumhi) (MOABITE STONE)	
884/3	–	SABBATIC YEAR		Urta-kibsi-usur (against Danabi)	
883/2	–			Urta-ilia (against Tabali – Tabal)	
882/1	–			Kurdi-Assur (against Melidi)	
881/0	–			Shepa-sharri (against Namri)	
880/9	–	JOASH(40)		Nergal-mudammik (to Que-Cilicia) Jehu's 7th year	
879/8	–			Iahalu (against Que-Cilicia)	
878/7	–			Ululaia gov.(to URARTU-ARMENIA – KING MENUAS)	
877/6	–	SABBATIC YEAR		Nishpati Bel gov. of Calah (to Unki)	
876/5	–			Nergal-ilia gov. of Arrapha (to Ulluba)	
875/4	–			Hubaia gov. of Mazamua Marduk-nadin-shum (Mannai)	
874/3	–			Ilu-mukin-ahe REVOLT	
873/2	–	(Shamshi-Adad V co-regent) –		Shulman-asharidu King of Assyria REVOLT – son	
		Ashur-danin-apla, son of Shalmaneser with		27 cities Assur, Nineveh, Erbil, & Arrapha-Kirkuk)	
872/1 B.C.	–		Sheshank III(39)	Daian-Assur field marshal	REVOLT
871/0	–			Assur-bunaia-usur cup-bearer	REVOLT
870/9	–	SABBATIC YEAR		Iahallu Abarakku	REVOLT
869/8	–			Bel-bunaia high chamberlain	REVOLT CRUSHED
868/7	–			Shamshi-Adad V(14) to Sikria	
867/6	–			Iahalu field marshal (against Madai)	
866/5	–			Bel-daian High Chamberlain (to Shumme)	
865/4	–			Urta-upahhir(Abarakku to Karne)Marduk-balatsu-iqbi	
864/3	–			Shamash-ilia (to Karne) Bau-arh-iddin ?	

JUDAH B.C.	ISRAEL	EGYPTIAN	ASSYRIAN	BABYLONIAN	URARTU (Armenian)
863/2 –	SABBATIC YEAR		Nergal-ilia gov. Arrapha (to Tilla)		
862/1 –	JUBILEE YEAR		Assur-bana-usur (cup bearer) (to Tilla)		
861/0 –			Nishpati-bel gov. Nasibina (to Zarate)		
860/9 –			Bel-balat gov. of Calah		
859/8 –			Mushiknish gov. of Kirruri (to Ahsana)		
858/7 –	23rd Joash–JEHOAHAZ(17)		Urta-asharid (against Chaldea)		
857/6 –			Shamash-kumua gov. Arrapha (against Babylon)		
856/5 –			Bel-kata-sabat gov. Mazamua.		(16)?
855/4 –	SABBATIC YEAR		Adad-Nirari III(29)(QUEEN SEMIRAMIS) to Madai		
854/3 –			Nergal-ilia F. Marshal(Guzana)		King Menuas of Urartu
853/2 –			Bel-daian chamberlain(to Mannai)		digs canal named it
852/1 –			Sil-bel cup-bearer (to Mannai)		"SHAMIRAM" after the
851/0 –			Assur-taklak(Abarakku Arpadda)		Queen Semiramis of
850/9 –			Shamash-ilia(" to Hazazi)		Assyria.
849/8 –			Nergal-eresh gov. Rasappa. (to Bali)		
848/7 –	SABBATIC YEAR		Assur-baltu-nishe gov. Arrapha (to the Sea)		
847/6 –			Urta-ilia gov. Ahi-Suhina (to Hubushkia)		
846/5 –			Shepa-Ishtar gov. Nasibina(to Madai)		
845/4 –			Marduk-ishme-ani gov. Amedi (against Madai)		
844/3 –	Joash's 37th Jehoash co-reg.–Israel II Kings 13:10		Mutakkil-Marduk Rabshaka (against Lusia)		
843/2 –			Bel-tarsi-iluma gov. Calah (against Namri)		
842/1 –	II K.13:1-25 – Benhadad II reigns		Assur-bel-usur gov. Kirruri (to Mansuate)		
841/0 –	SABBATIC JEHOASH(16)		Marduk-shaddua gov. Salmat (against Der		
840/9 –	AMAZIAH(29)		Kin-abua gov. Tashhan(to Der)		(8)?
839/8 –			Mannu-ki-assur gov. Guzana (against Madai)		
838/7 –			Mushallim-Urta gov. Tille (against Madai)		
837/6 –			Bel-ikishani gov. Mehinish (against Hubushkia)		
836/5 –		(Treaty with Babylon)	Shepa-Shamash gov. Isana (against Itu'a)		
835/4 –			Urta-mukin-ahi gov. Nineveh (against Madai)		
834/3 –	SABBATIC YEAR		Adad-Mushammir gov. Kakzi (against Madai)		
833/2 –		Pamay(6)	Sil-Ishtar gov. Arba-ilu(Nubu New Temple) Madai		
832/1 –			Balatu gov. Shibaniba Adad-shum-ibin		Argistis I(28)
831/0 –			Nabushar-usur gov. of Rimusi(to Kishi		
			Adad-uballit gov. Udnunna (to Hubushkia & Der)		

(Left out of chronology – Edwin R. Thiele, "Numbers of the Hebrew Kings", p.209.

JUDAH B.C.	ISRAEL	EGYPTIAN	ASSYRIAN	BABYLONIAN	URARTU
830/9 B.C.			Marduk-shar-usur governor	(to Hubushkia)	
829/8 –	Jehoash contemporary & submissive)		Ninurta-nasir gov. Mazamua	(against Utu'a)	
828/7 –	to King Shalmaneser IV.		Nabu-li gov. of Nasibina	(against Utu'a)	
827/6 –	SABBATIC YEAR	Sheshank IV(36)	Shalmaneser IV(11)	against URARTU	– – King Argistis I
826/5 –	15th Amaziah accession of Jeroboam		Shamshi-ilu field marshal	(against Urartu	
825/4 –	JEROBOAM(41)		Marduk-rimani bup bearer	(against Urartu	
824/3 –			Bel-lishir H. Chamberlain	(against Urartu	
823/2 –		Shamas-ishi-dia v.	Nabu-ishid-ukin	Shamas-Iva?	
822/1 –			Pan-Assur-lamur shaknu	(against Urartu	
821/0 –			Nergal-eresh gov. Rasappa	(against Erini	
820/9 –	SABBATIC YEAR		gov. Nasibina	(against Urartu & Namri)	
819/8 –			Mannu-ki-Adad gov. Salmat	(against Damascus	
818/7 –			Assur-bel-usur gov. Calah	(against Hatarika	
817/6 –			Ashurdan III(19) King of Assyria	(against Gananati	
816/5 –			Shamshi-ilu field marshal	(against Marrat	
815/4 –			Bel-ilia gov. Arrapha	(against Itu's	
814/3 –			Aplia gov. Mazamus	(in the land	
813/2 –	SABBATIC YEAR		Kurdi-Assur gov.Ahi-Suhina	(against Gananati	
812/1 –	JUBILEE Amaziah dies		Mushallim-Urta gov. Tille	(against Madai	
811/0 –	UZZIAH(52)		Urta-mukin-nishe gov. Kirruri	(to Hatarika-plague)	
810/9 –			Sidki-ilu gov. Tushhan Iva Lush(?)	(in the land)	
809/8 –		June 13 eclipse in 8th year Ashurdan)	Bur-Sagale gov. Guzana	(REVOLT IN CITY OF ASSUR)	
808/7 –		& 19 districts revolt to Uzziah.	Tab-bal gov Amedi	(REVOLT IN CITY OF ASSUR)	
807/6 –			Nabu-mukin-ani gov. Nineveh	(REVOLT IN ARRAPHA	
806/5 –			Lakipu gov. Kakzi	(REVOLT IN ARRAPHA	
805/4 –	SABBATIC YEAR	Plague –	Pan-Assur-lamur gov.Arbailu	(REVOLT IN GOZAN	
804/3 –			Bel-taklak gov. Isana	(against Gozan-PEACE)	
803/2 –		(In land)–	Urta-iddina gov. Kurban	(PEACE with Sarduros II)	
802/1 –			Bel-shadua gov. Parmunna	(in the land)	
801/0 –	(Uzziah looses Hamath area accord–)		Ikishu gov. Mehinish	(to Hatarika – Hamath area)	
800/9 –	(ing to II Kings 15:1-Eponyn List)		Urta-shezibani gov.Rimusi	(Hatarika-Hamath area)	
799/8 –	UZZIAH CO-REGENCY WITH ISRAEL		Ashur-Nirari V(9)	(in the land)	
798/7 –	SABBATIC YEAR		Shamshi-ilu field marshal	(in the land)	
797/6 –	Judah & Israel co-regency		Marduk-shallimani chamberlain	(in the land)	

B.C. | JUDAH | ISRAEL | EGYPTIAN | ASSYRIAN | BABYLONIAN | URARTU (Armenia)

796/5 – Judah & Israel co-regency — Bel-dan chief cup bearer — (in the land)

795/4 – JEROBOAM RECOVERS DAMASCUS & HAMATH) — Shamash-den-dugul Abarakku — (against Namri)

794/3 – TO JUDAH(Uzziah) – II Kings 14:25-28 — Adad-bel-ukin shaknu (against NAMRI-K. Sarduros II)

793/2 – Sin-shallimani gov. Rasappa — (in the land)

792/1 – FAMILY & ASHUR-NIRARI V KILLED (URARTU) Nergal-nasir gov.(REVOLT IN THE CITY OF CALAH)

791/0 – SABBATIC Eclipse Jun.24 Psdibulat(23)DYN XXIII ASSYRIA RULED BY URARTU-KING SARDUROS II –

790/9 – "About 790 B.C., 'For several years there was no king in the

789/8 – country'of Assyria confessed a chronicle'" – p. 251, "Ancient Iraq" – Georges Roux.

788/7 – Uzziah & Jeroboam co-regency for protection against Urartu (Armenia).

787/6 – Amos 1:1;7:11 prophecied 2 years before Uzziah's earthquake about destruction by Pul.

786/5 – Uzziah's 26th year & Jeroboam's 40th year.

785/4 – Uzziah's earthquake. Jeroboam dies by the sword in his 41st year – Amos 7:11;IIK.14:23.

784/3 – SABBATIC UZZIAH RULES BOTH KINGDOMS ALONE 10 YRS–1st year

783/2 – Uzziah – 2600 mighty men,307,500 soldiers, and – 2nd year – engines of war–II Chr.26:16

782/1 – Uzziah had vine dressers on Carmel(Israel)IIK.26. 3rd year — Ninurta apla-x(mos)

781/0 – "Thus, for 36 years (781-745 BC-added), 4th year — Marduk-bel-zeri

780/9 – Assyria was practically paralysed" – p. 251, 5th year — 46

779/8 – "Ancient Iraq" by Georges Roux. 6th year — years

778/7 – 7th year alone — of

777/6 – SABBATIC YEAR 8th year — Urartu

776/5 – Uzziah's 36th year – 1ST OLYMPIADS–Reign Psdibulat(9th yr) — rule

775/4 – Jotham co-reg. with Uzziah-leprous-IIChr.26:16. 10th year alone. — over

774/3 – Uzziah's 38th–ZACHARIAH(1) – because Jotham only 10 years old and Uzziah sick. — Assyria

773/2 – Uzziah's 39th–SHALUM(1 mo) – Pul recovers Hamath from Uzziah attacking from Arpad.

772/1 – MENAHEM(10) Pul gains area from Urartu in Assyria.

771/0 – Pul moves south into Israel & Judah.Marduk-apal-usur

770/9 – SABBATIC Uzziah & Menahem paid tribute to Pul (Tiglath Pileser).

769/8 – During the next 24 years Urartu gradually loosing control of Assyria to Pul

768/7 – Osorkon III(8) "

767/6 – "

766/5 – Uzziah's 46th year. "

765/4 – Eriba-marduk "

764/3 – "

763/2 – SABBATIC Eclipse, June 15th (Urartu) Sarduros III

232

B.C.	JUDAH	ISRAEL	EGYPTIAN	ASSYRIAN	BABYLONIAN	(URARTU Armenia)
762/1 –	JUBILEE	PEKAHIAH(2)–50th year Uzziah–IIK.15:23.			Nabu-shum-ukin I(15)	↑
760/9 –		Urartu gradually loosing control of Assyria to Pul.				46 years of Urartu rule over Assyria
760/9 –	52nd year–PEKAH(20)				"	
759/8 –	JOTHAM(16)		Psammus(10) – IIK.15:27		"	
758/7 –					"	
757/6 –	Jotham smote Ammon				"	
756/5 –					"	
755/4 –	SABBATIC YEAR				"	
754/3 –	Jotham built the wall of Ophal – II Chr. 27:3.				"	
753/2 –					"	
752/1 –					"	
751/0 –	Ahaz 12 years old conceives a son Hezekiah.				"	
750/9 –	Cushite & Sudanese – – Piankhy(16)(21) DYN XXIV				"	
749/8 –					"	
748/7 –	SABBATIC YEAR Feb. 27th of 747 B.C. Nisan year – – – King Nabonassar(15)				"	
747/6 –	Pul gains & Urartu's defeat begins while loosing control of Assyria to Pul.					↓
746/5 –					"	
745/4 –				Pul = Tiglath Pileser III(19)	Nabu-bel-usur gov. Arrapha	(against NAMRI)
				Bel-dan gov. Calah		
744/3 –	Accession of Ahaz			Tiglath in Arpadda–MASSACRE IN LAND OF URARTU		
743/2 –	AHAZ(16)			Nabu-daninani Field Marshal		(against Arpadda)
742/1 –	Ahaz wounded by Syria			Bel-harrah-bel-usuruh chamberlain(" ")		(against Arpadda)
741/0 –	SABBATIC YEAR					(against Arpadda)
740/9 –	20th of Jotham(prison)Hoshea rules–Tiglath Mabu-etirani					
739/8 –				Sin-taklak abarakku – Ulluba taken		
738/7 –	Ahaz subject to Assyria			Adad-bel-ukin shaknu Kulani (Calah) captured		
737/6 –				Bel-emurani gov. Rasappa		(against Madai)
736/5 –				Urta-ilia gov. Nasibina		(to Mount Nal)
735/4 –	Ahaz attacked by Philistines			Assur-shallimani gov. Arrapha(against URARTU)		
734/3 –	SABBATIC YEAR			Bel-dan gov. Calah	Nabu-nadin-zer(3)	(Philistines)
733/2 –			Bocchoris(3)	Assur-daninani	Nabu-shum-ukin II(mo)	Damascus
732/1 –	12th of Hoshea free from Assyria			Nabu-bel-usur gov.	Nabu-Ukin-zeri(4)	Sime(Damascus)
731/0 –			DYN XXV (SO) Shabalaka(13)	Nergal-uballit gov.	Ahi-suhina	(against Sapia)
730/9 –		HOSHEA(9)		Bel-ludalri gov. Tille		(in the land)
729/8 –				Naphar-ilu gov. Kirruri (took Bel)	Pulu (Pul)	

233

Year B.C.	JUDAH / ISRAEL / EGYPTIAN	ASSYRIAN	Babylonian
728/7 B.C.	Accession of Hezekiah	Dur-Assur gov. Tushhan(Took Bel)	Pulu (Pul)
727/6	SABBATIC HEZEKIAH(29)	Shalmaneser V(6)Bel-harran-bel-usur	Ulula(5) Damascus
726/5		Marduk-bel-usur gov. Amedi	(in the land
725/4		Mahde gov. Nineveh	(against Samaria)
724/3	4th year & Hoshea's 7th year	Assur-ishmeani gov. Kakze	(against Samaria)
723/2	5th year of Hezekiah	Shalmaneser King of Assyria	(against Samaria)
722/1	6th year was Hoshea's 9th year Samaria -	Sargon(19)Urta-ilia F.Marshal	Merodach-baladan II(14)
721/0	Eclipse Mar.19 (Co-regent General)	Nabutaria chamberlain	(New Temple Nabu)
720/9	SABBATIC YEAR (1st Battle)	- Assu-iska-danin F. Marshal	(to Tabala-Egypt)
719/8		Sargon King Assyria	(Nergal T.repair)
718/7	(Sibahki) Shebitku(7)	Zer-ibni gov. of Ra	(against Mannai)
717/6		Tab-shar-Assur Abarakku	Urartu - King Rusas I
716/5		Tab-sil-esharra gov. Assur	(Musasir-Haldia)
715/4	General co-regent - Tirhakak	Taklak-ana-bel gov. Namibina	(Great in Ellipa)
714/3	Lachish - Jerusalem (2nd Battle)	- Ishtar-duri gov. Arrapha	(New Temple Nergal)
713/2	SABBATIC Hezekiah Prayer(3rd Battle)	Assur-bani gov. Calah(to Musasir-Ashdod)	K.Midas of
712/1	JUBILEE(Meradach's V.)Tirhakak(39)	Sharru-emurani gov. Zamua	(in land) Phrygia.
711/0	Ashdod(Metinti) (4th Battle)	- Urta-alik-pani gov. Sima	(to Markasa-Ashdod)
710/9	Sargon II took Babylon - -	- Shamash-bel-usur gov. Arzuhina	(Sargon rules Babylon)
709/8		Mannu-ki-Assur-li gov. Tille	(Sargon II took Bel-5yrs)
708/7		Shamash-upahhir gov. Kirruri	(Kumuha captured - 2nd)
707/6		Sha-Assur-dubbi gov. Tushhan	(Sargon from Bel - 3rd)
706/5		Mutakkil-Assur gov. Guzana	(to Dur-Iakin - 4th)
705/4	SABBATIC YEAR	Nashu-Bel gov. Amedi	(Dur-Iakin destroyed - - 5th)
704/3		Sennacherib(24) Nabu-din-epush gov.	Nineveh-Interregnum
703/2		Kannunnai gov. Kakai(in Karalli)	Merodach-baladan(2)
702/1	Sennacherib - (5th Battle)	- Nab-li' gov. Arbailu - put on throne -	Bel-ibni(1)
701/0		Hananai gov. of bi. Rebellion of - -	Bel-ibni
700/9		Metunu gov. Isana. The Kings son-Assur-nadin-sum(7)	
699/8		Vel-sharani governor of Kurban	
698/7	SABBATIC YEAR	Shulmu-shar gov. of	

Chronological chart (rotated 90°). Columns: JUDAH | ISRAEL | EGYPTIAN | ASSYRIAN | BABYLONIAN

Year	JUDAH	ISRAEL	EGYPTIAN	ASSYRIAN	BABYLONIAN
697/6	MANASSEH(55)			Nabu-dur-usur	(Shagarakli-Shagiash seal after 600yrs)
696/5	(*;**;***; - Sidney Smith, "Early)			Shulmu-bel g. Rimusa	
695/4	(History of Assyria", page 355)			Assur-bel-usur	(Tiglath I defeated by)
694/3	Elam -		(6th Battle)	Ilu-ittia Damascus(Marduk-nadin-ahhe)	Nergal-ushezib(2)
693/2				Nadin-ahe gov.	(418 before Bel)Mushezib-Marduk(5)
692/1				Zazai gov. Arpadda	(or Babylon fell)
691/0	SABBATIC YEAR		(7th Battle)	Bel-emurani gov. Carchemish **	(Khalute)
690/9		Nabu-mukin-shi v.		Nabu-bel-usur gov. Samaria	
689/8			(8th Battle)	Gilhilu gov. Hatarika -	Sennacherib destroys Babylon
688/7	(Shalmaneser I built)			Nadin-ahe gov. of Simirra	2nd Interregnum(9)
687/6	(the Temple of Ashur)			Sennacherib King of both Assyria & Babylon - --	
686/5	(580 years before)			Bel-emuranni gov. of Calah	
685/4	(the reign of King)			Assur-daninanni gov. of	ub.
684/3	SABBATIC(Esarhaddon-Assyria)			Mannu-zirni gov. of Kullania	
683/2		****		Mannu-ki-Adad governor of Supite	
682/1		- - -(gov. Samalli)-Esarhaddon(14)Nabu-ah-aresh		Nabu-shar-usur governor of Markasi	
681/0				Assur-akh-iddin(10)	
680/9		Baal King of Tyre		Dananu gov. of Manaua	
679/8		Khansqabri King Edom		Iti-Adad-aninu gov. of Magidunu	
678/7		Mushuri King of Moab		Nergal-shar-usur cup-bearer	(600 years back to)
677/6	SABBATIC	Zilbel King of Gaza		Abi-rama high minister	(Rameses II and)
676/5		Mitinti King of Askalon		Banba second minister	(victory over the)
675/4		Ikasaun King of Ekron		Nabu-ahi-iddina governor	(Hittites-1273/2 B.C.)
674/3		Milliasap King of Byblos		Sharru-nuri gov. Barhalzi	
673/2		Matanbaal King of Arrad		Atar-ilu g. Lahiri(Attacked Egypt-Is.19:2;Ez.4:2)	
672/1			Tanutamon(3)	Nabu-bel-uaue g. Dur-Sharrukin(Egypt taken)	
671/0		Abibel King of Samsimuruna		Kanunai Sar-tinu (Babylon rebuilt in 3 years by)	
670/9	SABBATIC	Puduil King of Beth-Ammon		Shulmul-bel-lashme gov. of Der(King Esarhaddon)	
669/8	Ahimelech K. of Ashdod. Necho I(5)			Shamash-kashid-aibi gov. Ashdod-IIK.21;IIChr.33	
668/7		Jakinlu King of Arvad		Ashurbanipal(36)Mar-larim F.M. Shammash-shum-ukin	
667/6		Amminadab King of Beth-Ammon		Gabbar governor of	
666/5		Manasseh King of Judah		Kanunai governor of Bit-eshshi	
665/4				Mannu-ki-sharri prefect of the land	
664/663 B.C.	SABBATIC YEAR			Sharru-ludari governor of Dur-shur-rukia	
663/662 B.C.	JUBILEE	DYN XXVI Psammetiches I(54)		Bel-naid field-marshal -	they attacked Egypt

APPENDIX   K
TABLE VI
HEBREW MESSIANIC GENEALOGY

```
(Irano-Semitic)        /  ADAM THE SON OF GOD - Lk. 3:32-38 (Semitic - Gen. 5:3-32)
(Gen. 4:16f)          /    ADAM - expulsion from Eden - 4012 B.C.
          /                SETH
         / (Land of Nod)   ENOS
**                         CAINAN *
CAIN - - /                 MAHALALEEL
ENOCH
IRAD - inter-marriage - - JARED - means "to go down" (in sin) - Gen. 4:15;6:2 (3552-2590 B.C.)
MEHUJAEL      with        ENOCH - 7th from Adam - Jude 14                       (3390-3025 B.C.)
METHUSAEL  Irano-Semite   METHUSELAH - dies the year of Flood - 969 years old (2356-2355 B.C.)
LAMECH     results the    LAMECH - killed a man for wounding him
TUBAL-CAIN Sumarians.
NAAMAH(sister)            SHEM - 97th.yr.- NOAHIC WORLD WIDE FLOOD - - - - - - -(2356-2355 B.C.)
   ↓                      ARPHAXAD - born 2 years after the Flood - Gen. 11:10(2353 B.C. )
Married Ham? - **CAIN - - -(Cainan)*- inserted as identity of line - "Cainan" (Messianic), or"Cain's"
                          SALAH
         (Hebrew) - - - - EBER - Father of the Hebrew - lived 4 years longer then Abraham
                          PELEG - means "Division" of tribal lands and migration in his days
(2254-2015 B.C.) - - - -  REU - means "Friendship"
Moon worship in N.- Nahor-SERUG - means "Firmness" -Northern Moon Worship at city of Nahor
Joshua 24:2              NAHOR - means "Slayer" - Gutian's destroy Babel in 2169 B.C.
Moon worship in S.- Ur - -TERAH - means "wandering" - language confused - migration
Abram's call - 1928 B.C.--ABRAHAM means "Father of Faith" or of a multitude & born 2003 BC-1st Gen.
                          ISAAC                                                    2nd Gen.
                          JACOB        Matthew 1:17                                3rd Gen.
                          JUDAH by Tamar a Gentile (played a harlot) Gen. 38       4th Gen.
Days of Joseph- - - - - - PHARES - Israelites to Egypt in 1713 B.C.                5th Gen.
                          ESRON                                                    6th Gen.
                          ARAM                                                     7th Gen.
*Second Cainan added in   AMINADAB - EXODUS 1498 B.C. - 4th Gen. - Gen. 15:16      8th Gen.
Luke 3:36 simply to       NAASON                                                   9th Gen.
identify which line to    SALMON  - Rahab - entrance in land - 1458 B.C.          10th Gen.
take "Cain"** or          BOAZ    - Ruth                                          11th Gen.
"Cainan"* (Messianic).    OBED                                                    12th Gen.
                          JESSIE  - Matt. 1:6-16                                  13th Gen.
```

```
Lk. 3:28-31      / - DAVID--reign--began--in--1062 B.C.                              14th generation
NATHAN - /   I Chron. 3:5      1022 B.C. - SOLOMON                                   1st generation
MATTATHA                         982 B.C. - REHOBOAM                                 2nd generation
MENAN                                       ABIJAM                                   3rd generation
MELEA                                       ASA                                      4th generation
ELIAKIM                                     JEHOSHAPHAT                              5th generation
JONAN    EVIL - II Chr.21:5-7,12-15;22:2,9;23:17 - - JEHORAM (Evil)  (Wife Athaliah) 6th generation
JOSEPH   Curse to the 3rd and 4th generations - - - Ahaziah- Matt. 1:8 - Curse - Deut. 5:9 :
JUDA     Worshipper of Baal daughter of Ahab- - - Athaliah- daughter of Jezebel and Ahab - - -:
SIMEON   Curse - left out of genealogy- - - - - Joash   - Matt. 1:8 - Curse - Deut. 5:9
LEVI     Curse unto the 4th generation- - - - - Amaziah - Matt. 1:8 - Curse - Deut. 5:9
MATTHAT                                      UZZIAH                                   7th generation
JORIM                                        JOTHAM                                   8th generation
ELIEZER                                      AHAZ                                     9th generation
JOSE                                         HEZEKIAH                                 10th generation
ER                                           MANASSEH                                 11th generation
ELMODAM                                      AMON                                     12th generation
COSAM            (Only one brother chosen)   JOSIAH                                   13th generation
ADDI     Matt.1:11 "Brethren" Jehoahaz brother of JEHOIAKIM-Father of Jeconiah       14th generation
MELCHI   586 B.C. Babylon Captivity - Curse - Jer. 22:30----JECONIAH(Jehoiachin)---1st generation
NERI(his daughter)-Lk. 3:27 Neri father-in-law of - - - - -SHEALTIEL I Chr.3:19     2nd generation
     Zech. 4:6-10 HEADSTONE - ZERUBBABEL------------:Haggi 2:23-Signet               3rd generation
RHESI----JOANNA--------------▼      Luke 3:23-27          :-Seal of Royalty
JUDA     JOSEPH      STONE with 7 eyes (Spirits) Zechariah----ABIUD Matt. 1:14       4th generation
SEMEI    MATTATHIA   3:9; Rev.5:6;4:5;Isa. 11:2-Stone & Seven  ELIAKIM              5th generation
MAATH    NAGGE       Branch candlestick typifies Jesus Christ)  AZOR                6th generation
ESLI     NAUM                                                   SADOC               7th generation
AMOS     MATTATHIAS  Through Zerubbabel both Joseph and Mary    ACHIM               8th generation
JOSEPH   JANNA       were of the Seed Royal. Zerubbabel was     ELIUD               9th generation
MELCHI   LEVI        the HEADSTONE builder of a literal temple  ELEAZAR             10th generation
MATTHAT  HELI        & of the physical temple of Christ's       MATTHAN             11th generation
(Mary)------------daughter------------:        body.            JACOB               12th generation
         :-Mary made a Royal Seed by Zerubbabel :-Legal Step-father JOSEPH(Curse Jer.22) 13th generation
                    See Gen. 49:10*          :--JESUS CHRIST THE SON OF GOD---------14th generation
```

237

* "The sceptre (Kingly genealogy) shall not depart from Judah, nor a lawgiver from between his feet, until Shiloh (Peace Bringer) come; and unto Him shall the gathering of the people be".

APPROXIMATE POPULATION GROWTH OF THE WORLD
4012 B.C. - 2000 A.D.

B.C.
4012 - expulsion of Adam from Eden (Adam and Eve began to age)
```
        :
        :
   1656 years
        :                           Sumerian Period
        :
```
2356 - NOAHIC FLOOD - estimated 100,000,000 to 1,000,000,000 died
2355 - NOAHIC FLOOD ENDED -   8 souls(plus servants not counted)
2280 -   -   12,000 souls

	2254 B.C.			2254 BC Days of Peleg	
2225 -		-	1,366,000 souls	: M H :	"In his
	The			: i P e :	day
2185 -		-	2,015,000 souls	: g e r :	was
	Sumerian			: r r o :	the
2145 -		-	4,050,000 souls	: a i i :	earth
	Legendary			: t o c :	divided"
2105 -		-	7,891,750 souls	: i d :	or
	Period			: o A :	reinhabited
2065 -		-	12,218,500 souls	: n g :	Gen.
				: e :	10:5,25,32
2015 -	2015 B.C.	-	20,177,000 souls	: 2015 B.C. Peleg dies	

```
                                        (Abraham's days)
```
1500 -   -   52,315,500 souls
   (1498 BC - 603,550 fighting men in Israel - Num.1:45,46)(Moses' days)
1000 -   -   97,154,000 souls
   (1105 BC - 3,961,670 fighting men in Israel - Jud. 20)(Solomon's days)
 500 -   -   160,692,500 souls
```
                                             (Captivity ends)
```
   1 -   -   300,175,250 souls
```
                                             (Christ's days )
```
1650 A.D.   -   540,074,250 souls

1800 -   -   906,000,000 souls

1900 -   -   1,600,000,000 souls

1960 -   -   2,900,000,000 souls

1976*- (April 4th)   -   4,000,000,000 souls   (U.S.A. Bicentenial)

1985 A.D.(Feb)to Ju.1987   -   5,000,000,000 souls

1989/90 A.D.(Jubilee year)-   6,000,000,000 souls (6000 yrs-Prediction)

* Date the 4 billion mark was passed

BIBLIOGRAPHY

Adams, J. McKee, "Ancient Records and the Bible". Broadman Press,
    Nashville, Tennessee, 1946
Aharoni, Yohanan, "The Land of the Bible", Burns and Gales Limited,
    25 Ashley Place, London, S.W.1., 1968
Albright, William Foxwell, "From the Stone Age to Christianity", The
    Johns Hopkins Press, Baltimore, 1957
Albright, William Foxwell, "The Archaeology of Palestine", Pélican
    Books, Gibraltar House, Regent Rd., Sea Point, Cape Town, 1956
Albright, William Foxwell, & G.E. Mendenhall, translated "Empire in
    the Dust", from "Ancient Near Eastern Texts" edited by James B.
    Pritchard
Aldred, Cyril, "Egyptians" Published by Thames Hudson, London, 1961

Amplified New Testament, Zondervan Publishing House Grand Rapid
    Michigan 49503 (1415 Lake Drive S.E.)
Analytical Hebrew and Chaldee Lexicon by B. Davidson, Harper and
    Brothers, New York
Anderson, Bernard W., "The Living World of the Old Testament", Longman &
    Green, Longman House, Burnt Mill, Harlow, Essex CM 20 2 JE
Anderson, Sir Robert, "Coming Prince", Kregel Publications, Grand
    Rapids, Michigan 49503
Anstey, Dr. Martin, "The Romance of Bible Chronology", Marshall Bros.
    Ltd., London, 1913
Baron, David, "Visions and Prophecies of Zechariah" Hebrew Christian
    Testimony to Israel, 18 Whitechapel Road, E.1., 1951
Behwinkel, A.M., "The Flood", Concordia Publishing House, St. Louis,
    Missouri, 1951
Bibby, Geoffrey, "Four Thousand Years Ago", Collins, St. Jame's Place,
    London, 1962
Blackman, A.M., "Luxor and its Temples", A and C Bloch Ltd., 4-6 Soho
    Square, London, W.
Book of Enoch, Apocrypha book

Breasted, James Henry, "A History of Egypt", Seribner's, New York,
    1912; 1906, London, Hodder and Stoughton(J.H.Breasted,Univ.of Chicago)
Burney, Dr., "Israel's Settlement in Canaan" British Academy,London 1921
    Oxford University Press
Budge, Sir Ernest A. Wallis, "A History of the Egyptian People", J.M.
    Dent and Sons Limited, London, 1914
Cambridge Ancient History, Vol. II and Vol. VII, Macmillan, 1924, 2nd
    ed., New York(Camb.Univ.Press Bentley House 200 Euston Rd LdnNW1 ZDB)
Ceram, C.W., "Narrow Pass Black Mountain", Victor Gollancz Limited in
    association with Sidwich and Jackson limited, London made and
    printed in Great Britian by William Clowes and sons, Limited,
    London, and Beccles(1 Lavistock Chambers,Bloomsbury Way,Ldn,WCIA,2SG)
Christian Readers Digest, Pleasantville, N.Y.

Cleator, P.E., "The Past in Pieces", George Allen and Uniwin Ltd.,
    Ruskin House, Museum Street, London, 1957(P.O.Box18,ParkLane, Hemel
    Hempstead, Herts,HP2 4TE, England)

Coder, Maxwell S., "Archaeology Confirms the Old Testament",
    Moody Monthly, Jan., 1967, Moody Press, 820 N. LaSalle,
    Chicago, Illinois
Cole, D.I., "The Star of Bethlehem", monthly notes of the Astronomical
    Society of South Africa, Royal Observatory, Observatory, Cape, S.A.
Cottrell, Leonard, "The Mountains of Pharaoh", Trinity Press, London,
    1956; "The Anvil of Civilization"
Cottrell, Leonard, "The Lost Pharaoh's", Pan Books Ltd., 8 Headfort
    Place, London S.W.1,1961; "The Land of Shinar", Souvenir Press,
    London, 1965
Donnelly, Ignatius, "Atlantis: The Antediluvian World" quoted from
    Goodrich, "The Sea and Her Famous Sailors", London, 1859
Driver, S.R., "The book of Genesis", London, 1914

Elgood, Lieut-Colonel P.G. C.M.G. Basel, "Later Dynasties of Egypt",
    Blackwell and Holt, Oxford,England, 1951
Emery, W.B., "Archaic Egypt", Pelican Book Ltd., Harmondsworth,
    Middlesex, England
Encyclopaedia Britannica, 1942

Fausset's Bible Dictionary, Zondervan Publishing House, Grand Rapids,
    Michigan 49506   (1415 Lake Drive S.E.)
Feannerat, Pierre, "Flying to 3000 B.C."

Finagan, Jack, "Handbook of Biblical Chronology", Princeton New Jersey,
    Princeton University Press 1964
Finn, A.H., "The Creation, Fall & Deluge", Marshall Bros.,London, 1920

Frankfort, Henri, "The Birth of Civilization of the Near East",
    Published by Williams and Norgate Limited, Bouverie House, Fleet
    Street, London EC4
Free, Joseph P., "Archaeology and Bible History", Van Kempen Press,
    Wheaton, Illinois 1950
Gardiner, Sir Alan H., "Egypt of the Pharaoh's", Oxford University Press
    Walton St., Oxford OX26 DP England, UK   (UKO-55)
Garstang, J.B.E., "The Story of Jericho", Marshall, Morgan and Scott,Pub.
    (Revised Edition) London 1948,3 Beggarwood Lane,Basingstoke,Hants RG237
Glanville, S.R.K. (Edited), "The Legacy of Egypt", Oxford at Clarendon
    Press, Oxford University Press, Amen House, London E.C.4
Glueck, Nelson, "The River Jordon", The Westminister Press, Philadelphia
    1945
Griffith, J.S., "The Exodus in the Light of Archaeology", Robert Scott.
    Roxburghe House, Paternaster Row, London E.C.
Guinness, H. Grattan, "The Approaching End of the Age", Hadden and
    Stoughton, 47 Bedford Square, London NCIB 30P England
Halley, Henry H., "Halleys Bible Hand Book", Box 774, Chicago 90,
    Illinois
Hebrew Origins, by T.J. Meek, Harper and Row, Harper Torchbook ed.,
    1960, New York, VIII, IX, 7-24
Hrozny, Bedrich, "Ancient History of Western Asia, India, and Crete",
    Published by Philosophical Library Inc., 15 E. 40th Street, New
    York 16, N.Y.

Ipuwer's Papyrus, "The Admonitions of an Egyptian Sage" (Found in
    Memphis) Museum of Leidon, Netherlands Catalogue - Leiden 344,
    (1846), Pt.2, Face: Plates 105-113
"Journal of Near East studies", Vol.13, No. 4, Oct. 1954, University
    of Chicago Press, Chicago, Illinois
Kees, Hermann, "Ancient Egypt", Faber and Faber, 24 Russell Square,
    London
Keller, Werner, "The Bible as History", Hodder and Stoughton Limited,
    47 Bedford Square, London WCIB 3OP, England, by Richard Clay
    (The Chauar Press), Ltd., Bangay, Suffolk, England
Kramer, S.N., "Sumerian Historiography" in Israel Exploration Journal,
    Israel
Kramer, S.N., "History Begins at Sumer", Jarrold and sons, Ltd., Norwick,
    Britian
Lady Amherst, "A Sketch of Egypt History", Hackney, England
Layard,A.H."Nineveh & its Remains" Eyre & Methuen Vol II p.26
Lissner, Ivar, "The Living Past" translated by J. Maxwell Brownjohn,
    M.A., printed by A.P. Putnam's Sons, New York, 1957
Marston, Sir Charles, "The Bible Comes Alive", Eyre and Spottiswoods,
    London, 1950; 2 Serjeant's Inn,London,EC4y,ILU,UK 0-950 1996
Mauro, Philip, "The Chronology of the Bible", Mamilton Bros.,
    Scripture Truth Depot, 120 Fremont Street, Boston 9, Mass.
Miller, Madeleine S., Miller, J., "Harper's Bible Dictionary", Harper
    and Brothers Publishers, 49 East 33rd Street, New York 16,N.Y.
Moody Monthly, "Archaeology Confirms the Old Testament", Jan. 1967,
    Moody Press, 820 North LaSalle Street, Chicago, Illinois
Moscati, Sabatine, "The Semites in Ancient History", Cardiff, University
    of Wales Press, 1959("The Face of the Ancient Orient")*,6 Gwennyth St
Murry, Margaret A., "The Splendour that was Egypt", Sidgwick and
    Jackson Limited, London
Negev, Avraham, "Archaeology Encyclopedia of the Holy Land", The
    Jerusalem Publishing House, 2 Magnes Square, P.O. Box 7147 Jerusalem,
    Israel
New Bible Dictionary, Intervarsity Press, 38 De Montfort St.,Leicester LE1
                                                      7GP England
Noss, John B., "Man's Religions", Macmillan Co., New York, 1966

Oesterley, W.O.E., "The Legacy of Egypt", Edited by S.R.K. Granville,
    Oxford University Press, Walton street, Oxford OX2 6DP UK
Panin, Ivan, "Bible Numerics", Box 306, Aldershot, Onterio, Canada;"Bible
    Chronology", T.W. Cooks, 13 New Rd., North Walsham Norfolk, England
Petrie, W.M. Flinders, "A History of Egypt" (Methuen, 1924)Eyre Methuen,
    11 New Fetter Lane, London, E C4P 4EE, UK
Pfeiffer, Charles F., "Ra Shamra and the Bible", Baker Book House, Grand
    Rapids 6, Michigan, 1962

Rainburg, A.W., "The Day of Noah", The Keswick Week Journel, page 32,
    1970, England
Rawlinson, George, "Ancient Egypt", G.P. Putnams Sons, New York

* Vallentine, Mitchell Co., Ltd., 1960, 37 Furnival St., London, E.C.R.
  (Gainsborough House, 11 Gainsborough Rd., London E11 1RS)

Rea, John, "The Age of the Patriarchs in Palestine", Grace Theological
    Seminary notes, page 2,3
Rimmer, Henry, "Dead Men Tell Tales" Hart-Kolportasie-bekery,c/o
    Evangelik-Uitewers, P.O. Box 31636, Nraamfontein 2017
Roux, Georges, "Ancient Iraq", George Allen and Wnwin Ltd., Ruskin
    House, Museum Street, London, 1944
Saggs, H.W.F., "The Greatness That Was Babylon", Sidwick & Jackson Ltd.
    1 Lavistock Chambers, Bloomsbury Way, London, WCIA,2SG, 1962
Sale-Harrison, L., "Palestine God's Monument of Prophecy", c/o
    Evangelical Press, 3rd and Reily Street, Harrisburg Penn., 1933
Save-Soderbergh, Torgny, "Pharaohs and Mortals", The Bobbs-Merril Co.,
    Inc., a subsidary of Howard W. Sons and Co., Inc., Indianapolis,N.Y
Schonfield, Hugh J., "Secrets of the Dead Sea Scrolls", A.S. Barnes
    and Co., Inc., New York (A Petpetua Book)
Silverberg, Robert, "Empire in the Dust", Chilton Books, Chilton Co.
    Philadelphis, Pa.
Smith, Philip, "Ancient History of the East" by John Murry, Albemarle
    Street, London, 1876
Smith, Sidney, "Early History of Assyria", Chatte and Windus, London,
    1928
Smith and Fuller, Dictionary of the Bible", Vol. II, Fleming H. Revell
    Co., Old Tappan, New Jersey 07675
Smith, William, "Dictionary of the Bible", Fleming H. Revell Co.,Old Ta
    New York, New Jersey 07675
Thiele, Edwin R., "The Mystery Numbers of the Hebrew Kings", Eerdmans
    Publishing Company, Grand Rapids, Michigan, 1965
Toffteen, Olof A., "Ancient Chronology", The University of Chicago
    Press, Chicago Illinois, 1907
Unger, Merrill F., "Archaeology and the Old Testament", Zondervan
    Publishing House, Grand Rapids, Michigan
Unger, Merrill F., "Archaeology and the Bible", Zondervan Publishing
    House, Grand Rapids, Michigan, 1956
Velikovsky, Immanuel, "Ages in Chaos", (Publisher below)

Velikovsky, Immanuel, "Worlds in Collision", "Ages in Chaos", Sidwick &
    Jackson Ltd., 1 Lavistock Chambers, Bloomsburg Way,London WCIA, 2SG
Vos, Howard F., "And Introduction to Bible Archaeology", Moody Press,
    Chicago, Illinois, 1959
Vos, Howard F., "Genesis and Archaeology", Moody Press, 820 N. LaSalle,
    Chicago, Illinois 1963
Waddell, W.G., "Loeb Classical Library", Cambridge, Mass., 1941, Harvare
    University Press 79 Garden St., Cambride, Mass., 02138-9983
Westminster Historical Atlas, Westminster Press, Philadelphia, U.S.A.

White, J.E.M., "Ancient Egypt", Allan Wingate. 12 Beauchamp Place, S.W.
    3, London (Printed by W. Clowes & Son Limited, London & Beccles)
Whiston, William, "The Word of Josephus", The International Press, The
    John C. Winston Co., Philadelphia
Wilson, J.A., "The Burden of Egypt", University Chicago Press, Chicago,
    Illinois, 1951
Wilson, J.A., "The Culture of Ancient Egypt", Phoenix Books, University
    of Chicago Press, Chicago, Illinois, 1960
Witcomb, John C. & Morris, H.M., "The Genesis Flood", The Presbyterian
    & Reformed Publishing Company, Philadephia, Penn., 1961

J

Jermiah 13,15,16,19,21-24
Jeroboam 30,48-52,54
Jeroboam II 52
Jerome 200
Jertet-Meizoi 173
Jerubbaal 86
Jerusalem 12,16,29,30,31,60,63,76,
    98,177,183-185,188,189,192,196
Joash 47,48
Joktan 146
Joram 32,33
Joseph 9,68,70,102,119-120,122,127,
    129,131,134,135,167
Joseph (husband of Mary) 180,181-185
Josephus 16,27,80,99,115,123,142,158,
    161,177,188,199
Joshua 68,74,75,87,90,93,96,97
Jotham 53,54,57
Jubal 158,168
Jubilee 30,60,65,179,180,184-190
Jublius Africannus 69
Judah 13-24,26,30,32,33,34,49,51,53,
    54,57,58,61,87,99,177,188-189
Judas of Galilee 179
Jude 163
Judea 179,185

K

Kadesh Barnea 74,79,95
Kaiser-Frederick Museum 19
Kamose (Ahmose I) 123
Karkar 32,33
Kedme 116
Keller 107
Kenaiah 19
Kenez 81
Khafra 171
Kimuhu 15
Kirjath-Jearim 97,,98,100
Kish 138,141,142,150,175
Kohath 114
Kudur-mabug 105
Kummuha 54
Kurigalzi I 81
Kustaspi 54
Kyle, M.G. 130-131

L

Laban 105,119,124
Labaya 82
Lachish 60,61,82
Lagamur 105

Lagash 144
Lake Urmia 52
Lamech 158,167,168
Laou 144,175
Larkin, Clarence 195
Larsa 105
Layard 55
Leah 124,126
Lebanon 170,171
Levi 114
Leyden Museum 173
Libnah 60,61
Libyan 171,173
Libyia 141
Lipit-Ishtar 104
Lot 104,106,113,118
Louvre 33
Luash 49
Lucifer 152-153
Lugalzaggisi 138,150,175

M

Maaseiah 53
Magi 180,184,185
Mamre 16,18,113,122
Manasseh 28,62,79
Manetho 64,123,144,149,165
Marsimani 60
Mary 170,184,185,200
Masoudi 165
Massoretic text 161,169
Ma Tuan-lin 181
Mauro, Philip 47,102,169
McClain, A.J. Dr. 184
Medes 15,52,56
Megiddo 13,14,67,74,76,79,81,85,
    86,92
Melchisedic 105
Menahem 51,54
Menophis 69
Menuas 160
Merit-Amon 69
Merneptah 19,85,110
Merodack 51,61
Meroe 61
Mesha 33
Mesopotamia 14,79,81,90,148,150,160
Messiah 7,110,113,177,180,181,183
    184,190,197
Methaselah 108,138,159,164,167,168
    175
Metinti 61